Cluster Analysis for Corpus Linguistics

Quantitative Linguistics 66

De Gruyter Mouton

Cluster Analysis for Corpus Linguistics

by

Hermann Moisl

De Gruyter Mouton

ISBN 978-3-11-035025-8
e-ISBN (PDF) 978-3-11-036381-4
e-ISBN (EPUB) 978-3-11-039317-0
ISSN 0179-3616

Library of Congress Cataloging-in-Publication Data

A CIP catalog record for this book has been applied for at the Library of Congress.

Bibliographic information published by the Deutsche Nationalbibliothek

The Deutsche Nationalbibliothek lists this publication in the Deutsche Nationalbibliografie; detailed bibliographic data are available on the Internet at http://dnb.dnb.de.

Printing and binding: Hubert & Co. GmbH & Co. KG, Göttingen
♾ Printed on acid-free paper

Printed in Germany

www.degruyter.com

MIX
Papier aus verantwortungsvollen Quellen
FSC www.fsc.org
FSC® C003147

Preface

This book began as an interest and became an obsession. Thanks must, therefore, go first and foremost to my partner Liz for her graceful tolerance of long hours and blank stares. A close second for gratitude is my editor Professor Peter Grzybek of the University of Graz for his unfailing efficiency, patience, and good humour in what must have been a trying process for him. Sincere thanks go also to Peter and to the other two editors of the prestigious *Quantitative Linguistics* monograph series, Professor Gabriel Altmann and Professor Reinhard Köhler, for agreeing to publish this book, to Newcastle University for granting me leave to complete it, and to my academic colleagues for being friends as well as advisors.

H.M., Summer 2014

Contents

List of figures

1 Introduction

Linguistics is a science, and should therefore use scientific methodology. The currently dominant methodology is the hypothetico-deductive one associated in the philosophy of science with Karl Popper (1959, 1963), in which scientific research is conducted in a sequence of steps:

1. Some aspect of the natural world, that is, a domain of interest, is selected for study, and a research question that will substantially further scientific knowledge of the domain is posed.
2. A hypothesis that answers the research question is stated.
3. The hypothesis is tested by observation of the domain. If it is incompatible with observation the hypothesis must either be emended to make it compatible or, if this is not possible, must be abandoned. If it is compatible then the hypothesis is said to be supported but not proven; no scientific hypothesis is ever proven because it is always open to falsification by new evidence from observation.

On this model, the science of the selected aspect of the domain of interest at any given time is a collection of hypotheses that are valid with respect to observations of the domain made up to that time, or, in other words, a collection of best guesses about what that aspect of the natural world is like.

Because falsifiable hypotheses are central in science, it is natural to ask how they are generated. The consensus in philosophy of science is that hypothesis generation is non-algorithmic, that is, not reducible to a formula, but is rather driven by human intellectual creativity in response to a research question (Chalmers 1999; Gauch 2003; Machamer and Silberstein 2007; Psillos 2007; Psillos and Curd 2008). In principle any one of us, whatever our background, could wake up in the middle of the night with an utterly novel hypothesis that, say, unifies quantum mechanics and Einsteinian relativity, but this kind of inspiration is highly unlikely and must be exceedingly rare. In practice, hypothesis generation is a matter of becoming familiar with the domain of interest by observation of it, reading the associated research literature, formulating a research question which, if convincingly answered, will enhance scientific understanding, abstracting data relevant to the research question from the domain and drawing inferences from it, and on the basis of these inferences formulating a hypothesis which interestingly answers the research question. That hypothesis is then tested using data not used in the hypothesis formulation process.

In linguistics the research domain is text, where 'text' is understood generically as output from the human language faculty. In branches of linguistics concerned with chronological, geographical, and social language variation, text takes the form of collections of spoken and / or written language, or corpora. In those concerned with the structure of the language faculty, and in generative linguistics more specifically, the role of text is less obvious but no less fundamental: native speaker grammaticality judgment has historically been used as the basis for hypothesis generation and confirmation instead of evidence derived from corpora, but such judgment is based on example sentences generated by the researcher's language faculty, not on direct access to the language faculty within the human cognitive system itself, and therefore on text in the above sense. The use of corpora in Western linguistics began in the late eighteenth century with the postulation of an Indo-European protolanguage and its reconstruction based on examination of numerous living languages and of historical written documents; after two centuries of development the importance of corpora in linguistics research has increased to the extent that a subdiscipline has come into being, corpus linguistics, whose remit is to develop methodologies for creating collections of natural language speech and text, abstracting data from them, and analysing that data with the aim of generating or testing hypotheses about the structure of language and its use in the world.

Throughout the nineteenth and much of the twentieth centuries corpus linguistics was exclusively paper-based. The collections at the root of the discipline were in the form of hand-written or printed documents, and research using such collections involved reading through the documents, often repeatedly, creating data by noting features of interest on some paper medium such as index cards, inspecting the data directly, and on the basis of that inspection drawing conclusions that were published in printed books or journals. The advent of digital electronic technology in the second half of the twentieth century and its evolution since then have increasingly rendered this traditional methodology obsolete. On the one hand, the possibility of representing language electronically rather than as visual marks on paper, together with the development of electronic media, infrastructure, and computational tools for creation, emendation, storage, and transmission of electronic text have led to a rapid increase in the number and size of corpora available to the linguist, and these are now at or beyond the limit of what an individual researcher can efficiently use in the traditional way. On the other, data abstracted from very large corpora can themselves be so extensive and complex as to be impenetrable to understanding by direct inspection. Digital electronic technology has been a boon the corpus linguistics community, but in linguistics, as in life, it's possible to have too much of a good thing.

One response to digital electronic text and data overload is to use only corpora of tractable size or, equivalently, subsets of large corpora, but simply ignoring available information is not scientifically respectable. The alternative is to look to related research disciplines for help. The overload in corpus linguistics is symptomatic of a more general trend. Daily use of digital electronic information technology by many millions of people worldwide in their professional and personal lives has generated and continues to generate truly vast amounts of electronic speech and text, and abstraction of information from all but a tiny part of it by direct inspection is an intractable task not only for individuals but also in government and commerce – what, for example, are the prospects for finding a specific item of information by reading sequentially through the huge number of documents currently available on the World Wide Web? In response, research disciplines devoted to information abstraction from very large collections of electronic text have come into being. These go under a variety of names such as informatics, information science, information retrieval, text summarization, data mining, and natural language processing. They overlap to greater or lesser degrees but share an essentially identical remit: to make interpretation of large collections of digital text tractable. To achieve this they draw on concepts and methods from a range of other disciplines including mathematics, statistics, computer science, and artificial intelligence.

An increasingly important class of these concepts and methods is cluster analysis, which is used across a broad range of sciences for hypothesis generation based on identification of structure in data which are too large or complex, or both, to be interpretable by direct inspection. The aim of the present discussion is to show how cluster analysis can be used for corpus-based hypothesis generation in linguistics by applying it to a case study of a dialect corpus, the *Diachronic Electronic Corpus of Tyneside English*, and thereby to contribute to the development of an empirically-based quantitative methodology for hypothesis generation in the linguistics community.

This aim is realized by presenting the relevant material so as to make it accessible to that community. The implication is that accessibility is a problem, and in the author's view it is, for several reasons.

- The number of available clustering methods is so large that even specifically dedicated surveys of them are selective, more are continually being proposed, and the associated technical literature is correspondingly extensive and growing. Selection of the method or methods appropriate to any given research application requires engagement with this literature, and for the non-specialist the magnitude of the task can be a substantial obstacle to informed use. The present discussion addresses this obstacle by selectivity. There is no prospect of being able to cover all or even very many of the available clustering methods at a level of description sufficient to convey understanding, and as such no attempt is made at comprehensiveness; surveys exist already, and no obvious purpose would be served by a précis of them here. What is offered instead are detailed descriptions of a relatively small selection of methods which are likely to be of most use to corpus linguists, where usefulness is judged on criteria of intuitive accessibility, theoretically and empirically demonstrated effectiveness, and availability of software implementations for practical application. These detailed descriptions are supplemented by references to variations on and alternatives to the methods in question, thereby providing the reader with pointers to the broader range of methods in which the selected ones are embedded. Needless to say, there is a subjective element in this, and a different writer might well have made different choices.
- Understanding of the nature, creation, representation, and properties of data is fundamental to successful cluster analysis. The clustering literature typically assumes familiarity with these topics, however, and consequently tends to deal with them in relatively cursory fashion; anyone lacking the requisite familiarity must acquire it by engaging with the relevant and also very extensive mathematical, statistical, and data processing literatures. To forestall the need for this, the account of the selected cluster analytical methods that follows is preceded by a detailed discussion of data issues.
- Cluster analysis and data processing are based on concepts from mathematics, statistics, and computer science, and discussions of them in the above-mentioned literatures are, in general, quite technical. This can be a serious obstacle in that, though it has become less pronounced, the arts / science divide is still with us, and many professional linguists have little or no background in and sometimes even an antipathy to mathematics and statistics; for further discussion of this assertion see

Chapter 6 below. Understanding of these concepts and associated formalisms is, however, a prerequisite for informed application of cluster analysis, and so introductory-level explanations of them are provided. No a priori knowledge of them is assumed, and all are explained before any use is made of them. They are, moreover, introduced in the course of discussion as they are needed, so that the requisite knowledge is built up gradually. This approach itself presents a problem. Some linguists are as mathematically and statistically sophisticated as any in the traditional 'hard' sciences, and to such readers intuitive explanations can seem tedious. The choice, therefore, is between puzzling some readers and probably putting them off the discussion entirely, and boring others. There is no obvious solution. In my experience more linguists need the explanations than not; every effort is made to avoid long-windedness, but where there is a choice between brevity and intuitive clarity, the latter wins every time.

The discussion is in six main chapters. The first chapter is the present Introduction. The second chapter motivates the use of of cluster analysis for hypothesis generation in linguistics. The third deals with data creation: the nature of data, its abstraction from text corpora, its representation in a mathematical format suitable for cluster analysis, and transformation of that representation so as to optimize its interpretability. The fourth chapter describes a range of cluster analysis methods and exemplifies their application to data created in Chapter 3. The fifth shows how these methods can serve as the basis for generation of linguistic hypotheses, and the sixth reviews existing applications of cluster analysis in corpus linguistics. In addition, there is an Appendix that identifies software implementations which make the clustering methods described in Chapter 4 available for practical application. Exemplification throughout the discussion is based on data abstracted from the *Diachronic Electronic Corpus of Tyneside English* .

This book is one of a series entitled *Quantitative Linguistics*, whose remit is to cover 'all aspects of quantitative methods and models in linguistics, text analysis, and related research fields', and more specifically 'the whole spectrum of theoretical and empirical research, ultimately striving for an exact mathematical formulation and empirical testing of hypotheses: observation and description of linguistic data, application of methods and models, discussion of methodological and epistemological issues, modelling of language and text phenomena'. In addressing methodological issues in linguistic hypothesis generation using mathematically-based data creation and cluster analysis methods, the author's belief is that it satisfies this remit.

2 Motivation

As noted in the Introduction, cluster analysis is a tool for hypothesis generation. It identifies structure latent in data, and awareness of such structure can be used to draw the inferences on the basis of which a hypothesis is formulated. To see how this works, let us assume that the domain of interest is a speech community and that one wants to understand the relationship between phonetic usage and social structure within it; for concreteness, that community will be assumed to be Tyneside in north-east England, shown in Figure 2.1.

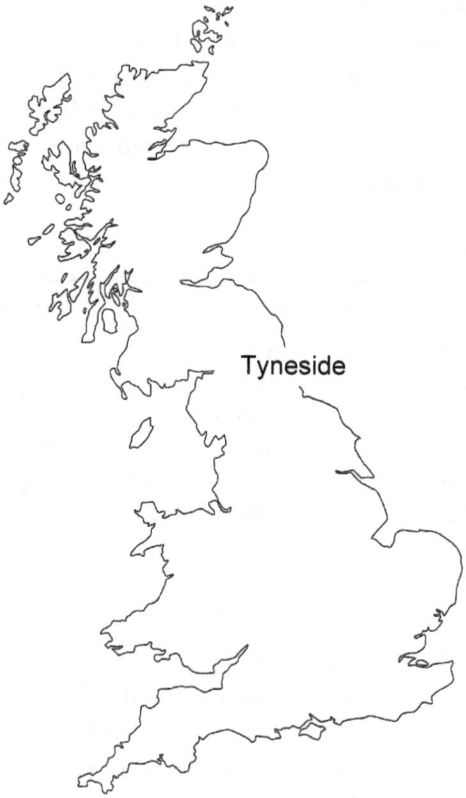

Figure 2.1: The Tyneside area of Great Britain

The research question is:

Is there systematic phonetic variation in the Tyneside speech community, and, if so, does that variation correlate systematically with social factors?

To answer the question, a representative sample of Tyneside speech is collected and relevant data is abstracted from it. The sample used here is the *Diachronic Electronic Corpus of Tyneside English* (henceforth DECTE), a collection of interviews with Tyneside English speakers that will be fully described in due course.

A group of 24 speakers is selected at random and a set of phonetic variables descriptive of Tyneside pronunciation is defined. The number of times each speaker uses the phonetic variable or variables of interest is recorded, thereby building up a body of data. To start, each speaker's speech is described by a single variable, the phonetic segment $ə_1$; the labels in the *Speaker* column of Table 2.1 designate speakers in DECTE, and the values in the variable column are the frequencies with which each of the corresponding 24 speakers uses that segment.

Table 2.1: Frequency data for $ə_1$

Speaker	$ə_1$	Speaker	$ə_1$	Speaker	$ə_1$
tlsg01	3	tlsg07	5	tlsg52	11
tlsg02	8	tlsg08	3	tlsg53	6
tlsg03	3	tlsg09	5	tlsn05	145
tlsn01	100	tlsg10	6	tlsn06	109
tlsg04	15	tlsn03	142	tlsg54	3
tlsg05	14	tlsn04	110	tlsg55	7
tlsg06	5	tlsg11	3	tlsg56	12
tlsn02	103	tlsg12	2	tlsn07	104

It is easy to see by direct inspection of the data that the speakers fall into two groups: those that use $ə_1$ relatively frequently and those that use it infrequently. Based on this result, the obvious hypothesis is that there is systematic variation in phonetic usage with respect to in the speech community. For two variables, shown in Table 2.2, direct inspection again shows two groups: those that use both $ə_1$ and $ə_2$ relatively frequently and those that do not, and the hypothesis is analogous to the one just stated.

Table 2.2: Frequency data for $ə_1$ and $ə_2$

Speaker	$ə_1$	$ə_2$	Speaker	$ə_1$	$ə_2$	Speaker	$ə_1$	$ə_2$
tlsg01	3	1	tlsg07	5	0	tlsg52	11	1
tlsg02	8	0	tlsg08	3	0	tlsg53	6	0
tlsg03	3	1	tlsg09	5	0	tlsn05	145	102
tlsn01	100	116	tlsg10	6	0	tlsn06	109	107
tlsg04	15	0	tlsn03	142	107	tlsg54	3	0
tlsg05	14	6	tlsn04	110	120	tlsg55	7	0
tlsg06	5	0	tlsg11	3	0	tlsg56	12	0
tlsn02	103	93	tlsg12	2	0	tlsn07	104	93

There is no theoretical limit on the number of variables. As the number of variables and speakers grows, so does the difficulty of generating hypotheses from direct inspection of the data. In the present case, the selection of $ə_1$ and $ə_2$ in Tables 2.1 and 2.2 was arbitrary, and the speakers could have been described using more phonetic segment variables. Table 2.3 shows twelve.

Table 2.3: Frequency data for a range of phonetic segments

Speaker	$ə_1$	$ə_2$	o:	$ə_3$	ɪ	eɪ	n	$a:_1$	$a:_2$	aɪ	r	w
tlsg01	3	1	55	101	33	26	193	64	1	8	54	96
tlsg02	8	0	11	82	31	44	205	54	64	8	83	88
tlsg03	4	1	52	109	38	25	193	60	15	3	59	101
tlsn01	100	116	5	17	75	0	179	64	0	19	46	62
tlsg04	15	0	12	75	21	23	186	57	6	12	32	97
tlsg05	14	6	45	70	49	0	188	40	0	45	72	79
tlsg06	5	0	40	70	32	22	183	46	0	2	37	117
tlsn02	103	93	7	5	87	27	241	52	0	1	19	72
tlsg07	5	0	11	58	44	31	195	87	12	4	28	93
tlsg08	3	0	44	63	31	44	140	47	0	5	43	106
tlsg09	5	0	30	103	68	10	177	35	0	33	52	96
tlsg10	6	0	89	61	20	33	177	37	0	4	63	97
tlsn03	142	107	2	15	94	0	234	15	0	25	28	118
tlsn04	110	120	0	21	100	0	237	4	0	61	21	62
tlsg11	3	0	61	55	27	19	205	88	0	4	47	94
tlsg12	2	0	9	42	43	41	213	39	31	5	68	124
tlsg52	11	1	29	75	34	22	206	46	0	29	34	93
tlsg53	6	0	49	66	41	32	177	52	9	1	68	74
tlsn05	145	102	4	6	100	0	208	51	0	22	61	104
tlsn06	109	107	0	7	111	0	220	38	0	26	19	70
tlsg54	3	0	8	81	22	27	239	30	32	8	80	116
tlsg55	7	0	12	57	37	20	187	77	41	4	58	101
tlsg56	12	0	21	59	31	40	164	52	17	6	45	103
tlsn07	104	93	0	11	108	0	194	5	0	66	33	69

What hypothesis would one formulate from inspection of the data in Table 2.3, taking into account all the variables? And what about, say, 100 speakers and 150 variables? These questions are clearly rhetorical, and there is a straightforward moral: human cognitive makeup is unsuited to seeing regularities in anything but the smallest collections of numerical data. To see the regularities we need help, and that is what cluster analysis provides.

As noted in the Introduction, cluster analysis is a family of computational methods for identification and graphical display of structure in data when the data are too large either in terms of the number of variables or of the number of objects described, or both, to be readily interpretable by direct inspection. All the members of the family work by partitioning a set of objects in the domain of interest into disjoint subsets in accordance with how relatively similar those objects are in terms of the variables that describe them. The objects of interest in Tables 2.1, 2.2, and 2.3 are speakers, and each speaker's phonetic usage is described by a set of variables. Any two speakers' usage will be more or less similar depending on how similar their respective variable values are: if the values are identical then so are the speakers in terms of their usage, and the greater the divergence in values the greater the differences. Cluster analysis of the data in Table 2.3 groups the 24 speakers in accordance with how similar their frequency of usage of 12 phonetic segments is. There are various kinds of cluster analysis; Figure 2.2 shows the result from application of one of the most frequently-used of them, hierarchical clustering, to the data.

This and other varieties of cluster analysis are described in detail later in the discussion; the aim at this stage is to give an initial impression of how they can be used in linguistic analysis. Figure 2.2 shows the cluster structure of the speaker data as a hierarchical tree. To interpret the tree correctly one has to understand how it is constructed, so a short intuitive account is given here. The labels at the leaves of the tree are the speaker-identifiers corresponding to those of the data in Table 2.3, abbreviated so as to make them more amenable to graphical display. These labels are partitioned into clusters in a sequence of steps. Initially, each speaker is interpreted as a cluster on his or her own. At the first step Table 2.3 is searched to determine which two speakers are most similar in terms of their frequency of usage of the 12 phonetic variables; in practice this involves comparison of the rows of the data table to determine which two rows are numerically the most similar, as described in due course, but for present expository purposes visual inspection of the cluster tree will suffice. When the two most similar speakers are found, they are joined into a superordinate cluster in which their degree of similarity is graphically repre-

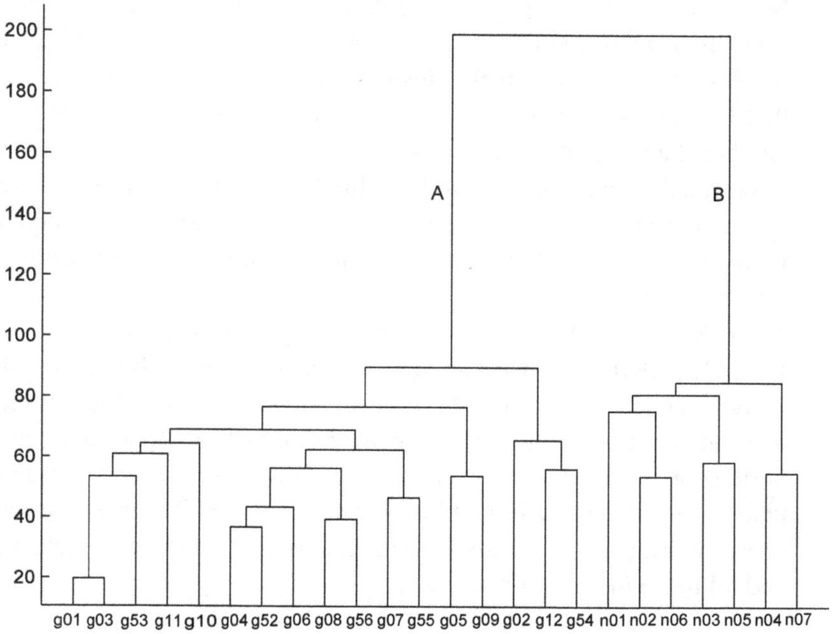

Figure 2.2: Hierarchical cluster analysis of the data in Table 2.3

sented by the length of the vertical branches joining the subclusters. To judge by the relative shortness of the branches in the tree, the singleton clusters g01 and g03 at the very left are the most similar. These are joined into a composite cluster (g01 g03). At the second step the data is searched again to determine the next-most-similar pair of clusters. Visual inspection indicates that these are g04 and g52 about a quarter of the way from the left of the tree, and these are joined into a composite cluster (g04 g52). At step 3, the next-most-similar pair (g04 g52) and g06 are joined into a composite cluster ((g04 g52) g06), and so on. The sequence of steps continues in this way, combining the most similar pair of clusters at each step, and stops when there is only one cluster remaining which contains all the subclusters.

Once the structure of the data has been identified by the above procedure it can be used for generation of a hypothesis in response to the research question.

- *Is there systematic phonetic variation in the Tyneside speech community?*

Since the relative lengths of the branches joining subclusters represents their relative similarity, the speakers included in the analysis can be seen to fall into two main clusters, labeled *A* and *B* in the tree, such that the speakers in cluster *A* are relatively much more similar to one another than any of them are to speakers in cluster *B*, and vice versa. A reasonable hypothesis based on this finding would be that there is systematic phonetic variation in the Tyneside speech community, and more specifically that the speakers who constitute that community fall into two main groups.

– *Does that variation correlate systematically with social factors?*
DECTE includes a range of social information for each speaker, such as age, gender, educational level, occupation, and so on. Also included is an indication of whether the speaker comes from Newcastle on the north shore of the river Tyne or Gateshead on the south side; correlating place of residence with the cluster tree in Figure 2.2 gives the result shown in Figure 2.3, where all the speakers in *A* come from Gateshead and all those in *B* from Newcastle.

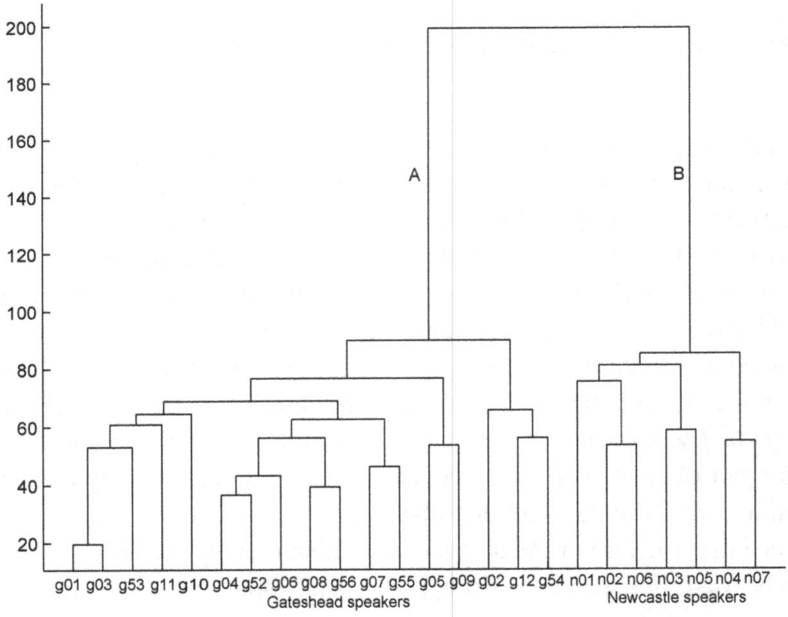

Figure 2.3: Hierarchical cluster analysis of the data in Table 2.3 with social information added

This result supports the hypothesis that there is a systematic correlation between phonetic variation and social factors, and more specifically with place of residence: phonetic variation among speakers in these two areas of Tyneside is relatively small compared to the relatively much larger difference between the areas.

This hypothesis can be refined on the one hand by correlating the internal structures of clusters *A* and *B* with a larger number of social factors, and on the other by identifying the phonetic segments which are most important as determinants of the cluster structure. The former is analogous to what has already been described and does not need to be made explicit at this stage, though subsequent discussion will do so. One approach to the latter is to create summary descriptions of the phonetic characteristics of the two main clusters *A* and *B* and then to compare them. This is done by taking the mean of variable values for the speakers in each cluster, as in Table 2.4.

All the speakers whom the cluster tree assigns to *A* are collected in the cluster *A* sub-table of Table 2.4. The mean of each column in cluster *A* is calculated and shown at the bottom of the sub-table, and the list of 12 values then represents the average phonetic characteristics of the speakers in *A*. The same is done for *B*. The means for *A* and *B* can now be compared; the bar plot in Figure 2.4 shows the result graphically.

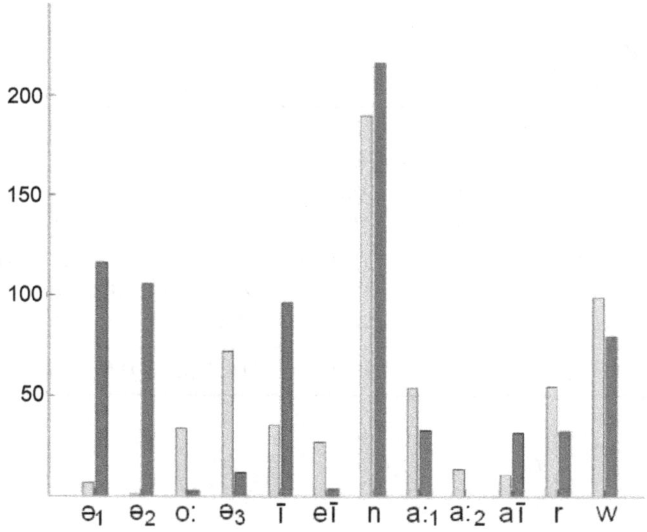

Figure 2.4: Bar-plot of mean *A* and mean *B* from Table 2.4

Table 2.4: Rows of Figure 2.3 grouped by cluster and corresponding means

Cluster A

Speaker	ə$_1$	ə$_2$	o:	ə$_3$	ī	eī	n	a:$_1$	a:$_2$	aī	r	w
tlsg01	3	1	55	101	33	26	193	64	1	8	54	96
tlsg02	8	0	11	82	31	44	205	54	64	8	83	88
tlsg03	4	1	52	109	38	25	193	60	15	3	59	101
tlsg04	15	0	12	75	21	23	186	57	6	12	32	97
tlsg05	14	6	45	70	49	0	188	40	0	45	72	79
tlsg06	5	0	40	70	32	22	183	46	0	2	37	117
tlsg07	5	0	11	58	44	31	195	87	12	4	28	93
tlsg12	2	0	9	42	43	41	213	39	31	5	68	124
tlsg55	7	0	12	57	37	20	187	77	41	4	58	101
tlsg08	3	0	44	63	31	44	140	47	0	5	43	106
tlsg09	5	0	30	103	68	10	177	35	0	33	52	96
tlsg10	6	0	89	61	20	33	177	37	0	4	63	97
tlsg11	3	0	61	55	27	19	205	88	0	4	47	94
tlsg52	11	1	29	75	34	22	206	46	0	29	34	93
tlsg53	6	0	49	66	41	32	177	52	9	1	68	74
tlsg54	3	0	8	81	22	27	239	30	32	8	80	116
tlsg55	7	0	12	57	37	20	187	77	41	4	58	101
tlsg56	12	0	21	59	31	40	164	52	17	6	45	103
Mean (*A*)	6.6	0.5	34.0	72.2	35.4	27.0	189.9	53.6	13.4	10.7	54.3	98.5

Cluster B

Speaker	ə$_1$	ə$_2$	o:	ə$_3$	ī	eī	n	a:$_1$	a:$_2$	aī	r	w
tlsn01	100	116	5	17	75	0	179	64	0	19	46	62
tlsn02	103	93	7	5	87	27	241	52	0	1	19	72
tlsn03	142	107	2	15	94	0	234	15	0	25	28	118
tlsn04	110	120	0	21	100	0	237	4	0	61	21	62
tlsn05	145	102	4	6	100	0	208	51	0	22	61	104
tlsn06	109	107	0	7	111	0	220	38	0	26	19	70
tlsn07	104	93	0	11	108	0	194	5	0	66	33	69
Mean (*B*)	116.1	105.4	2.6	11.7	96.4	3.9	216.1	32.7	0.0	31.4	32.4	79.6

The relative degrees of disparity in phonetic usage are shown by the differences in the heights of the bars representing *A* and *B*; clearly, the two varieties of ə are the most important differentiators between clusters.

Cluster analysis can be applied to hypothesis generation in any research where the data consists of objects described by variables; since most research uses data of this kind, it is very widely applicable. It can *usefully* be applied where the number of objects and variables is so large that the data cannot easily be interpreted by direct inspection, as above. The foregoing discussion has sketched one sort of application to linguistic analysis; a few other random possibilities are, briefly:

- A historical linguist might want to infer phonetic or phonological structure in a legacy corpus on the basis of spelling by cluster analyzing alphabetic *n*-grams for different magnitudes 2, 3, 4 ... of *n*.
- A generative linguist might want to infer syntactic structures in a little-known or endangered language by clustering lexical *n*-grams for different magnitudes of *n*.
- A philologist might want to use cluster analysis of alphabetic *n*-grams to see if a collection of historical literary texts can be classified chronologically of geographically on the basis of their spelling.

Further applications to linguistic analysis are given in Chapter 6, the literature review.

3 Data

'Data' is the plural of 'datum', the past participle of Latin 'dare', *to give*, and means *things that are given*. A datum is therefore something to be accepted at face value, a true statement about the world. What is a true statement about the world? That question has been debated in philosophical metaphysics since Antiquity and probably before, and, in our own time, has been intensively studied by the disciplines that comprise cognitive science (Audi 2010). The issues are complex, controversy abounds, and the associated academic literatures are vast – saying what a true statement about the world might be is anything but straightforward. We can't go into all this, and so will adopt the attitude prevalent in most areas of science: data are abstractions of what we perceive using our senses, often with the aid of instruments.

Data are ontologically different from the world. The world is as it is; data are an interpretation of it for the purpose of scientific study. The weather is not the meteorologist's data – measurements of such things as air temperature are. A text corpus is not the linguist's data – measurements of such things as lexical frequency are. Data are constructed from observation of things in the world, and the process of construction raises a range of issues that determine the amenability of the data to analysis and the interpretability of the analytical results. The importance to cluster analysis of understanding such data issues can hardly be overstated (Jain 2010). On the one hand, nothing can be discovered that is beyond the limits of what the data say about the world. On the other, failure to understand and where necessary to emend relevant characteristics of data can lead to results and interpretations that are distorted or even worthless. For these reasons, a detailed account of data issues is given before moving on to discussion of cluster analytical methods. For general discussions of data see (Izenman 2008; Tan, Steinbach, and Kumar 2006).

The discussion is in three main sections. The first section deals with data creation, the second presents a geometrical interpretation of data on which all subsequent discussion is based, and the third describes several ways of transforming data prior to cluster analysis in terms of that geometrical interpretation.

3.1 Data creation

3.1.1 Research domain

To state the obvious, data creation presupposes a research domain from which the data are to be abstracted. In corpus-based linguistics the research domain is some collection of natural language utterances. In the present case the domain is the English spoken in the Tyneside region of north-east England (Beal, Burbano-Elzondo, and Llamas 2012; Wales 2006), and the sample from that domain is the *Diachronic Electronic Corpus of Tyneside English* (Corrigan, Mearns, and Moisl 2012), henceforth referred to as DECTE. This section first briefly describes the corpus and then gives a detailed account of the DECTE phonetic transcriptions on which the cluster analyses in the remainder of the discussion are based.

3.1.1.1 The DECTE corpus

DECTE contains samples of the Tyneside English speech variety dating from the later 20th and the early 21st centuries collected from residents of Tyneside and surrounding areas of North-East England. The main locations in this area represented in the corpus are the cities of Newcastle upon Tyne on the north side of the river Tyne and Gatesheadon the south side, but its geographical reach is currently being extended to include speakers from other regions in the North-East such as County Durham, Northumberland, and Sunderland. It updates the existing *Newcastle Electronic Corpus of Tyneside English* (NECTE), which was created between 2000 and 2005 and consists of two pre-existing corpora of audio-recorded Tyneside speech (Allen et al. 2007).

The earlier of the two components of NECTE, the *Tyneside Linguistic Survey* (TLS), was created in the late 1960s (Jones-Sargent 1983; Pellowe and Jones 1978; Pellowe et al. 1972; Strang 1968), and consists of audio-taped interviews of about 30 minutes' duration with Tyneside speakers who were encouraged to talk freely about their lives but were also asked for judgements on various linguistic features and constructions. These interviews were then orthographically transcribed in their entirety and the first ten minutes or so of each interview phonetically transcribed. Detailed social data for each speaker were also recorded. The TLS project was never satisfactorily concluded, and the materials it produced were archived and largely forgotten until Joan Beal and Karen Corrigan of Newcastle University undertook to recover them and make them available to the corpus linguistics community, which led to their incorporation into the NECTE corpus. It is presently unclear how many interviews were conducted and associated transcriptions produced by the TLS project. The NECTE project was able to identify components relating to 114 interviews, of which only 37 are complete sets consisting of audio interview, orthographic and phonetic transcription, and social data.

The other component of NECTE is the *Phonological Variation and Change in Contemporary Spoken English* (PVC) project (Docherty and Foulkes 1999; Milroy et al. 1999; Watt and Milroy 1999) was collected between 1991 and 1994, and, as its name indicates, it investigated patterns of phonological variation and change in Tyneside English. The core of its materials consists of 18 digital audio-taped interviews of up to one hour's duration with self-selected dyads of friends or relatives, matched in terms of age and social class, who had freedom to converse on a wide range of subjects with minimal interference from the fieldworker. Only selective phonetic transcriptions of lexical items of interest were produced, and records of social data were limited to the gender, age and broadly defined socio-economic class of the participants.

NECTE amalgamated the TLS and PVC materials into a single Text Encoding Initiative (TEI)-conformant XML-encoded corpus and made them available online in a variety of aligned formats: digitized audio, standard orthographic transcription, phonetic transcription, and part-of-speech tagged.

In 2011–12 the DECTE project combined NECTE with the NECTE2 corpus, which was begun in 2007 and is ongoing. NECTE2 consists of digitized audio recordings and orthographic transcriptions of dyadic interviews together with records of informant social details and other supplementary material, collected by undergraduate and postgraduate students and researchers at Newcastle University. The interviews record the language use of a variety of local informants from a range of social groups and extend the geographical domain covered in the earlier collections to include other parts of the North East of England. Successive cohorts of students add their own interviews to NECTE2.

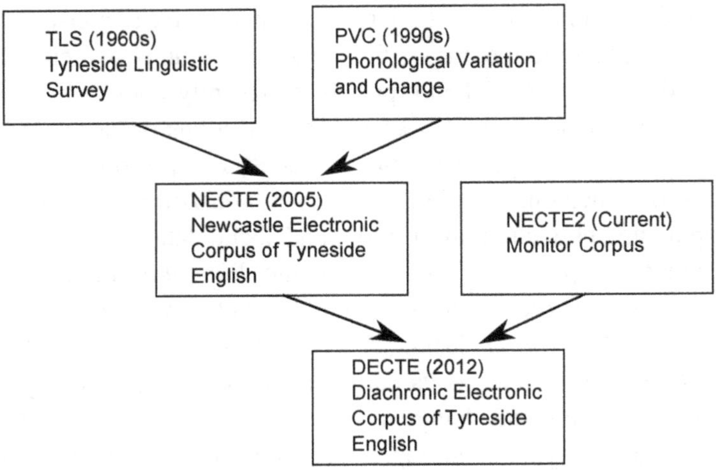

Figure 3.1: Structure of the DECTE corpus

The components of DECTE and their interrelationship are shown schematically in Figure 3.1. The resulting DECTE corpus imposes a uniform structure on its components and is conformant with the current P5 Text Encoding Initiative guidelines.

3.1.1.2 The DECTE phonetic transcriptions

The main motivation for the TLS project was to see whether systematic phonetic variation among Tyneside speakers of the period could be interestingly correlated with variation in their social characteristics. To this end the project developed a methodology that was radical at the time and remains so today: in contrast to the then-universal and still-dominant theory driven approach, where social and linguistic factors are selected by the analyst on the basis of some combination of an independently-specified theoretical framework, existing case studies, and personal experience of the domain of inquiry, the TLS proposed a fundamentally empirical approach in which salient factors are extracted from the data itself and then serve as the basis for model construction. To realize this research aim using its empirical methodology, the TLS had to compare the audio interviews it had collected at the phonetic level of representation. This required the analog speech signal to be discretized into phonetic segment sequences, or, in other words, to be phonetically transcribed.

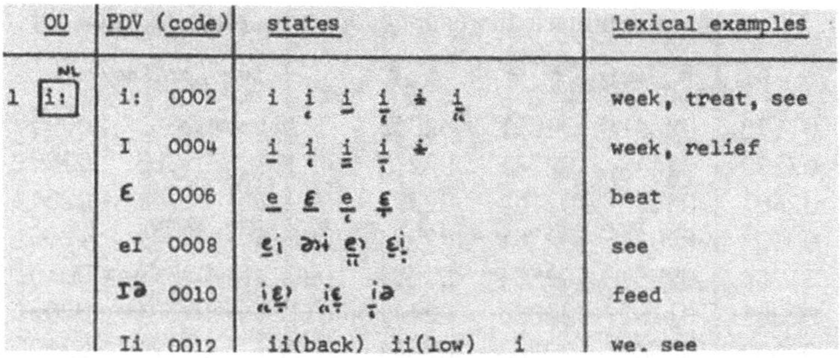

OU	PDV (code)	states	lexical examples
1 i:	i: 0002	i i i i i i	week, treat, see
	I 0004	i i i i i	week, relief
	ɛ 0006	e ɛ e ɛ	beat
	eI 0008	ɛi əi ɛi ɛi	see
	Iə 0010	iɛ iɛ iə	feed
	Ii 0012	ii(back) ii(low) i	we, see

Figure 3.2: Sample page from the Tyneside Linguistic Survey's phonetic transcription scheme

Phonetic transcriptions for 63 speakers survive and have been incorporated into DECTE; these are the basis for the case study developed in this book.

The TLS project wanted a very detailed phonetic transcription of its audio files, and the standard International Phonetic Alphabet scheme was not detailed enough. It therefore developed an extended version of the IPA scheme; the sample from the TLS encoding manual in Figure 3.2 shows what this was like.

There are four columns in the encoding table. The first three give symbols for increasingly fine-grained transcription, and the fourth examples of what speech sounds the symbols represent: the OU column lists phonological segments, the PDV ('Putative Diasystemic Variable') column lists IPA-level phonetic segments, and the State column lists the TLS's detailed elaboration of the IPA scheme.

The graphical symbols had to be numerically encoded for computational processing. For this encoding, each PDV symbol was assigned a unique four-digit code. A fifth digit was then added to any given PDV code to give a five-digit State code. This fifth digit was the State symbol's position in the left-to-right sequence in the State column. For example, the PDV code for [iː] is 0002; the State code for, say, [i̠] is 00024, because [i̠] is fourth in the left-to-right State symbol sequence. Using this encoding scheme, a PDV-level transcription of any given speaker interview audio file is a sequence of four-digit codes, and a State-level transcription is a sequence of five-digit codes. Reference to the DECTE transcriptions will henceforth be to the PDV level.

Associated with each of the 63 DECTE speakers is a file representing a transcription of about the first 10 minutes of the audio interview and containing a sequence of numerical State-level codes of which only the first four

Table 3.1: Fragment of a numerically encoded DECTE phonetic transcription

```
 <u who="informantTlsg01">
02441  01123  02301  02621  02363  02741  02881  00906  02081
02301  02322  01443  02741  02201  01284  02383  02801  00421
02421  02501  00342  02164  02721  02021  02741  02642  04321
02621  00503  02825  02301  02721  00246  02341  12601  02642
02541  01284  02561  02881  01641  02941  02781  00161  02561
02363  02301  01181  02825  02441  01123  02301  02621  02365
02721  00903  02561  02363  02541  02861  02721  02605  01822
02263  00906  00241  02825  02621  02083  02421  02621  02263
02861  00023  02301  02442  01041  02301  02621  02364  02606
00343  02301  02621  02621  00823  02741  02041  02741  02363
02861  01164  02364  02541  02861  00501  02721  02605  01822
02263  02781  00501  02221  02801  02123  02701  00144  02822
02741  00504  02301  00161  02621  02701  01424  02621  00823
02606  01942  02704  01443  02621  02383  02861  00504  02621
...    ... etc.
 </u>
```

PDV-level digits are used in what follows. A fragment of one of these is shown in Table 3.1, with XML markup.

The 63 phonetic transcription files and the social data associated with them constitute the corpus from which the data used in the remainder of this discussion were abstracted. For ease of cross-reference, abbreviated versions of the DECTE naming conventions for speakers together with the phonetic symbols and corresponding numerical codes are used throughout.

A final note: earlier work (Moisl and Maguire 2008; Moisl, Maguire, and Allen 2006) on the TLS phonetic transcriptions refers to 64 of them, but in the course of writing this book one of these, *tlsg57*, was found to be a doublet of *tlsg50* and has been omitted. This needs to be kept in mind when referring to the earlier work.

3.1.2 Research question

Any aspect of the world can be described in an arbitrary number of ways and to arbitrary degrees of precision. A desktop computer can, for example, be described in terms of its role in an the administrative structure of an organization, its physical appearance, its hardware components, the functionality of the software installed on it, the programs which implement that functionality, the design of the chips on the circuit board, or the atomic and subatomic characteristics of the transistors on the chips, not to speak of its connectivity to the internet or its social and economic impact on the world at large. Which description is best? That depends on why one wants the description. A software developer wants a clear definition of the required functionality but doesn't care about the details of chip design; the chip designer doesn't care about the physical appearance of the machines in which her devices are installed but a marketing manager does; an academic interested in the sociology of computers doesn't care about chip design either, or about circuit boards, or about programs. In general, how one describes a thing depends on what one wants to know about it, or, in other words, on the question one has asked.

The implications of this go straight to the heart of the debate on the nature of science and scientific theories in philosophy of science (Chalmers 1999; Gauch 2003; Machamer and Silberstein 2007; Psillos 2007; Psillos and Curd 2008), but to avoid being drawn into that debate this discussion adopts the position that is pretty much standard in scientific practice: the view that there is no theory-free observation of the world. In essence, this means that there is no such thing as objective observation in science; entities in a domain of inquiry only become relevant to observation in terms of a research question framed using the axioms and ontology of a theory about the domain. For example, in linguistic analysis variables are selected in terms of the discipline of linguistics broadly defined, which includes the division into subdisciplines such as sociolinguistics and dialectology, the subcategorization within subdisciplines such as phonetics through syntax to semantics and pragmatics in formal grammar, and theoretical entities within each subcategory such as constituency structures and movement in syntax. Claims, occasionally seen, that the variables used to describe a corpus are 'theoretically neutral' are naive: even word categories like 'noun' and 'verb' are interpretative constructs that imply a certain view of how language works, and they only appear to be theory-neutral because of familiarity with long-established tradition.

In a scientific context, the question one has asked is the research question component of the hypothetico-deductive model outlined earlier. Given a domain of interest, how is a good research question formulated? That, of course, is the central question in science. Asking the right questions is what leads to scientific breakthroughs and makes reputations, and, beyond a thorough knowledge of the research area and possession of a creative intelligence, there is no known guaranteed route to the right questions. What is clear from the preceding paragraph, though, is that a well-defined question is the key precondition to the conduct of research, and more particularly to the creation of the data that will support hypothesis formulation. The research question provides an interpretative orientation; without such an orientation, how does one know what to observe in the domain, what is important, and what is not? A linguist's domain is natural language, but syntacticians want to know different things about it than semanticists, and they ask commensurately different questions. In the present case we will be interested in sociophonetics, and the research question is the one stated earlier:

> *Is there systematic phonetic variation in the Tyneside speech community, and, if so, does that variation correlate systematically with social variables?*

3.1.3 Variable selection

Given that data are an interpretation of some domain of interest, what does such an interpretation look like? It is a description of objects in the domain in terms of variables. A variable is a symbol, that is, a physical entity to which a meaning is assigned by human interpreters; the physical shape A in the English spelling system means the phoneme /a/, for example, because all users of the system agree that it does. The variables chosen to describe a domain constitute the conceptual template in terms of which the domain is interpreted and on which the proposed analysis is based. If the analysis is to be valid with respect to the domain, therefore, it is crucial that the set of selected variables be adequate in relation to the research question, where adequacy is understood as follows:

- The variables should represent all and only those aspects of the domain which are relevant to the research question, that is, relevant aspects of the domain should not be unrepresented in the set of variables, and irrelevant aspects should not be represented. Failure to include relevant aspects in the data renders the description of the domain incomplete and thereby self-evidently compromises the validity of analysis based on it; inclusion of irrelevant aspects is less serious but introduces potentially confounding factors into an analysis.
- Each variable should be independent of all the others in terms of what it represents in the domain, that is, the variables should not overlap

with one another in what they describe in the domain because such overlap describes the same thing multiple times and can thereby skew the analysis by overemphasizing the importance of some aspects of the domain over others.

In general, adequacy so defined cannot be guaranteed in any given research application because neither relevance nor independence is always obvious. Any domain can be described by an essentially arbitrary number of finite sets of variables, as the foregoing example of computer description makes clear. Selection of one particular set can only be done on the basis of personal knowledge of the domain and of the body of scientific theory associated with it, tempered by personal discretion. In other words, there is no algorithm for choosing an adequate set of variables.

The research question defined on DECTE involves phonetic analysis, and that implies phonetic transcription of the audio speaker interviews: a set of variables is defined each of which represents a characteristic of the speech signal taken to be phonetically significant, and these are then used to interpret the continuous signal as a sequence of discrete symbols. The standard way of doing this is to use the symbols defined by the International Phonetic Alphabet (IPA), but the TLS researchers felt that the IPA was too restrictive in the sense that it did not capture phonetic features which they considered to be of interest, and so they invented their own transcription scheme, described earlier. The remainder of this discussion refers to data abstracted from these TLS transcriptions, but it has to be understood that the 156 variables in that scheme are not necessarily optimal or even adequate relative to our research question. They only constitute one view of what is important in the phonetics of Tyneside speech. In fact, as we shall see, many of them have no particular relevance to the research question.

3.1.4 Variable value assignment

Once variables have been selected, a value is assigned to each of them for each of the objects of interest in the domain. This value assignment is what makes the link between the researcher's conceptualization of the domain in terms of the variables s/he has chosen and the actual state of the world, and allows the resulting data to be taken as a valid representation of the domain. The type of value assigned to any given variable depends on its meaning. The fundamental distinction of types is between quantitative, that is, numerical values, and qualitative ones such as binary 'yes / no' or categorial 'poor / adequate / good / excellent' (Gan, Ma, and Wu 2007; Jain and Dubes 1988; Jain, Murty, and Flynn 1999; Kaufman and Rousseeuw 1990; Xu and Wunsch 2009). This discussion concentrates on quantitative variables because the vast majority of clustering applications are defined relative to them, but the literature cited in the course of discussion also describes provisions for clustering of qualitative variables.

The objects of interest in DECTE are the 63 speakers, each of whom is described by the values for each of the 156 phonetic variables. What kind of value should be assigned? One possibility is to use qualitative binary ones: the value of any given variable is 'yes' if the speaker in question uses the corresponding phonetic segment in his or her interview, and 'no' if not. Another, and the one adopted here, is to use quantitative values which represent the number of times the speaker uses each of the phonetic segments.

3.1.5 Data representation

If they are to be analyzed using mathematically-based computational methods like cluster analysis, the descriptions of the entities in the domain of interest in terms of the selected variables must be mathematically represented. A widely used way of doing this in, for example, information retrieval (Manning, Raghavan, and Schütze 2008; Manning and Schütze 1999), and the one used throughout the remainder of the discussion, is based on the concept of the vector. A vector is a sequence of n numbers each of which is indexed by its position in the sequence. Table 3.2 shows $n = 6$ real-valued numbers, where the first number v_1 is 2.1, the second v_2 is 5.1, and so on.

Table 3.2: A vector

$v =$	2.5	5.1	9.3	0.2	1.5	6.8
	1	2	3	4	5	6

Vectors are a standard data structure in computer science and are extensively used in numerical computation. Applied to the DECTE data, each speaker profile can be represented as a 156-component vector in which every component represents a different PDV variable, and the value in any given vector component is the frequency with which the speaker uses the associated PDV variable. Table 3.3 shows such a vector, including phonetic segment symbols and the corresponding PDV codes.

Table 3.3: Vector representation of phonetic profile for a single DECTE speaker

Column number	1	2	3	...	156
Phonetic segment symbol	g	i	t	...	ɜː
Phonetic segment code	0244	0112	0230	...	0184
Phonetic profile for tlsg01	31	28	123	...	0

Speaker *tlsg01* uses phonetic segment 0244 31 times, 0112 28 times, and so on. The set of 63 DECTE speaker vectors is assembled into a matrix MDECTEin which the rows i (for $i = 1..n$, where n is the number of speakers) represent the speakers, the columns j (for $j = 1..156$) represent the PDV variables, and the value at $MDECTE_{ij}$ is the number of times speaker i uses the phonetic segment j. A fragment of this 63×156 matrix is shown in Table 3.4. MDECTEis the basis for all subsequent DECTE-based examples.

Table 3.4: The matrix MDECTE of the 63 DECTE phonetic profile vectors

Column number	1	2	3	...	156
Phonetic segment symbol	g	i	t	...	ɜː
Phonetic segment code	0244	0112	0230	...	0184
1. tlsg01	31	28	123	...	0
2. tlsg02	22	8	124	...	0
...	...	...	...	...	...
63. tlsn07	19	3	73	...	0

3.1.6 Data validation

The most basic characteristic of data is that they be complete and accurate, where ‘complete’ means that all variables for all cases in the data have values associated with them, and ‘accurate’ that all values assigned to variables faithfully reflect the reality they represent. These are stringent requirements: most datasets large enough to have cluster analysis usefully applied to them probably contain error, known as ‘noise’, to greater or lesser degrees. Measurement error arises in numerous ways – tolerances in measuring instruments, human inaccuracy in the use of the instruments, corruption at one or more points in data transmission, and so on.

Because error in data distorts analytical results, it has to be eliminated as much as possible. This is a two-step process, the first step of which is to determine the amount and nature of the error, and the second to mitigate or remove it. Methods for error identification and correction are standardly discussed in statistics and multivariate analysis textbooks, for example by Tabachnick and Fidell (2007) and Hair et al. (2010), including those on cluster analysis cited in this and the subsequent chapter.

The DECTE data are generated by counting the frequency of phonetic segments in interviews, so completeness and accuracy should not be issues if the survey is carefully done using a reliable procedure; manual counting of features in physical text by direct observation is in general far less accurate than the software equivalent for electronic text.

3.2 Data geometry

Data matrices have a geometrical interpretation, and the remainder of the discussion is based on it. This section first presents a few relevant mathematical and geometrical concepts and then shows how data can be represented and interpreted in terms of them.

3.2.1 Space

In colloquial usage, the word ‘space’ denotes a fundamental aspect of how humans understand their world: that we live our lives in a three-dimensional space, that there are directions in that space, that distances along those directions can be measured, that relative distances between and among objects in the space can be compared, that objects in the space themselves have size and shape which can be measured and described. The earliest geometries were attempts to define these intuitive notions of space, direction, distance, size, and shape in terms of abstract principles which could, on the one hand, be applied to scientific understanding of physical reality, and on the other to practical problems like construction and navigation. Basing their ideas on the first attempts in ancient Mesopotamia and Egypt, Greek philosophers from the sixth century BCE onwards developed such abstract principles systematically, and their work culminated in the geometrical system attributed to Euclid (floruit ca. 300 BCE), which remained the standard for more than two millennia thereafter (Tabak 2004).

In the nineteenth century CE the validity of Euclidean geometry was questioned for the first time both intrinsically and as a description of physical reality. It was realized that the Euclidean was not the only possible geometry, and alternative ones were proposed in which, for example, there are no parallel lines and the angles inside a triangle always sum to less than 180 degrees. Since the nineteenth century these alternative geometries have continued to be developed without reference to their utility as descriptions of physical reality, and as part of this development 'space' has come to have an entirely abstract meaning which has nothing obvious to do with the one rooted in our intuitions about physical reality. A space under this construal is a mathematical set on which one or more mathematical structures are defined, and is thus a mathematical object rather than a humanly-perceived physical phenomenon [Lee 2010]. The present discussion uses 'space' in the abstract sense; the physical meaning is often useful as a metaphor for conceptualizing the abstract one, though it can easily lead one astray, as we shall see.

3.2.1.1 Cartesian product

Given two sets A and B, the Cartesian product (Whitehead and Towers 2002) of A and B is the set of all possible unique ordered pairings of members of A with members of B, that is, $A \times B = \{(a,b) | a \in A \text{ and } b \in B\}$, where

1. $\times$ denotes multiplication
2. $|$ is read as 'such that'
3. $\in$ is read as 'belongs to'
4. the paired brackets $\{\ldots\}$ denote a set

The expression $A \times B = \{(ab) | a \in A \text{ and } b \in B\}$ therefore reads: 'The Cartesian product of set A and set B is the set of all pairs (ab) such that a belongs to set A and b belongs to set B'. If, for example, $A = \{vw\}$ and $B = \{xyz\}$, then $A \times B = \{(vx)(vy)(vz)(wx)(wy)(wz)\}$. The Cartesian product of three sets $A \times B \times C$ is the set of all possible ordered triples of members of A, B, and C, that is, $A \times B \times C = \{(abc) | a \in A \text{ and } b \in B \text{ and } c \in C\}$, the Cartesian product $A \times B \times C \times D$ is the set of all possible quadruples, and so on for any number n of sets. Note that these n-tuples are ordered: $A \times B$ generates all possible pairs $(a \in A, b \in B)$, and $B \times A$ all possible pairs $(b \in B, a \in A)$. The sets multiplied by Cartesian product need not be different; the same set can be multiplied by itself any number of times. N-fold multiplication of A, for example, generates the set of all possible unique n-tuples of the the components of A; $A \times A$ generates the set of pairs $\{(aa)(ab)(ba)(bb)\}$, $A \times A \times A$ generates the set of triples $\{(aaa)(aab)(aba)(abb)(baa)(bab)(bba)(bbb)\}$, and so on.

3.2.1.2 Vector space

If the mathematical structures of addition and scalar multiplication, that is, multiplication by a single number, are defined on the n-tuples of a Cartesian product X, then X together with these two structures is a vector space V subject to a range of conditions which are for present purposes assumed to apply (Lay 2010). The n-tuples in a vector space are referred to as vectors, and n is the dimension both of the vectors and of the space itself.

Given an n-dimensional vector space V, n vectors can be selected from the space to constitute a basis for it, and the set of these n vectors is so called because all other vectors in V can be generated from it using the operations of addition and scalar multiplication, as described below. Selection of basis vectors is constrained by certain conditions explained in any and every linear algebra textbook, but understanding of these constraints is unnecessary for present purposes because we shall be using orthogonal bases, and such bases automatically satisfy the conditions. Orthogonal basis vectors have the property that their inner product is zero; the inner product of n-dimensional vectors v and w, also called the dot product and written $v.w$, is defined in Equation (3.1).

$$v.w = v_1w_1 + v_2w_2 + \ldots + v_nw_n \tag{3.1}$$

that is, corresponding components of v and w are multiplied and all the products summed. For $n = 2$, the inner product of, say, $v = [2.2, 3.5]$ and $w = [1.9, 6.0]$ is $(2.2 \times 1.9) + (3.5 \times 6.0) = 25.18$, and so v and w are not orthogonal, but the inner product of $v = [12.89, 0]$ and $w = [0, 3.8]$ is 0, and in this case they are.

For orthogonal basis vectors $v_1, v_2, \ldots v_n$ and scalars $s_1, s_2, \ldots s_n$, a linear combination of the $v_1 \ldots v_n$ generates a new vector x of the same dimensionality, as given in Equation (3.2):

$$x = [(s_1v_1) + (s_2v_2) + \ldots + (s_nv_n)] \tag{3.2}$$

where, in multiplication of a vector by a scalar, each component of the vector is multiplied by the scalar, and in vector addition corresponding components are added. For example, take V to be based on a twofold Cartesian product $A = R \times R$ of the set of real numbers $\mathbb{R}$. Select any two orthogonal vectors from V, say $v_1 = [12.89, 0]$ and $v_2 = [0, 3.8]$, adopting the convention that vectors are shown between square brackets and components are comma-separated. Table 3.5 then shows some linear combinations of v_1 and v_2, using randomly selected s-values.

Table 3.5: Examples of linear combinations

$s_1 = 5.0, s_2 = 2.2$	x = [(5.0 x [12.89, 0]) + (2.2 x [0, 3.8])] = [64.45, 8.36]
$s_1 = 1.2, s_2 = 3.4$	x = [(1.2 x [12.89, 0]) + (3.4 x [0, 3.8])] = [15.47, 12.92]
$s_1 = 1.3, s_2 = 0.9$	x = [(1.3 x [12.89, 0]) + (0.9 x [0, 3.8])] = [16.76, 3.42]
... etc	... etc

It should be clear from Table 3.5 that every different combination of scalars s_1 and s_2 results in a different vector generated from v_1 and v_2, and, because there is no constraint on the choice of scalars, there is correspondingly no constraint on the number of vectors than can be generated in this way.

Vector spaces have a geometrical interpretation. Under this interpretation, orthogonal vectors are perpendicular to one another, as in Figure 3.3.

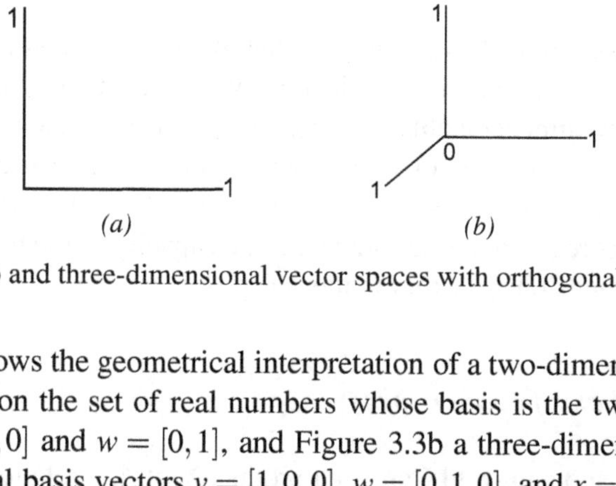

Figure 3.3: Two and three-dimensional vector spaces with orthogonal bases

Figure 3.3a shows the geometrical interpretation of a two-dimensional vector space defined on the set of real numbers whose basis is the two orthogonal vectors $v = [1, 0]$ and $w = [0, 1]$, and Figure 3.3b a three-dimensional space with orthogonal basis vectors $v = [1, 0, 0]$, $w = [0, 1, 0]$, and $x = [0, 0, 1]$. This idea extends to any dimensionality n. For $n = 4$ the analogy between mathematical and physical space breaks down in that four and higher dimensional spaces can neither be conceptualized nor represented in terms of physical space as in Figure 3.3, except in the special case where the fourth dimension is time and the physical representation can be animated to show the temporal evolution of the space. Mathematically and geometrically, however, higher-dimensional spaces are defined in the same way as the foregoing lower-dimensional ones; for $n = 2$ and $n = 3$ the basis vectors are the familiar Cartesian coordinates.

Vectors generated by linear combination of the basis vectors of an n-dimensional space are conceptualized as coordinates of points in the space. Thus, for the linear combination of the basis vectors $v = [1,0]$ and $w = [0,1]$ with randomly selected scalars $s_1 = 0.9$ and $s_2 = 0.8$, the resulting vector is a point at coordinates $[0.9, 0.8]$, as in Figure 3.4. This idea extends, again, to any dimensionality.

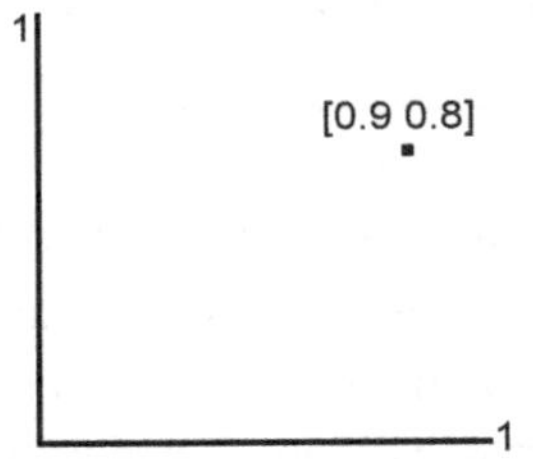

Figure 3.4: Linear combination of orthogonal basis vectors $v = [1,0]$ and $w = [0,1]$ with scalars $s_1 = 0.9$ and $s_2 = 0.8$ as coordinates in a two-dimensional vector space: $x = (0.9 \times [1,0]) + (0.8 \times [0,1]) = [0.9, 0.8]$

Finally, note in the above examples that the non-zero components of basis vectors can be any real number. It is, however, convenient to have a standard basis for every dimensionality $n = 1, 2, 3 \ldots$. These standard bases, called orthonormal bases, consist of vectors whose values are restricted to 0 and 1, so that the basis for $n = 2$ is $(1,0), (0,1)$, for $n = 3$ $(1,0,0), (0,1,0), (0,0,1)$ and so on. This restriction does not affect the capacity of the bases to generate all other n-dimensional vectors in the space.

3.2.1.3 *Manifolds in vector space*

Given a set A and an n-fold Cartesian product X of A, a relation is a subset of X. The subset may be random or made on the basis of some explicit criterion. In the latter case, for example, if A is the set of all people in some city, then $X = A \times A$ is the set of all possible pairings of people in the city. If one now defines a selection criterion, say 'loves', then the subset of pairs which satisfy the criterion constitute the relation: it is the set of all pairs of people one of whom loves the other.

In a vector space, a relation defined on an n-fold Cartesian product is a subset of vectors in the space. Geometrically, such a subset is a manifold (Lee 2010) whose constituent points define a shape in the space. The data in Figure 3.5 shows a collection of two-dimensional vectors and Figure 3.5 illustrates the corresponding locations of those vectors in the two-dimensional space.

$v_1 = [2,5]$	$v_2 = [3,5]$
$v_3 = [4,5]$	$v_4 = [5,5]$
$v_5 = [2,4]$	$v_6 = [3,4]$
$v_7 = [4,4]$	$v_8 = [5,4]$
$v_9 = [2,3]$	$v_{10} = [3,3]$
$v_{11} = [4,3]$	$v_{12} = [5,3]$
$v_{13} = [2,2]$	$v_{14} = [3,2]$
$v_{15} = [4,2]$	$v_{16} = [5,2]$

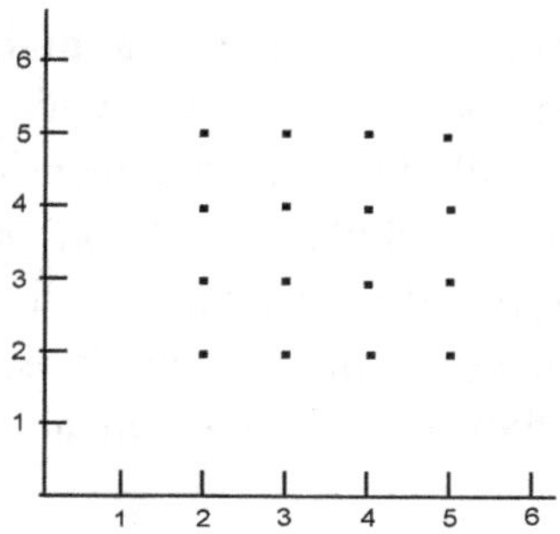

Figure 3.5: A manifold in two-dimensional space

The shape here is a square plane, but many other shapes are possible; Figure 3.6 gives a few examples.

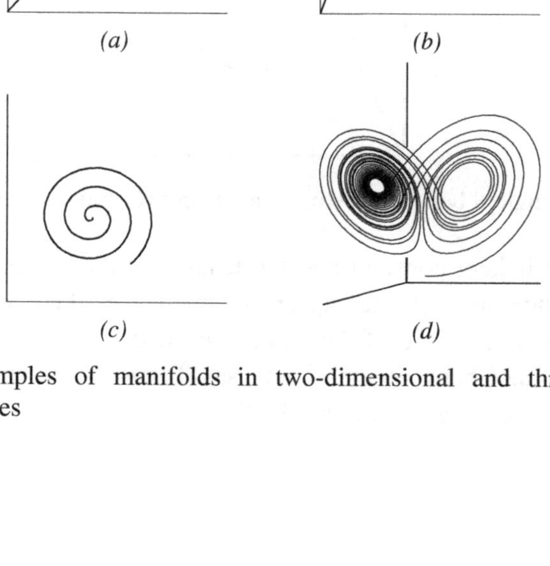

Figure 3.6: Examples of manifolds in two-dimensional and three-dimensional spaces

Figures 3.5 and 3.6a–3.6d exemplify a fundamental distinction between two types of manifold shape, linear and nonlinear, which will be important in subsequent discussion because it reflects a corresponding distinction in the characteristics of natural processes and the data that describe them. As their names indicate, linear manifolds are straight lines, as in Figure 3.6a, and planes, as in 3.5, and nonlinear ones are curved lines and surfaces, as in 3.6b–3.6d. In the nonlinear case the complexity of curvature can range from simple curved lines and planes to highly convoluted fractal shapes.

What has been said about manifolds in two-dimensional space applies straightforwardly to arbitrary dimensionality n; for $n > 3$ lines are referred to as hypercurves and planes and nonlinear surfaces as hyperplanes and hypersurfaces. Hyper-objects cannot be directly visualized or even conceptualized except by analogy with two and three dimensionalshapes, but as mathematical objects they are unproblematical.

3.2.1.4 Proximity in vector space

The geometrical proximity of two vectors v and w in a vector space V is determined by a combination of the size of the angle between the lines joining them to the origin of the space's basis, and by the lengths of those lines. Assume that v and w have identical lengths and are separated by an angle θ, as in Figure 3.7.

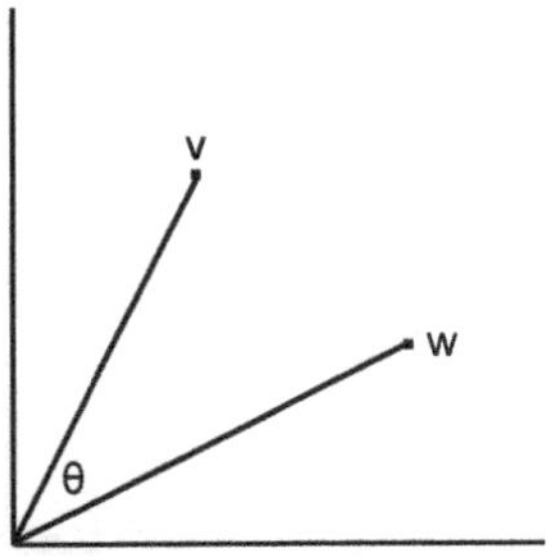

Figure 3.7: An angle θ between vectors v and w in two-dimensional space

If the angle is kept constant and the lengths of the vectors are made unequal by lengthening or shortening one of them, then the distance increases, as in Figures 3.8a and 3.8b; if the lengths are kept equal but the angle is increased the distance between them increases (Figure 3.8c), and if the angle is decreased so is the distance (Figure 3.8d).

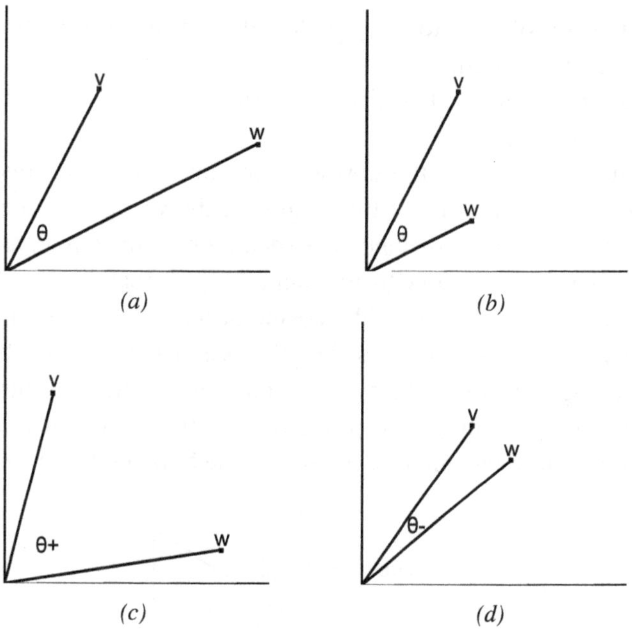

Figure 3.8: Increasing and decreasing the distance between v and w

The proximity of v and w in this two-dimensional space, and generally of any two vectors in n-dimensional space, can be found either by measuring the distance between them directly or by measuring the angle between them. These are dealt with separately, beginning with angle.

The angle between two vectors v and w can be found by first finding its cosine and then translating that into the corresponding angle using standard trigonometric tables. The cosine is found using the formula given in Equation (3.3):

$$cosine(\theta) = \frac{v}{|v|} \cdot \frac{w}{|w|} \tag{3.3}$$

where:

1. θ is the unknown angle between v and w.
2. $|v|$ and $|w|$ are the lengths or 'norms' of v and w, that is, the lengths of the lines connecting them to the origin in the basis, as in Figure 3.3 (p. 31). The norm of an n-dimensional vector $v = [v_1, v_2 \ldots v_n]$ is defined by Equation (3.4)

$$|v| = \sqrt{v_1^2 + v_2^2 + \ldots + v_n^2} \tag{3.4}$$

3. The division of a vector v by its length $|v|$ is always a unit vector, that is, a vector of length 1.
4. The dot between the two division terms designates the inner product, as described earlier.

The formula for finding the cosine of the angle between two vectors is based on the observation that, if the lengths of the vectors are the same, then the sole determinant of the distance between them is the angle, as noted. The formula rescales both vectors to the same length, that is, 1, and the inner product of the rescaled vectors is the cosine of the angle between them.

To see why the inner product of length-normalized vectors should be the cosine of the angle between them, recall that the cosine of either one of the non-right angles in a right-angled triangle is defined as the ratio of the length of the side adjacent to the angle of interest to the hypotenuse, as in Figure 3.9.

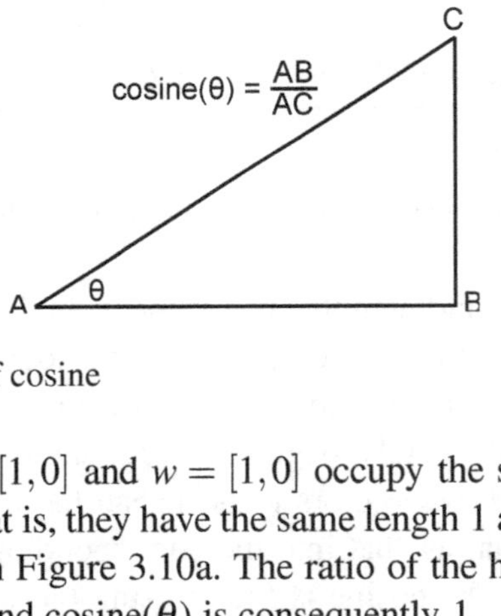

Figure 3.9: Definition of cosine

Two unit vectors $v = [1,0]$ and $w = [1,0]$ occupy the same location in two-dimensional space, that is, they have the same length 1 and the angle between them is 0, as shown in Figure 3.10a. The ratio of the hypotenuse to the side adjacent to θ is $1/1$, and cosine(θ) is consequently 1.

The vector v is rotated 30 degrees about the origin keeping its length constant; the new coordinates are $[0.87, 0.50]$ as in Figure 3.10b. The vectors v and w have moved apart in the two-dimensional space and so are no longer identical, but they do retain some similarity. The degree of similarity can be determined by projecting v onto w, that is, by drawing a line from v perpendicular to a line drawn the origin to w, as in Figure 3.10b. The length of the line segment from the origin to where the perpendicular meets the origin-to-w line is given by the inner product of v and w, and is the degree to which v and w are similar. It is also the length of the side of a right-angled triangle adjacent to θ, so that cosine(θ) can be calculated. Now rotate v a further 30 degrees as in Figure 3.10c, again keeping its length constant: the vectors have

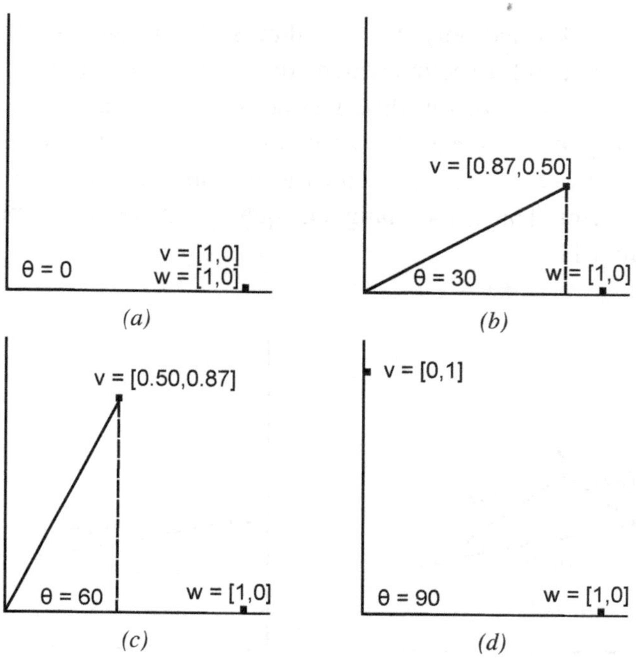

Figure 3.10: Vector projections for different angles
(a): v and w identical; $v.w = (1\times 1) + (0\times 0) = 1$; cosine($\theta$) = $1/1 = 1$
(b): v rotated 30 degrees; $v.w = (0.87\times 1) + (0.5\times 0) = 0.87$; cosine($\theta$) = $0.87/1 = 0.87$
(c): v rotated 60 degrees; $v.w = (0.5\times 1) + (0.87\times 0) = 0.5$; cosine($\theta$) = $0.5/1 = 0.5$
(d): v rotated 90 degrees; $v.w = (0\times 1) + (1\times 0) = 0$; cosine($\theta$) = $0/1 =$ 0

moved further apart in the space, and their similarity and the cosine have decreased commensurately. And, finally, v is rotated a further 30 degrees as in Figure 3.10d, so that the distance between v and w increases yet further, the projection of v on w, that is, their degree of similarity, is 0, and cosine(θ) is also 0.

Moving on to proximity measurement by distance, the distance between two vectors v and w in a vector space V can be measured in terms of a metric. Given a set X, a metric (cf. Deza and Deza (2009) and O Searcoid (2006)) is a function $d : X \times X \rightarrow \mathbb{R}$ if, for all $x, y, z \in X$, the following properties hold:

1. $d(x,y) \geq 0$, that is, the distance between any two vectors in the space is non-negative.

2. $d(x,y) = 0$ if and only if $x = y$, that is, the distance from a vector to itself is 0, and for vectors which are not identical is greater than 0.
3. $d(x,y) = d(y,x)$, that is, distances are symmetrical.
4. $d(x,y) \leq d(x,y) + d(y,z)$, that is, the distance between any two vectors is always less than or equal to the distance between them and a third vector. This is the triangle inequality, shown diagrammatically in Figure 3.11.

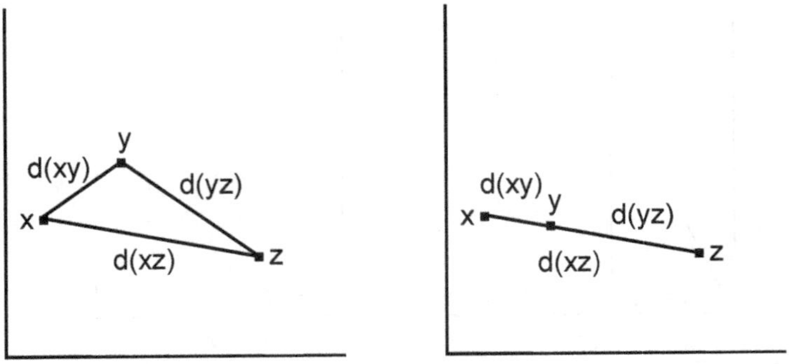

(a) $d(x,z) < d(x,y) + d(y,z)$ *(b)* $d(x,z) = d(x,y) + d(y,z)$

Figure 3.11: Triangle inequality for distances among vectors in metric space

A metric space $M(V,d)$ is a vector space V on which a metric d is defined in terms of which the distance between any two points in the space can be measured. Numerous distance metrics exist (Deza and Deza 2009: Chs. 17, 19). For present purposes these are divided into two types:

1. Linear metrics, where the distance between two points in a manifold is taken to be the length of the straight line joining the points, or some approximation to it, without reference to the shape of the manifold.
2. Nonlinear metrics, where the distance between the two points is the length of the shortest line joining them along the surface of the manifold and where this line can but need not be straight.

This categorization is motivated by the earlier observation that manifolds can have shapes which range from perfectly flat to various degrees of curvature. Where the manifold is flat, as in Figure 3.12a, linear and nonlinear measures are identical. Where it is curved, however, linear and nonlinear measurements can differ to varying degrees depending on the nature of the curvature, as shown in Figures 3.12b and 3.12c.

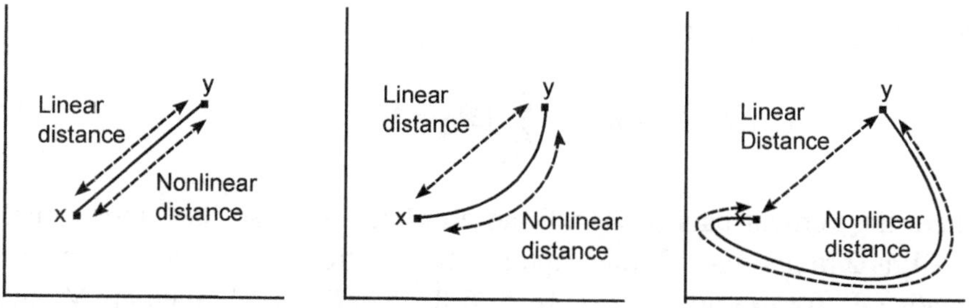

(a) (b) (c)

Figure 3.12: Linear and nonlinear distances on flat and curved manifolds

Distance in vector space will figure prominently in the discussion from this point onwards, and as such it is discussed in some detail; see further Deza and Deza (2009), Everitt et al. (2011), Gan, Ma, and Wu (2007), Jain, Murty, and Flynn (1999), and Xu and Wunsch (2009).

The most commonly used linear metric is the Minkowski, given in Equation (3.5); for others see Deza and Deza (2009).

$$d_{i,j} = \sqrt[p]{\sum_{k=1}^{n} |M_{ik} - M_{jk}|^p} \tag{3.5}$$

where

1. M is a matrix each of whose n-dimensional row vectors specifies a point in n-dimensional metric space.
2. i and j index any two row vectors in M.
3. p is a real-valued user-defined parameter in the range $1 \ldots \infty$.
4. $|M_{i,j} - M_{j,k}|$ is the absolute difference between the coordinates of i and j in the space.
5. $\sum_{k=1..n} |M_{i,j} - M_{j,k}|$ generalizes to n dimensions the Pythagorean theorem for the length of the hypotenuse of a right angled triangle in two dimensions, which is the shortest distance between any two 2 dimensional vectors.

Three parameterizations of Minkowski distance are normally used in clustering contexts. Where $p = 1$ the result is the Manhattan or city-block distance in Equation (3.6).

$$d_{i,j} = \sqrt[1]{\sum_{k=1}^{n} |M_{ik} - M_{jk}|^1} \tag{3.6}$$

which simplifies to Equation (3.7)

$$d_{i,j} = \sum_{k=1}^{n} |M_{ik} - M_{jk}| \tag{3.7}$$

where the vertical bars $|\ldots|$ indicate absolute value, so that, for example, $|2-4|$ is 2 rather than -2; this captures the intuition that distances cannot be negative. If, for example, $n=2$, $M_i=[6,8]$ and $M_j=[2,4]$, then the Manhattan distance $d(M_i,M_j)$ is calculated as in Figure 3.13.

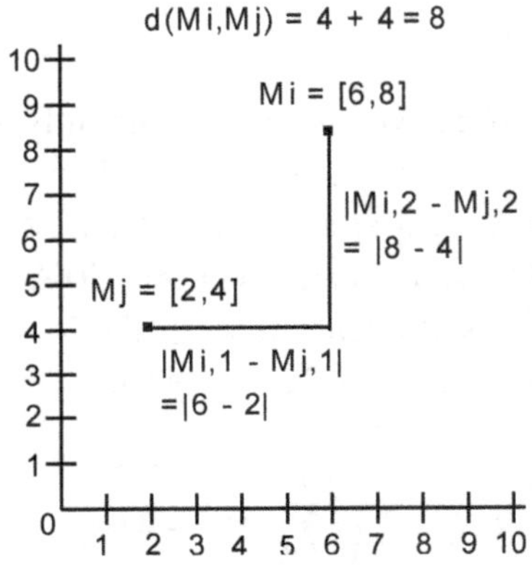

Figure 3.13: Calculation of Manhattan distance between row vectors i and j of matrix M

Figure 3.13 demonstrates the reason for the names 'Manhattan' and 'city block': to get from one corner of a city block to the diagonally opposite one it is necessary to walk around the block, and the distance walked is the sum of the lengths of its sides.

By far the most often used parameterization of the Minkowski metric is $p=2$, the Euclidean distance, shown in Equation (3.8).

$$d_{i,j} = \sqrt[2]{\sum_{k=1}^{n} |M_{ik} - M_{jk}|^2} \tag{3.8}$$

This is just the Pythagorean rule known, one hopes, to all schoolchildren, that the length of the hypotenuse of a right-angled triangle is the square root of the sum of the squares of the lengths of the other two sides. Again for $n=2$, $M_i=[6,8]$ and $M_j=[2,4]$, the Euclidean distance $d(M_i,M_j)$ is calculated as in Figure 3.14, and it is the shortest distance between the two points.

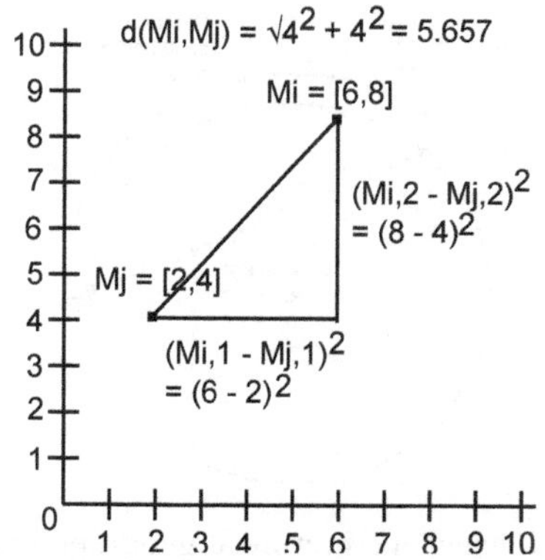

Figure 3.14: Calculation of Euclidean distance between row vectors i and j of matrix M

Finally, for $p = \infty$, the result is the Chebyshev distance given in Equation (3.9)

$$d_{i,j} = \sqrt[\infty]{\sum_{k=1}^{n} |M_{ik} - M_{jk}|^{\infty}} \tag{3.9}$$

which simplifies to Equation (3.10)

$$d_{i,j} = \max_{k=1..n} |M_{ik} - M_{jk}| \tag{3.10}$$

The parameter p can be any positive real-number value: where $1 < p < 2$, the Manhattan approximation approaches the Euclidean distance, and where $p > 2$ the approximation moves away from the Euclidean and approaches the Chebyshev. Like the Euclidean distance, the Manhattan distance is based on the differences between any pair of row vectors M_i and M_j in their n dimensions, but it merely sums the absolute values of the differences without using them to calculate the shortest distance, and is thereby an approximation to the shortest distance; Chebyshev takes the distance between any pair of row vectors M_i and M_j to be the maximum absolute difference across all n dimensions and, like Manhattan, is therefore an approximation to linear distance.

There are many nonlinear metrics (Deza and Deza 2009), the most useful of which for present purposes is the geodesic. The word 'geodesy' comes from Greek *geodaisia*, 'division of the earth'; geodesic distance is the shortest distance between any two points on the Earth measured along its curved surface as opposed to the linear shortest distance, as in Figure 3.15.

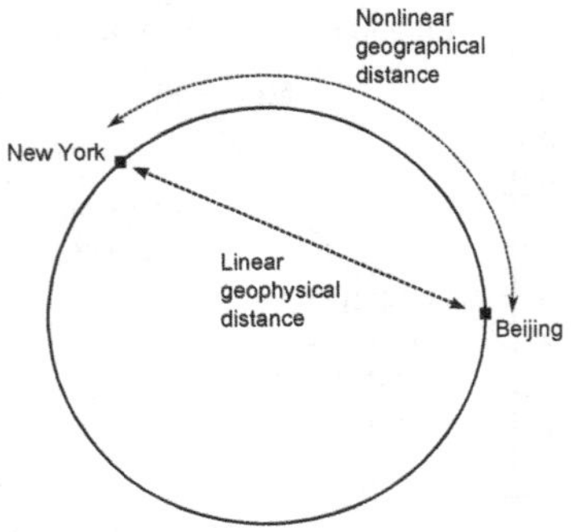

Figure 3.15: Linear geophysical and nonlinear geographical distance between points on the Earth's surface

Mathematically, geodesic distance is a generalization of linear to nonlinear distance measurement in a space: the geodesic distance $g_{x,y}$ is the shortest distance between two points x and y on a manifold measured along its possibly-curved surface (Deza and Deza 2009). One approach to measurement of geodesic distance on manifolds is to approximate it using graph distance (Lee and Verleysen 2007), and that is the approach taken here. To see how this approximation works, we begin with the small nonlinear data matrix M and scatterplot showing the corresponding nonlinear manifold shape in Figure 3.16.

	v_1	v_2
1	0.27	0.20
2	0.32	0.50
2	0.40	0.70
4	0.50	0.80
5	0.60	0.70
6	0.68	0.50
7	0.73	0.20

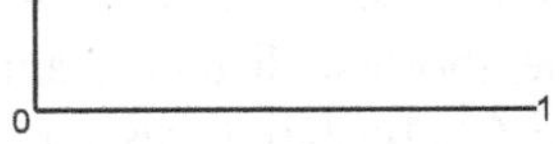

Figure 3.16: Nonlinear data matrix and corresponding scatterplot

Given a matrix M with m rows and n columns, a Euclidean distance matrix D is an $m \times m$ matrix each of whose values d_{ij} (for $i, j = 1..m$) is the Euclidean distance from row vector i to row vector j of M in n-dimensional space. Figure 3.17a shows all the distances d for the M of Figure 3.16 together with a graphical representation of them.

	1	2	3	4	5	6	7
1	0	0.30	0.52	0.64	0.60	0.51	0.46
2	0.30	0	0.22	0.35	0.34	0.36	0.51
3	0.52	0.22	0	0.14	0.20	0.34	0.60
4	0.64	0.35	0.14	0	0.14	0.35	0.64
5	0.60	0.34	0.20	0.14	0	0.22	0.52
6	0.51	0.36	0.34	0.35	0.22	0	0.30
7	0.46	0.51	0.60	0.64	0.52	0.30	0

(a) Euclidean distance matrix *D* for *M*

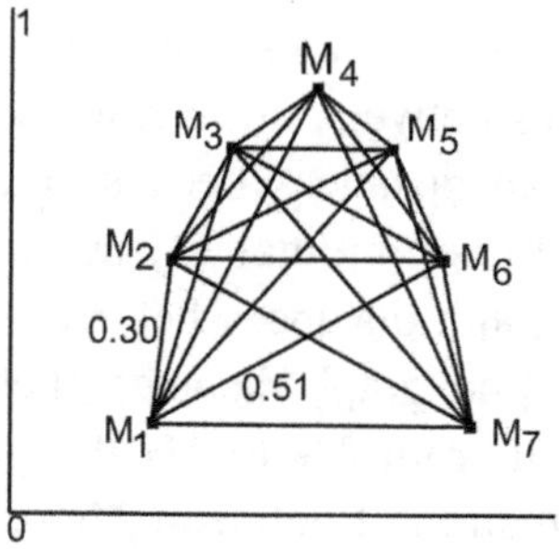

(b) M interpreted as a graph *G*

Figure 3.17: Euclidean distance matrix for the data in Fig. 3.16 and interpretation of manifold in Fig. 3.16 as a connected graph with Euclidean distances as arc labels

M is interpretable as a connected graph *G* each of whose arcs from *i* to *j* is labelled with the Euclidean distance between G_i and G_j, as shown in Figure 3.17b; the distance between node 1 and node 2, for example, is given in the table as 0.30, between 1 and 6 as 0.51, and so on; only two arcs are explicitly labelled with distances in Figure 3.17b to avoid clutter.

A spanning tree for *G* is an acyclic subgraph of *G* which contains all the nodes in *G* and some subset of the arcs of *G* (Gross and Yellen 2006: 72ff.). A minimum spanning tree of *G*, as its name indicates, is a spanning tree which contains the minimum number of arcs required to connect all the nodes in *G*, or, if the arcs have weights, the smallest sum of weights (ibid.: 176ff.). The minimum spanning tree for *G* in Figure 3.17b is shown in Figure 3.18, with the arcs comprising the tree emboldened.

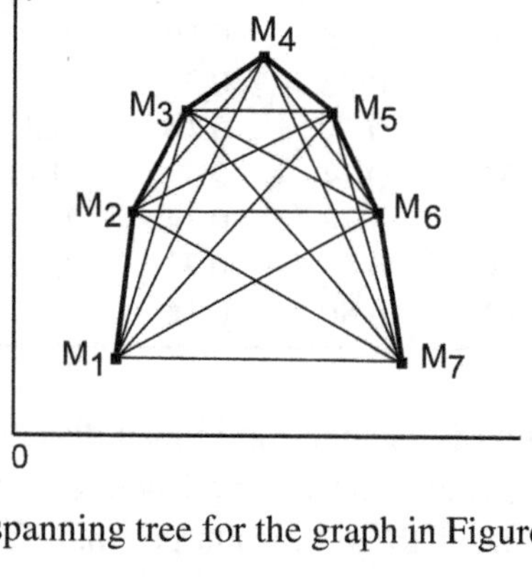

Figure 3.18: Minimum spanning tree for the graph in Figure 3.17b

Such a minimum spanning tree can be used to approximate the geodesic distances between all m row vectors of M in n dimensional space using the Euclidean distances because the distance between any two nodes is guaranteed to be minimal. By summing the shortest paths between nodes, a table of graph distances between all data vectors can be constructed: the Euclidean and graph distances between M_1 and M_2 in Figure 3.18 are identical, but from M_1 to M_3 the graph distance is $(M_1 \to M_2) + (M_2 \to M_3)$ rather than the Euclidean $M_1 \to M_3$, from M_1 to M_4 the graph distance is is $(M_1 \to M_2) + (M_2 \to M_3) + (M_3 \to M_4)$ rather than the Euclidean $M_1 \to M_4$, and so on. The graph distance table and the Euclidean one from which it was derived in this way are shown in Tables 3.6 and 3.7.

Table 3.6: Euclidean distance matrix for M: Sum of all distances: 16.27; Mean distance: 0.39; Distance $M_1 \to M_7$: 0.46

	M_1	M_2	M_3	M_4	M_5	M_6	M_7
M_1	0	0.30	0.52	0.64	0.60	0.51	0.46
M_2	0.30	0	0.22	0.35	0.34	0.36	0.51
M_3	0.52	0.22	0	0.14	0.20	0.34	0.60
M_4	0.64	0.35	0.14	0	0.14	0.35	0.64
M_5	0.60	0.34	0.20	0.14	0	0.22	0.52
M_6	0.51	0.36	0.34	0.35	0.22	0	0.30
M_7	0.46	0.51	0.60	0.64	0.52	0.30	0

Table 3.7: Graph distance matrix for M: Sum of all distances: 22.53; Mean distance: 0.54; Distance $M_1 \to M_7$: 1.32

	M_1	M_2	M_3	M_4	M_5	M_6	M_7
M_1	0	0.30	0.53	0.66	0.80	1.00	1.32
M_2	0.30	0	0.22	0.35	0.49	0.70	1.00
M_3	0.53	0.22	0	0.14	0.28	0.49	0.79
M_4	0.66	0.35	0.14	0	0.14	0.35	0.66
M_5	0.80	0.49	0.28	0.14	0	0.22	0.52
M_6	1.01	0.70	0.49	0.35	0.22	0	0.30
M_7	1.32	1.01	0.80	0.66	0.52	0.30	0

As expected, the sum of distances and mean distance for the graph matrix are both substantially greater than for the Euclidean one, and the graph distance between M_1 and M_7 is three times larger than for the Euclidean, which Figure 3.18 confirms visually.

Figure 3.19 gives an example of the approach to geodesic distance approximation just described which is less tidy and contrived than the one just discussed: a version of the the so-called 'Swiss roll' often used in discussions of nonlinearity such as that of Lee and Verleysen (2007).

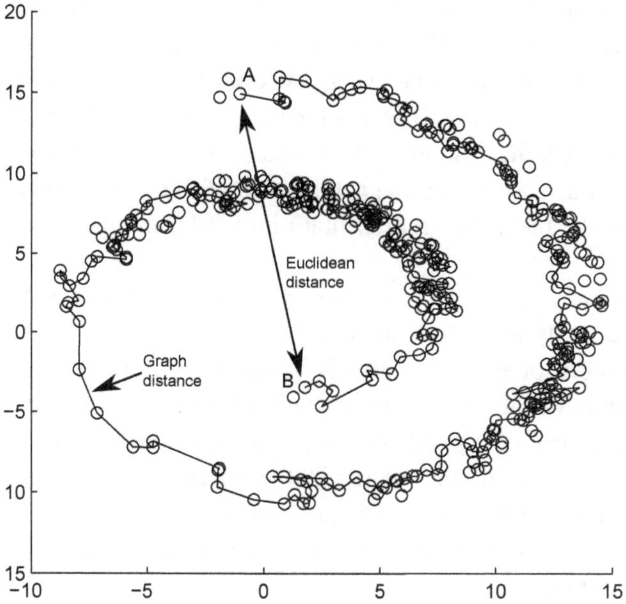

Figure 3.19: Euclidean and graph distance in an empirically derived data manifold. Mean Euclidean distance: 11.38; Mean geodesic distance: 41.91; Ratio mean geodesic / mean Euclidean: 3.68; Euclidean distance $A \to B$: 18.66; Geodesic distance $A \to B$: 191.73; Ratio geodesic / Euclidean $A \to B$: 6.60

As before there is a disparity, reflected in the given ratios, between Euclidean and graph distances both in terms of means across the distances between all data vectors and between the sample vectors *A* and *B*; these ratios, particularly the one for the *A* to *B* distance, are readily confirmed visually.

In general, the graph approximation of geodesic distance is constrained to follow the shape of the manifold by the need to visit its nodes in the course of minimum spanning tree traversal. Intuitively, this corresponds to approximating the geodesic distance between any two cities on the surface of the Earth, say from New York to Beijing in Figure 3.15, by stopping off at intervening airports, say New York → London → Istanbul → Dehli → Beijing.

3.2.1.5 Geometrical interpretation of data

Data are a description of objects from a domain of interest in terms of a set of variables such that each variable is assigned a value for each of the objects. We have seen that, given m objects described by n variables, a standard representation of data for computational analysis is a matrix M in which each of the m rows represents a different object, each of the n columns represents a different variable, and the value at M_{ij} describes object i in terms of variable j, for $i = 1 \ldots m, j = 1 \ldots n$. The matrix thereby makes the link between the researcher's conceptualization of the domain in terms of the semantics of the variables s/he has chosen and the actual state of the world, and allows the resulting data to be taken as a representation of the domain based on empirical observation.

Once data are represented as a matrix M, the foregoing geometrical concepts apply directly to it. Specifically:

- The dimensionality of M, that is, the number n of columns representing the n data variables, defines an n-dimensional data space.
- The sequence of n numbers comprising each row vector of M specifies the coordinates of the vector in the space, and the vector itself is a point at the specified coordinates; because the row vectors represent the objects in the research domain, each object has a specified location in the data space.
- The set of all data vectors in the space constitutes a manifold; the shape of the manifold is the shape of the data.
- Distance between the data vectors comprising the manifold can be measured linearly or nonlinearly.

The issue of whether the data manifold is linear or nonlinear will be prominent in the discussion to follow because it reflects a corresponding distinction in the characteristics of the natural process that the data describe. Linear processes have a constant proportionality between cause and effect. If kicking a ball is a linear system and kicking it x hard makes it go y distance, then a $2x$ kick will make it go $2y$ distance, a $3x$ kick $3y$ distance and so on for nx and ny. Experience tells us that reality is not like this, however: air and rolling resistance become significant factors as the ball is kicked harder and harder, so that for a $5x$ kick it only goes, say, $4.9y$, for $6x$ $5.7y$, and again so on until it bursts and goes hardly any distance at all. This is nonlinear behaviour: the breakdown of strict proportionality between cause and effect. Such nonlinear effects pervade the natural world, giving rise to a wide variety of complex and often unexpected, including chaotic, behaviours (Bertuglia and Vaio 2005). Depending on the choice of variables used to describe a nonlinear natural process, the data manifold may or may not capture the nonlinearity as curvature in its shape, and, if it does, the researcher must judge whether or not to take the nonlinearity into account during analysis.

Conceptualizing data as a manifold in n-dimensional space is fundamental to the discussion of clustering that follows for two main reasons. On the one hand, it becomes possible to visualize the degrees of similarity of data vectors, that is, the rows of a data matrix, as clusters in a geometrical space, thereby greatly enhancing intuitive understanding of structure in data. And, on the other, the degrees of similarity among data vectors can be quantified in terms of relative distance between them, and this quantification is the basis for most of the clustering methods presented later on.

3.2.2 Data transformation

Once a data matrix has been constructed, it can be transformed in a variety of ways prior to cluster analysis. In some cases such transformation is desirable in that it enhances the quality of the data and thereby of the analysis. In others the transformation is not only desirable but necessary to mitigate or eliminate characteristics in the matrix that would compromise the quality of the analysis or even render it valueless. The present section describes various types of transformation and the motivations for using them.

3.2.2.1 Variable scaling

The variables selected for a research project involving cluster analysis may require measurement on different scales. This is not an issue with respect to MDECTEbecause all its variables measure phonetic segment frequency and are thus on the same scale, but it is not difficult to think of cases in corpus-based linguistics where it can be. In sociolinguistics, for example, speakers might be described by a set of variables one of which represents the frequency of occurrence of some phonetic segment in interviews, another one speaker age, and a third income. Because these variables represent different kinds of thing in the world, they are measured in numerical units and ranges appropriate to them: phonetic frequency in the integer range, say, 1..1000, age in the integer range 20..100, and income in some currency in the real-valued range 0..50000. Humans understand that one can't compare apples and oranges and, faced with different scales, use the variable semantics to interpret their values sensibly. But cluster analysis methods don't have common sense. Given an $m \times n$ data matrix M in which the m rows represent the m objects to be clustered, the n columns represent the n variables, and the entry at M_{ij} (for $i = 1 \ldots m, j = 1 \ldots n$) represents a numerical measure of object i in terms of variable j, a clustering method has no idea what the values in the matrix mean and calculates the degrees of similarity between the row vectors purely on the basis of the relative numerical magnitudes of the variable values, as we shall see. As a consequence, variables whose scales permit relatively larger magnitudes can have a greater influence on the cluster analysis than those whose scales restrict them to relatively small values, and this can compromise the reliability of the analysis, as has often been noted – cf., for example, Kaufman and Rousseeuw (1990: 4ff.); Gnanadesikan (1997: 102ff.); Kettenring (2006); Tan, Steinbach, and Kumar (2006: 64f., 81); Gan, Ma, and Wu (2007: Ch. 4.1); Chu, Holliday, and Willett (2009); Xu and Wunsch (2009: 22); Hair et al. (2010: Ch. 8). This section first examines the nature of the problem, and then describes a resolution.

Table 3.8 shows three variants of a matrix that describes a dozen speakers in terms of three variables and, for each variant, a cluster analysis of the matrix rows; all the cluster trees were generated using squared Euclidean distance and Ward's Method, but for present purposes any other combination of distance measure and clustering method chosen from among those described later on in the discussion would have done just as well.

In Table 3.8a the first variable represents the frequency of speakers' usage of some phonetic segment of interest, the second the age of the speakers in years, and the third speaker annual income in Euros. In 3.8b frequency and age are as in 3.8a but income is now expressed as the number of thousands of Euros (K), and 3.8c both retains the income scale of 3.8b and also expresses age in terms of days rather than years.

Table 3.8: Versions of a data matrix with different variable scales

a

	Frequency	Age	Income
1	100	20	30000
2	105	21	30250
3	110	22	30500
4	115	30	30750
5	200	31	31000
6	205	32	31250
7	210	40	35000
8	215	41	35250
9	300	42	35500
10	305	50	35750
11	310	51	36000
12	315	52	36250

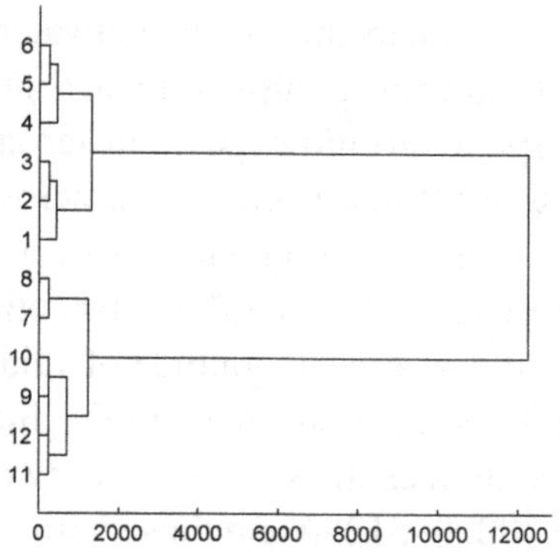

b

	Frequency	Age	Income (k)
1	100	20	30.00
2	105	21	30.25
3	110	22	30.50
4	115	30	30.75
5	200	31	31.00
6	205	32	31.25
7	210	40	35.00
8	215	41	35.25
9	300	42	35.50
10	305	50	35.75
11	310	51	36.00
12	315	52	36.25

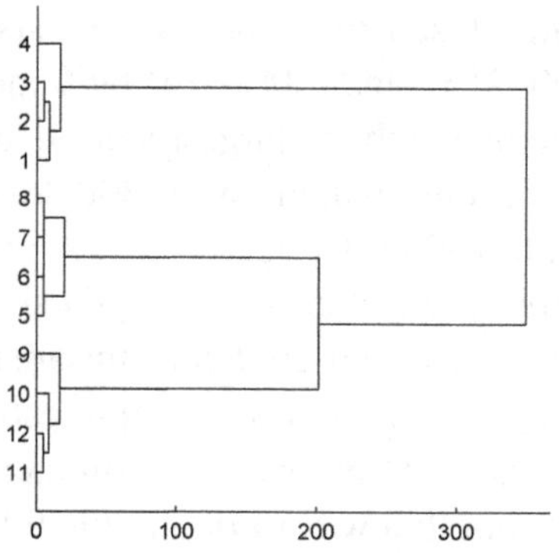

c

	Frequency	Age	Income (k)
1	100	7300	30.00
2	105	7665	30.25
3	110	8030	30.50
4	115	10950	30.75
5	200	11315	31.00
6	205	11680	31.25
7	210	14600	35.00
8	215	14965	35.25
9	300	15330	35.50
10	305	18250	35.75
11	310	18615	36.00
12	315	18980	36.25

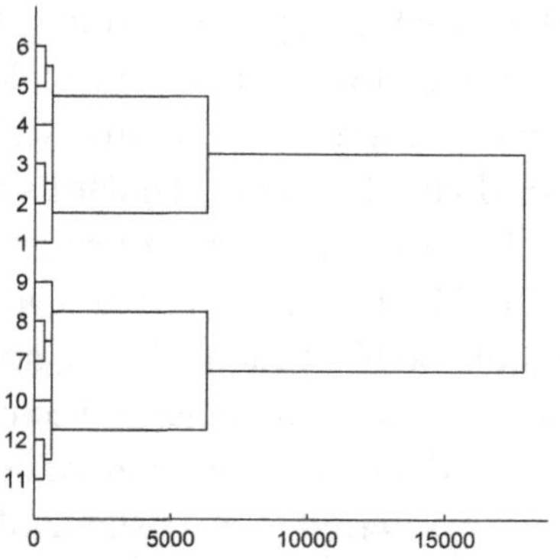

Using the variable semantics, a human interpreter would see from direct inspection of the matrices that, irrespective of variation in scale, the descriptions of the speakers are in fact equivalent and that the speakers fall into three phonetic frequency groups, four age groups, and two income groups. That same interpreter would expect cluster analysis to use these groupings as the basis for a result that is consistent across all three matrices and independent of the variation in scaling, but it does not.

The trees in Table 3.8 differ substantially, and they cluster the speakers according to the relative magnitude of values in the matrix columns. The largest values in Table 3.8a are those in the Income column and the corresponding cluster tree divides the speakers into two main groups, those with incomes in the range 30000–31250 and those with incomes in the range 35000–36250. In Table 3.8b the largest values are those in the Frequency column, and the corresponding cluster tree classifies the speakers into three main groups (100–115), (200–215), and (300–315) by frequency; and in Table 3.8c the Age column is the one with the largest values, and, predictably, the speakers are now divided into four main groups (7300–8030), (10950–11680), (14600–15330), and (18250–18980) by age.

That the result of cluster analysis should be contingent on the vagaries of scale selection is self-evidently unsatisfactory both in the present case and also more generally in research applications where variables are measured on different scales. Some way of eliminating scale as a factor in such applications is required; a way of doing this follows.

Relative to a data matrix M, a solution to the above problem is to standardize the variables by transforming the values in the column vectors of M in such a way that variation in scale among them is removed: if all the variables are measured on the same scale, none can dominate. The textbook method for doing this is via standard score, also known as z-score and autoscaling – cf., for example, Kaufman and Rousseeuw (1990: 6f.), Everitt and Dunn (2001: 51), Hair et al. (2010: Ch. 8), Kettenring (2006), Boslaugh and Watters (2008: 369f.), Chu, Holliday, and Willett (2009)) –, which transforms the original values in any column vector M_j into ones which say how many standard deviations those original values are from the vector mean; in what follows, this is referred to as 'z-standardization'.

For the ith value in any given vector x, the z-standardization is defined as in Equation (3.11)

$$zscore(x_i) = \frac{x_i - \mu(x)}{\delta(x)} \tag{3.11}$$

where

- $\mu(x)$ is the mean of the values in the vector. Given a variable x whose values are represented as a vector of n numerical values distributed across some range, the mean or average of those values is the value at the centre of the distribution. The values in Table 3.9 have been sorted by magnitude for ease of interpretation.

Table 3.9: An $n = 10$ dimensional vector x

	1	2	3	4	5	6	7	8	9	10
x	2	4	6	8	10	12	15	16	18	20

Direct inspection suggests that the value at the centre of this distribution is around 10 or 12. A more precise indication is given by Equation (3.12).

$$mu(x) = \frac{\sum_{i=1..n} x_i}{n} \tag{3.12}$$

where μ is the conventional symbol for 'mean', $\sum$ denotes summation, and n is the number of values in x: the mean of a set of n values is their sum divided by n. In the case of Table 3.9 this is $(2+4+\ldots+20 = 110)/10 = 11$.

- $\delta(x)$ is the standard deviation. The mean hides important information about the distribution of values in a vector. Consider, for example, these two (fictitious) runs of student marks A and B on a percentage scale in Table 3.10:

Table 3.10: Two fictitious student mark vectors

	1	2	3	4	5	6	7	8	9	10	Mean
A	40	58	92	31	27	85	67	77	73	30	58
B	55	60	56	64	59	58	57	54	58	59	58

The means in Table 3.10 are identical, but the variations across the mark scale differ strikingly: student A is capable of both failure and excellence, and student B is remarkably consistent. Knowing only the averages one could not make the distinction. Both the average and an indication of the spread of marks across the range are required in order to do proper justice to these students. Assessing the spread can be problematic in practice, however. Where the number of marks is few, as in the above example, visual inspection is sufficient, but what about longer runs? Visual inspection quickly fails; some quantitative measure that summarizes the spread of marks is required. That measure is variance.

Given a variable x whose values are represented as a vector of n values $[x_1, x_2 \ldots x_n]$, variance is calculated as follows.

- The mean of the values μ is $(x_1 + x_2 + \ldots + x_n)/n$.
- The amount by which any given value x_i differs from μ is then $x_i - \mu$.
- The average difference from μ across all values is therefore $\sum_{i=1}^{n} x_i - \mu / n$.
- This average difference of variable values from their mean almost corresponds to the definition of variance. One more step is necessary, and it is technical rather than conceptual. Because μ is an average, some of the variable values will be greater than μ and some will be less. Consequently, some of the differences $(x_i - \mu)$ will be positive and some negative. When all the $(x_i - \mu)$ are added up, as above, they will cancel each other out. To prevent this, the $(x_i - \mu)$ are squared.
- The definition of variance for n values $x = [x_1, x_2 \ldots x_n]$ is given in Equation (3.13)

$$variance(x) = \frac{\sum_{i=1..n}(x_i - \mu(x))^2}{n} \tag{3.13}$$

Thus, in Table 3.10, the variance for A is $(40-58)2+(30-58)2+\ldots+30-58)2/10 = 594.44$. Doing the same calculation for student B, the variance works out as 8.00. Comparing the two variances, it is clear that the variability in A's run of marks is much greater than B's.

The variance of a set of values can be difficult to interpret because it is the average of the square of distances of the values from their mean. In the above example it is clear enough that 594.44 is much larger than 8, and that the amount of variability in the A distribution is therefore much larger than that of B. What if the variance of A is taken on its own without comparing it to any other variance, however? Relative to the values in A, how is one to interpret the observation that the variance is 594.44? Is that a large variance, or a small one, or somewhere in between? It would be intuitively much clearer to express the variability in terms of the range of values in A. This is done by taking the square root of the variance, as in Equation (3.14).

$$\delta(x) = \sqrt{variance(x)} \tag{3.14}$$

where δ is the standard deviation. The standard deviation of A is the square root of 594.44 = 24.38, which tells one that, on average, the marks for A vary by that amount, and by 2.83 for B, both of which are readily interpretable in terms of their respective runs of marks.

The z-standardization of an arbitrary vector x is shown in Table 3.11.

Table 3.11: z-standardization of an arbitrary vector x

	Original x	z-standardized x
Values	1 0 0 1 1 1 1 6 10 19 13 88 90 157 91 141 199 331	-0.701 -0.712 -0.712 -0.701 -0.701 -0.701 -0.701 -0.645 -0.600 -0.500 -0.567 0.269 0.2909 1.037 0.302 0.859 1.505 2.976
Mean	63.89	0
Std. dev.	89.76	1

Figures 3.20a and 3.20b illustrate the result of z-standardization.

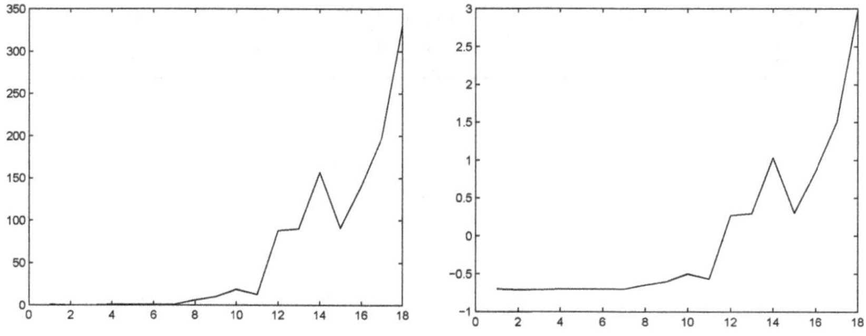

(a) Original x *(b)* z-standardized x

Figure 3.20: Plots of original and z-standardized vector x of Table 3.11

Application of z-standardization transforms any vector into one having a mean of 0 and a standard deviation of 1, and, because division by a constant is a linear operation, the shape of the distribution of the original values is preserved, as is shown by the pre- and post-standardization plots in Table 3.11. Only the scale changes: $0 \ldots 331$ for original x, and $-0.7006 \ldots 2.9759$ for transformed x.

When z-standardization is applied to each of the column vectors of a matrix, any variation in scale across those variables disappears because all the variables are now expressed in terms of the number of standard deviations from their respective means. Tables 3.12a–c show, for example, z-standardization of the matrices in Table 3.8.

Table 3.12: Comparison of matrices in Table 3.8 and their z-standardized versions

Table 3.12a

	Freq	Age(yrs)	Income	Freq	Age(yrs)	Income
1	100	20	30000	-1.31	-1.42	-1.23
2	105	21	30250	-1.25	-1.33	-1.13
3	110	22	30500	-1.19	-1.24	-1.03
4	115	30	30750	-1.13	-0.53	-0.93
5	200	31	31000	-0.09	-0.44	-0.83
6	205	32	31250	-0.03	-0.35	-0.73
7	210	40	35000	0.03	0.35	0.73
8	215	41	35250	0.09	0.44	0.83
9	300	42	35500	1.13	0.53	0.93
10	305	50	35750	1.19	1.24	1.03
11	310	51	36000	1.25	1.33	1.13
12	315	52	36250	1.31	1.42	1.23
Mean	207.5	36	33125	0	0	0
Std. dev.	81.84	11.21	2536.20	1	1	1

Table 3.12b

	Freq	Age(yrs)	Income(K)	Freq	Age(yrs)	Income(K)
1	100	20	30.00	-1.31	-1.42	-1.23
2	105	21	30.25	-1.25	-1.33	-1.13
3	110	22	30.50	-1.19	-1.24	-1.03
4	115	30	30.75	-1.13	-0.53	-0.93
5	200	31	31.00	-0.09	-0.44	-0.83
6	205	32	31.25	-0.03	-0.35	-0.73
7	210	40	35.00	0.03	0.35	0.73
8	215	41	35.25	0.09	0.44	0.83
9	300	42	35.50	1.13	0.53	0.93
10	305	50	35.75	1.19	1.24	1.03
11	310	51	36.00	1.25	1.33	1.13
12	315	52	36.25	1.31	1.42	1.23
Mean	207.5	26	33.13	0	0	0
Std dev	81.84	11.21	2536.20	1	1	1

Table 3.12c

	Freq	Age(days)	Income(K)	Freq	Age(days)	Income(K)
1	100	7300	30.00	-1.31	-1.42	-1.23
2	105	7665	30.25	-1.25	-1.33	-1.13
3	110	8030	30.50	-1.19	-1.24	-1.03
4	115	10950	30.75	-1.13	-0.53	-0.93
5	200	11315	31.00	-0.09	-0.44	-0.83
6	205	11680	31.25	-0.03	-0.35	-0.73
7	210	14600	35.00	0.03	0.35	0.73
8	215	14965	35.25	0.09	0.44	0.83
9	300	15330	35.50	1.13	0.53	0.93
10	305	18250	35.75	1.19	1.24	1.03
11	310	18615	36.00	1.25	1.33	1.13
12	315	18980	36.25	1.31	1.42	1.23
Mean	207.5	13140	33.13	0	0	0
Std. dev.	81.84	4091.69	2.54	1	1	1

Despite the variation of scale in the matrices in the left-hand columns of Tables 3.12a–c, the *z*-standardized versions in the right-hand columns are identical. Cluster analysis of the rows of this standardized matrix, moreover, generates a tree, shown in Figure 3.21, that differs from any of those in Table 3.8; it was generated using squared Euclidean distance and Ward's Method, as before, and this combination is used throughout the remainder of the discussion to maintain comparability among analyses.

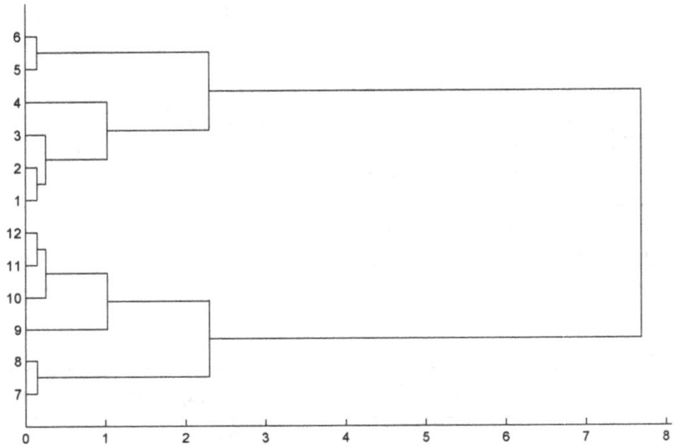

Figure 3.21: Cluster analysis of the *z*-standardized matrix in Table 3.12

No one variable is dominant by virtue of the magnitudes of its values relative to the magnitudes of the others. Instead, all three variables play an equal part in determining the cluster structure, resulting in a symmetrical tree which reflects the symmetry of the standardized matrix's row vectors in Table 3.12: vectors 1 and 12 are numerically identical but with opposite signs, as are 2 and 11, 3 and 10, and so on.

Application of *z*-standardization appears to be a good general solution to the problem of variation in scaling among data variables, and it is in fact widely used for that purpose. It is, however, arguable that, for cluster analysis, *z*-standardization should be used with caution or not at all, again as others have observed (Chu, Holliday, and Willett 2009; Gnanandesikan, Tsao, and Kettenring 1995; Kettenring 2006; Milligan and Cooper 1988). The remainder of this section first presents the argument against *z*-standardization, then proposes an alternative standardization method, and finally assesses the alternative method relative to some others proposed in the literature.

The argument against z-standardization for cluster analysis depends on making a distinction between three properties of a variable:

- The absolute magnitude of values of a variable is the numerical size of its values, and can for present purposes be taken as the absolute maximum of those values. For *Frequency* in Table 3.12, for example, it is 315 on that criterion.
- The absolute magnitude of variability is the amount of variation in the values of a variable expressed in terms of the scale of those values, and is measured by the standard deviation; in Table 3.12 the absolute magnitude of variability of the non-z-standardized *Frequency* column is 81.84.
- The intrinsic variability is the amount of variability in the values of a variable expressed independently of the scale of those values. This is measured in statistics by the coefficient of variation – see for example Boslaugh and Watters (2008: 62) –, which is defined with respect to a variable x as the ratio of x's standard deviation to its mean, as in Equation (3.15).

$$\textit{Coefficient of Variation}(x) = \frac{\delta(x)}{\mu(x)} \tag{3.15}$$

The intuition gained from direct inspection of the matrices in Table 3.12 is that there is much more variability in the non-z-standardized values of the *Frequency* and *Age* columns than there is for those in the *Income* column regardless of the variation in their absolute magnitudes and absolute magnitudes of variability. The coefficient of variation captures this intuition: for *Frequency* it is 0.383, for *Age* almost as much at 0.311, and for *Income* much less at 0.141. Because the coefficient of variation is scale-independent it can be used as a general way of comparing the degrees of variability of variables measured on different scales. Tables 3.13 and 3.14 exemplify the interrelationship of these three properties via a sequence of two-dimensional matrices together with standard deviations and coefficients of variation of the column vectors and with cluster analyses of the row vectors; the subtables show a two-dimensional matrix with the standard deviations and coefficients of variation of its column vectors together with a cluster analysis of the row vectors, and the values of v_1 are altered in various ways in the sequence while those of v_2 are held constant.

	v_1	v_2
1	1000	100
2	1000	100
3	1000	100
4	1000	200
5	1000	200
6	1000	200
7	1000	300
8	1000	300
9	1000	300
10	1000	400
11	1000	400
12	1000	400
Std. dev.	0	111.80
Coeff. var.	0	0.447

(a)

	v_1	v_2
1	10000	100
2	10000	100
3	10000	100
4	10000	200
5	10000	200
6	10000	200
7	10000	300
8	10000	300
9	10000	300
10	10000	400
11	10000	400
12	10000	400
Std. dev.	0	111.80
Coeff. var.	0	0.447

(b)

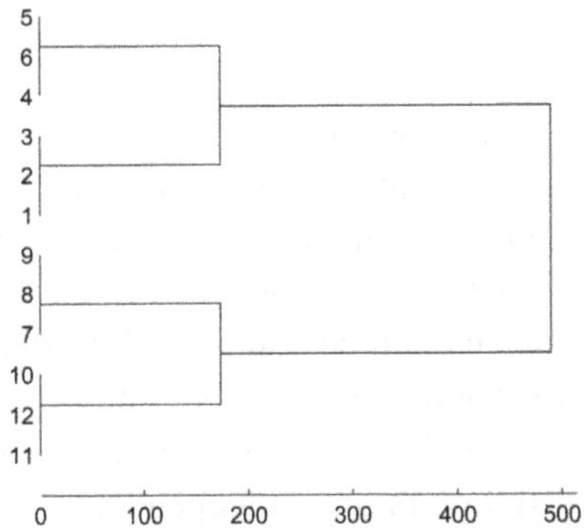

(c)

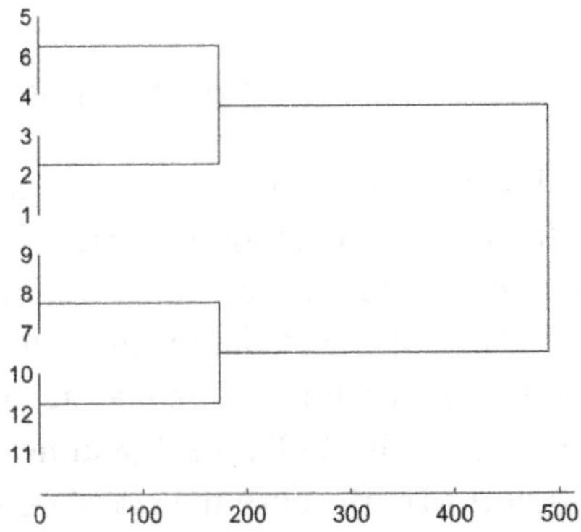

(d)

Table 3.13: Interrelationship of absolute magnitude, absolute magnitude of variability, and intrinsic variability

(a)

	v_1	v_2
1	1000	100
2	1000	100
3	1000	100
4	1000	200
5	1050	200
6	1050	200
7	1050	300
8	1050	300
9	1100	300
10	1100	400
11	1100	400
12	1100	400
Std. dev.	40.82	111.80
Coeff. var.	0.039	0.447

(b)

	v_1	v_2
1	1000	100
2	1000	100
3	1000	100
4	1000	200
5	1200	200
6	1200	200
7	1200	300
8	1200	300
9	1400	300
10	1400	400
11	1400	400
12	1400	400
Std. dev.	151.84	111.80
Coeff. var.	0.136	0.447

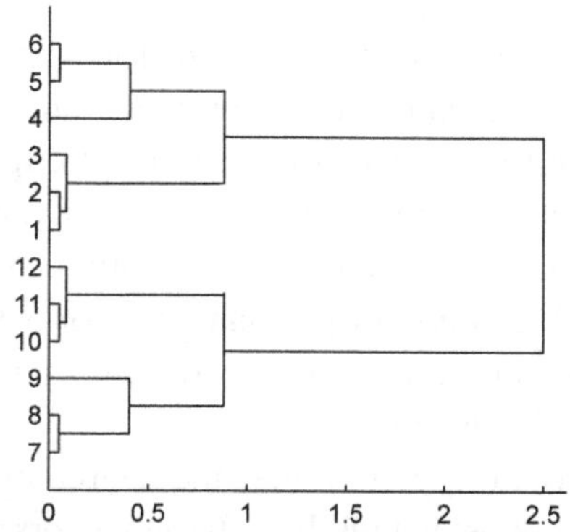

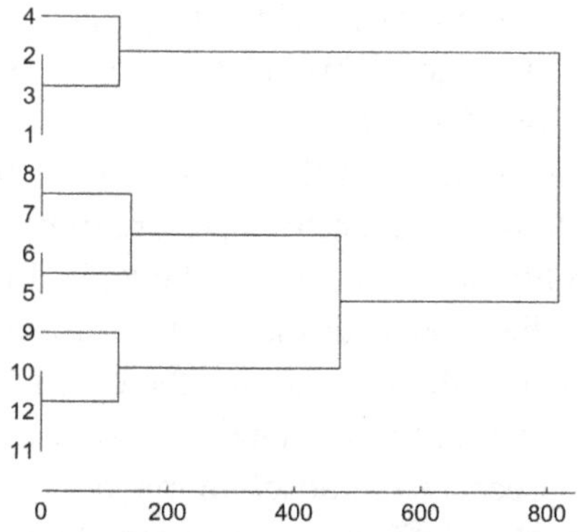

Table 3.14: Interrelationship of absolute magnitude of values, absolute magnitude of variability, and intrinsic variability

In Table 3.13a there is no variability in the values of v_1, the coefficient of variation and standard deviation are commensurately 0, and, even though the absolute magnitude of the values in v_1 is much greater than that in v_2, clustering is determined entirely by the variation in the values of v_2: there are four primary clusters corresponding to the value-groups (100–120), (200–220), (300–320), (400–420). In Table 3.13b the absolute magnitude of v_1 is substantially increased but the increase is uniform so that there is still no variability and the standard deviation and coefficient of variation remain 0;

the cluster analysis is again determined by the variability in v_2 and is identical to the one in 3.13a. In Table 3.14a a relatively small amount of variability is introduced into the values of v_1, which results in nonzero standard deviation and coefficient of variation, though both of these are smaller than those of v_2; the cluster tree differs from the ones in Tables 3.13a and 3.13b in that the same four primary clusters remain, but the pattern of variability across rows 4–6 and 7–9 is now different from that in rows 1–3 and 10–12, and this is expressed in the internal structures of the corresponding clusters. Finally, the amount of variability in v_1 is increased still further in 3.14b, and this is reflected in a higher standard deviation and coefficient of variation. For the first time, however, the standard deviation of v_1 is greater than that of v_2, and, even though the coefficient of variation is still smaller than that of v_2, there are now three rather than the previous four primary clusters corresponding to the v_1 value groups 1000, 1200, and 1400. It is neither the absolute magnitude of values nor the intrinsic variability of a variable's values that determine clustering, but their absolute magnitude of variability: the larger the standard deviation of a variable, the greater its effect on clustering.

How does this relate to the use of z-standardization of data for cluster analysis? It is a general property of every z-standardized vector, noted above, that its standard deviation is 1. Application of z-standardization to multiple columns of a matrix therefore imposes a uniform absolute magnitude of variability on them. This is shown in Table 3.15; the coefficient of variation cannot be shown for the z-standardized variables on the right-hand side of 3.15 because the formula for the coefficient of variation involves division by the mean and, for a z-standardized vector, this is always 0.

Because the absolute magnitude of variability determines the degree of a variable's effect on clustering, the implication is that all the column vectors in a z-standardized matrix have an equal influence; we have already seen an example of this above. This obviously eliminates any possibility of dominance by variables with relatively high absolute magnitudes of variability, but there is a price, and that price might be felt to be too high in any given research application. Intuitively, real-world objects can be distinguished from one another in proportion to the degree to which they differ: identical objects cannot be distinguished, objects that differ moderately from one another are moderately easy to distinguish, and so on. Data variables used to describe real-world objects to be clustered are therefore useful in proportion to the variability in their values: a variable with no variability says that the objects are identical with respect to the characteristic it describes and can therefore contribute

Table 3.15: Unstandardized and z-standardized versions of a matrix

	v_1	v_2	v_3	v_1	v_2	v_3
1	100	20	1000.10	-1.39	-1.38	-0.42
2	110	21	1000.11	-1.27	-1.34	-0.36
3	120	22	1000.08	-1.15	-1.29	-0.54
4	130	40	1000.01	-1.03	0.49	-0.94
5	200	41	1000.20	-0.18	0.44	0.15
6	210	42	1000.07	-0.06	0.40	-0.60
7	220	60	1000.23	0.06	0.40	0.32
8	230	61	1000.30	0.18	0.44	0.73
9	300	62	1000.03	1.03	0.49	-0.83
10	310	80	1000.14	1.15	1.29	-0.19
11	320	81	1000.13	1.27	1.34	-0.25
12	330	82	1000.68	1.39	1.38	2.94
Std. dev.	82.41	22.38	1	1	1	1
Coeff. var.	0.38	0.44	0.0002			

nothing as a clustering criterion, a variable with moderate variability says that the corresponding objects are moderately distinguishable with respect to the associated characteristic and is therefore moderately useful as a clustering criterion, and again so on. Variables v_1 and v_2 in Table 3.15 have high intrinsic variabilities relative to v_3 and are therefore more useful clustering criteria than v_3; in fact, the variability of v_3 is so small that it could be the result of random observational noise with respect to a characteristic that is constant across the objects to be clustered. To equate v_3 with v_1 and v_2 in terms of its influence on clustering, as z-standardization does, cannot be right. Rescaling data values so that all variables have an identical absolute magnitude of variability diminishes the distinguishing power of high-variability variables and enhances the power of low-variability ones relative to what is warranted by observed reality. In other words, z-standardization can distort the validity of data as an accurate description of reality, and this is the reason why it should be used with caution or not at all in data preparation for cluster analysis.

For multivariate data whose variables are measured on different scales, what is required is a standardization method that, like z-standardization, eliminates the distorting effect of disparity of variable scale on clustering but, unlike z-standardization, also preserves the relativities of size of the pre-standardization intrinsic variabilities in the post-standardization absolute magni-

Table 3.16: Unstandardized and mean-standardized versions of a matrix

	v_1	v_2	v_3	v_1	v_2	v_3
1	100	20	1000.10	0.46	0.39	0.9999
2	110	21	1000.11	0.51	0.41	0.9999
3	120	22	1000.08	0.55	0.43	0.9999
4	130	40	1000.01	0.60	0.78	0.9998
5	200	41	1000.20	0.93	0.80	1.0000
6	210	42	1000.07	0.97	0.82	0.9999
7	220	60	1000.23	1.02	1.17	1.0001
8	230	61	1000.30	1.06	1.19	1.0001
9	300	62	1000.03	1.39	1.21	0.9998
10	310	80	1000.14	1.44	1.56	0.9999
11	320	81	1000.13	1.48	1.58	0.9999
12	330	82	1000.68	1.53	1.60	1.0005
Std. dev.	82.41	22.38	0.17	0.38	0.44	0.00018
Coeff. var.	0.38	0.44	0.0002	0.38	0.44	0.00018

tudes of variability. In other words, what is required is a method that generates standardized variable vectors such that the ratios of their absolute magnitudes of variability are identical to those of the intrinsic variabilities of the unstandardized ones. In this way the standardized variables can influence the clustering in proportion to the real-world distinguishability of the objects they describe. Such a method follows.

The literature (Chu, Holliday, and Willett 2009; Gnanandesikan, Tsao, and Kettenring 1995; Milligan and Cooper 1988) contains a variety of alternatives to z-standardization, but, relative to the desiderata just stated, one of them seems the obvious choice: mean-standardization, which was first proposed by (Anderberg 1973). This standardization involves division of the values of a numerical vector v by their mean μ_v, as in Equation (3.16).

$$v_{std} = \frac{v}{\mu_v} \tag{3.16}$$

The right-hand side of Table 3.16 shows the application of mean-standardization to the column vectors of the unstandardized matrix on the left. Note that mean-standardization has preserved the coefficients of variation of the unstandardized variables. This is because division by a scalar – here the column vector mean – is a linear operation that alters the scale while preserving the shape of the original value distribution, as shown in Figure 3.22.

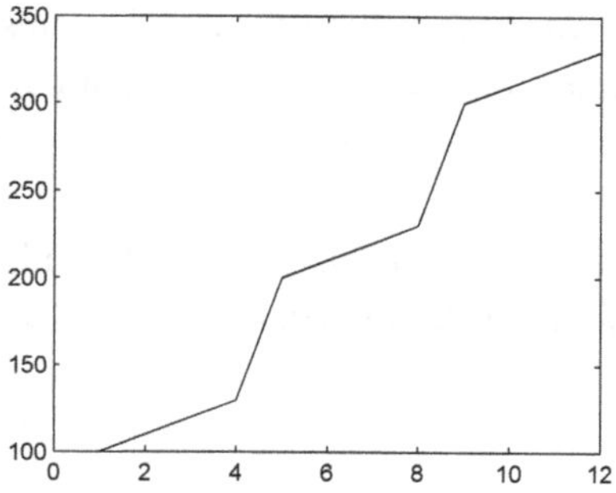

(*a*) a: Plot of unstandardized v_1 from Table 3.16

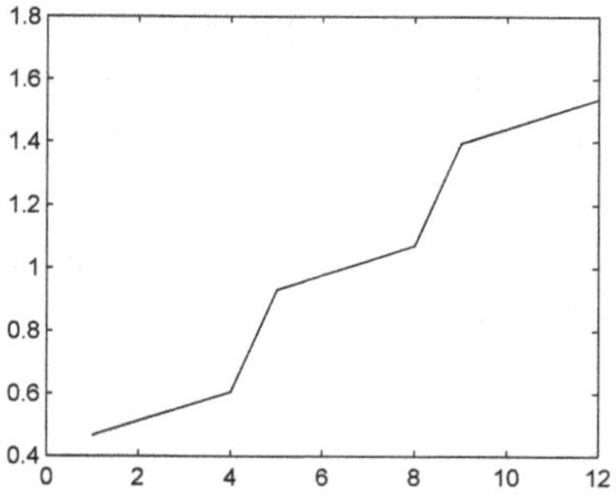

(*b*) b: Plot of mean-standardized v_1 from Table 3.16

Figure 3.22: Preservation of distribution shape by linear transformation

Note also that the mean-standardized standard deviations of $v_1 - v_3$ in Table 3.16 are identical to the corresponding coefficients of variation. This is because, for any vector v, it is always the case that its coefficient of variation is identical to the standard deviation of the mean-standardized version of v.

Table 3.17: Calculation of coefficient of variation and standard deviation of a vector v

Coefficient of variation of v	$CoeffVar(v) = 1/\mu(StdDev(v))$
Std. dev. of mean-standardized v	$StdDev(v_{mean-std}) = StdDev(1/\mu(v)$

Table 3.17 shows that for the coefficient of variation of v the standard deviation is calculated first and then multiplied by the inverse of the mean, and for the standard deviation of the mean-standardized version of v, v is first divided by its mean and the standard deviation of the result then calculated. But one of the properties of the standard deviation is that, for a vector v and a constant c, $StdDev(cv) = cStdDev(v)$, or, in other words, the two are mathematically equivalent.

Since, therefore, (i) the coefficient of variation is a scale-independent measure of variability, and (ii) the standard deviation of a mean-standardized variable is always identical to the coefficient of variation of the unstandardized variable, and (iii) the standard deviation of a variable is what measures its absolute magnitude of variability, mean-standardization fulfils the above-stated requirements for a general standardization method: that it eliminate the distorting effect of disparity of variable scale on clustering while preserving the ratios of the intrinsic variabilities of the unstandardized variables in the ratios of the absolute magnitudes of variation of the standardized ones. The absolute magnitudes of variation of mean-standardized variables are identical to the intrinsic variabilities of the unstandardized ones, and hence so are the ratios.

Figures 3.23a–3.23c compare the cluster trees for the unstandardized, z-standardized, and mean-standardized versions of the matrix in Table 3.16.

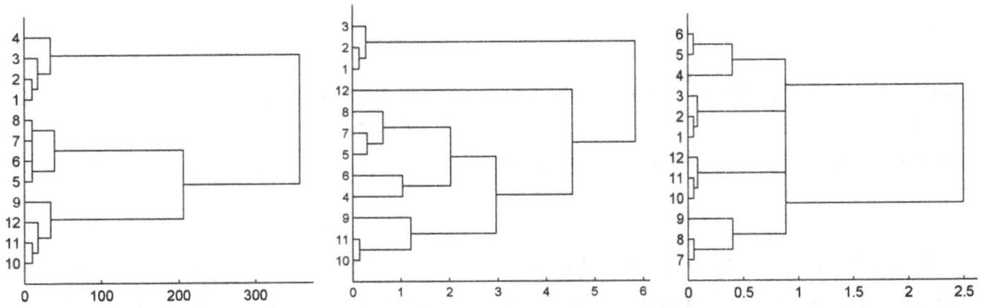

(a) Unstandardized *(b)* z-standardized *(c)* Mean-standardized

Figure 3.23: Cluster analyses of unstandardized, z-standardized, and mean-standardized versions of the matrix in Table 3.16

Direct inspection of the unstandardized matrix in Figures 3.23a–3.23c reveals three value-groups for v_1, four groups for v_2, and small random variations on a constant for v_3. The primary clustering in Figure3.23a is by v_1 because it has the highest absolute magnitude of variability and subclustering within the three primary clusters is by v_2, with the effect of v_3 invisible, all as expected. The cluster tree for the z-standardized matrix is much more complex, and any sense of the groups observable in v_1 and v_2 is lost as the clustering algorithm takes account of the numerically much-enhanced random variation in v_3 generated by z-standardization; the tree in Figure 3.23b bears no obvious relationship to any reasonable intuition about structure in the unstandardized matrix. The mean-standardized tree, however, captures these intuitions very well: there are four primary clusters corresponding to the four numerical groups in v_2, which has the highest intrinsic variability and therefore represents the characteristic that most strongly distinguishes objects 1-12 from one another in the real world; the effect of v_2, the variable with the next-highest intrinsic variability, is seen in the internal structures of the primary clusters, so that, for example, the flat subtree for objects 1–3 corresponds to very similar row vectors in the unstandardized matrix, the segregation of 4 from 5 and 6 in the subtree corresponds to the anomalously-low value of 130 in row 4 of the unstandardized matrix, and similarly for the remaining two groups 7–9 and 10–12; the influence of v_3, with its very low intrinsic variability, is invisible.

Most statistics, data processing, and cluster analysis textbooks say something about standardization. The z-standardization procedure is always mentioned and, when different methods are cited or proposed, there is typically little discussion of the relative merits of the alternatives, though, as noted earlier, quite a few express reservations about z-standardization. The relatively few studies that are devoted specifically to the issue are empirical, that is, they assess various methods' effectiveness in allowing clustering algorithms to recover clusters known a priori to exist in specific data sets, and their conclusions are inconsistent with one another and with the results of the present discussion. Milligan and Cooper (1988) compared eight methods and concluded that standardization using the range of the variables works best; Gnanandesikan, Tsao, and Kettenring (1995) proposed and favoured one that uses estimates of within-cluster and between-cluster variability, though also noted that "much more research is needed before one attempts to cull out the best approaches"; Chu, Holliday, and Willett (2009) concluded that, for the data they used, "there is no consistent performance benefit that is likely to be obtained from the use of any particular standardization method". For a principled comparison of the various standardization methods which demonstrates the superiority of mean-standardization, see Moisl (2010). In applications where preservation of the intrinsic variabilities of data variables is felt to be important for reliable cluster analysis, therefore, mean-standardization should be used.

3.2.2.2 Normalization

This section deals with a problem that arises when clustering is based on frequency data abstracted from multi-document corpora and there is substantial variation in the lengths of the documents. The discussion is in three main parts. It first shows why variation in document length can be a problem for frequency-based clustering, then goes on to describe a matrix transformation or 'normalization' designed to deal with the problem, and finally shows that such normalization is ineffective where documents are too short to provide reliable probability estimates for data variables. The 63 interviews that comprise the DECTE corpus differ substantially in length and so, consequently, do the phonetic transcriptions of them. Figure 3.24 shows the relative lengths of the transcriptions.

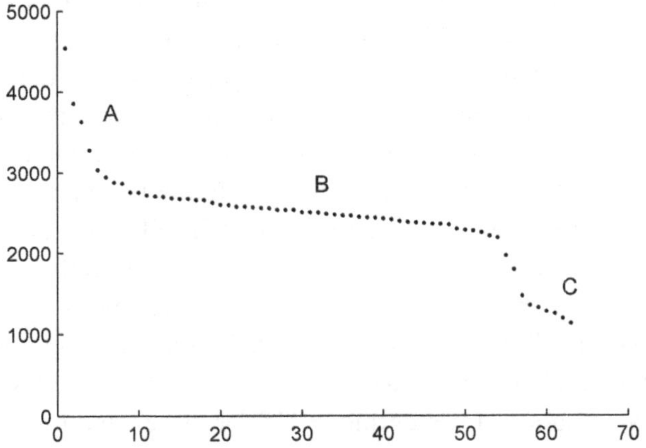

Figure 3.24: Lengths of the DECTE phonetic transcriptions in Kb

Most of the transcriptions in Figure 3.24, labelled B, are in the range $\approx 15-20$ Kb, but a few (A) are substantially longer and a few (C) substantially shorter. The rows of the MDECTE abstracted from the transcriptions were cluster analyzed, and the result is shown in Figure 3.25.

The labels (A)–(C) in Figure 3.25 correspond to those in Figure 3.24, and the correspondence exemplifies a general rule often observed in the data processing literature: that, for frequency data abstracted from multi-document corpora, variation in document length will affect clustering to greater or lesser degrees depending on the degree of variation. In the present case relatively long, relatively short, and intermediate-length transcriptions cluster irrespective of the phonetic usage of the speakers; as we shall see, elimination of interview length variation as a factor in the data yields a very different clustering result than the one in Figure 3.25.

The reason for this effect is easy to see. Whatever set of linguistic features one is counting, be they phonetic, phonological, morphological, lexical, syntactic, or semantic, it is in general probable that a longer document will contain more instances of those features than a shorter one: a newspaper will, for example, contain many more instances of, say, the word 'the' than an average-length email. If frequency profiles for varying-length documents are constructed, as here for the phonetic usage of the DECTE speakers, then the

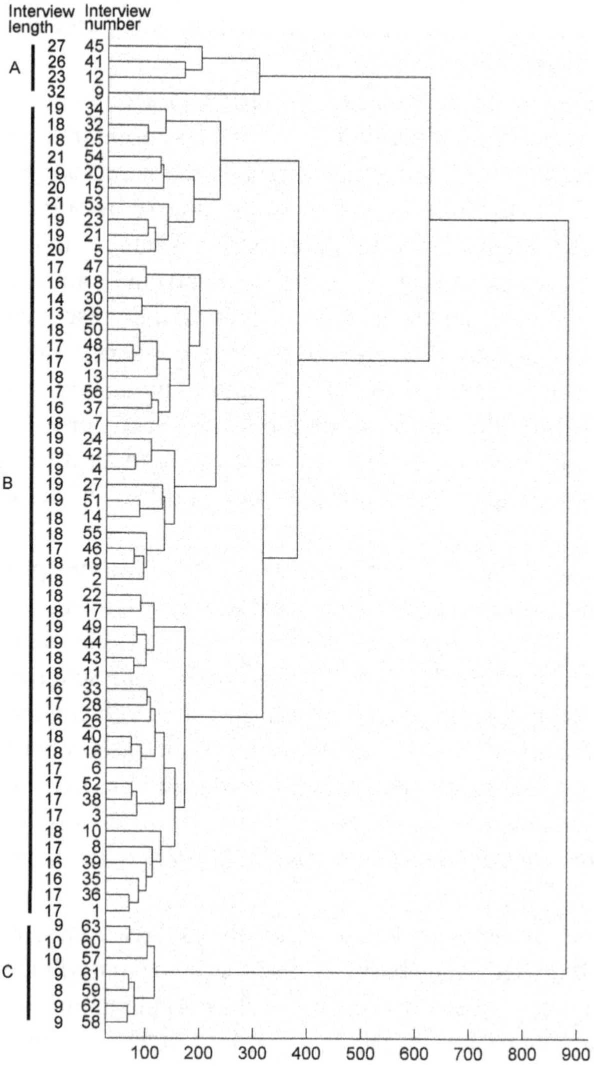

Figure 3.25: Cluster analysis of the MDECTE matrix

profiles for the longer documents will, in general, have relatively high values and those for the shorter documents relatively low ones. The preceding discussion of scaling has already observed that clustering is strongly affected by the relative magnitudes of variable values. When, therefore, the rows of a frequency matrix are clustered, the profiles are grouped according to relative frequency magnitude, and the grouping will thus be strongly influenced by document length.

The solution to the problem of clustering in accordance with document length is to transform or 'normalize' the values in the data matrix in such a way as to mitigate or eliminate the effect of the variation. Normalization is accomplished by dividing the values in the matrix by some constant factor which reflects the terms in which the analyst wants to understand the data; a statistician, for example, might want to understand data variation in terms of standard deviation, and so divides by that. In the present case the normalization factor is document length, so that the frequency values representing any given document are divided by its length or by the mean length of all the documents in the collection to which it belongs. Such normalization is an important issue in Information Retrieval because, without it, longer documents in general have a higher probability of retrieval than shorter ones relative to any given query.

The associated literature consequently contains various proposals for how such normalization should be done – for example Greengrass (2001: 20ff.), Singhal et al. (1996), Singhal, Buckley, and Mitra (1996), Spärck Jones, Walker, and Robertson (2000), Manning, Raghavan, and Schütze (2008: Ch. 6). These normalizations are judged in terms of their effectiveness for retrieval of relevant documents and exclusion of irrelevant ones rather than for cluster analysis, and the cluster analysis literature has little to say on the subject, so it is presently unclear what the best document length normalization method for cluster analysis might be among those currently in the literature, or indeed what the criteria for 'best' are.

Normalization by mean document length (Spärck Jones, Walker, and Robertson 2000) is used as the basis for discussion in what follows because of its intuitive simplicity. Mean document length normalization involves transformation of the row vectors of the data matrix in relation to the average length of documents in the corpus being used, and, in the present case, transformation of the row vectors of MDECTE in relation to the average length of the $m = 63$ DECTE phonetic transcriptions, as in Equation (3.17).

$$M_i = M_i(\frac{\mu}{length(T_i)}) \tag{3.17}$$

where M_i is the matrix row representing the frequency profile of i'th DECTE transcription T_i, $length(T_i)$ is the total number of phonetic segments in T_i, and μ is the mean number of phonetic segments across all transcriptions T in DECTE, as in Equation (3.18).

$$\mu = \sum_{i=1..m} \frac{length(T_i)}{m} \tag{3.18}$$

The values in each row vector M_i are multiplied by the ratio of the mean number of segments per transcription across the set of transcriptions T to the number of segments in transcription T_i. The longer the document the numerically smaller the ratio, and vice versa; the effect is to decrease the values in the vectors that represent long documents, and increase them in vectors that represent short ones, relative to average document length.

MDECTEwas normalized by mean document length and then cluster analyzed using the same clustering method as for Figure 3.25, and the result is shown in Figure 3.26.

The tree in Figure 3.26 differs substantially from the one in 3.25. The constituents of cluster C have remained unchanged, but as subsequent discussion will show, there is good phonetic reason for this. Note, however, that the relatively long transcriptions comprising cluster *A* in Figure 3.25 have now been distributed among those in cluster *B*.

Caveat emptor, however. Mean document length normalization has eliminated variation in transcription length as a factor in clustering of the DECTE speakers. There is a limit to the effectiveness of normalization, however, and it has to do with the probabilities with which the linguistic features of interest occur in the corpus. Given a population E of n events, the empirical interpretation of probability (Milton and Arnold 2003: Ch. 1) says that the probability $p(e_i)$ of $e_i \in E$ (for $i = 1..n$) is the ratio $frequency(e_i)/n$, that is, the proportion of the number of times e_i occurs relative to the total number of occurrences of events in E. A sample of E can be used to estimate $p(e_i)$, as is done with, for example, human populations in social surveys. The Law of Large Numbers (ibid.: 227f.) in probability theory says that, as sample size increases, so does the likelihood that the sample estimate of an event's population probability is accurate: a small sample might give an accurate estimate but is less likely to do so than a larger one, and for this reason larger samples are preferred.

Applying these observations to the present case, each of the constituent transcriptions of T is taken to be a sample of the population of all Tyneside speakers. The longer the transcription the more likely it is that its estimate of the population probabilities of the 156 phonetic segment types in T will

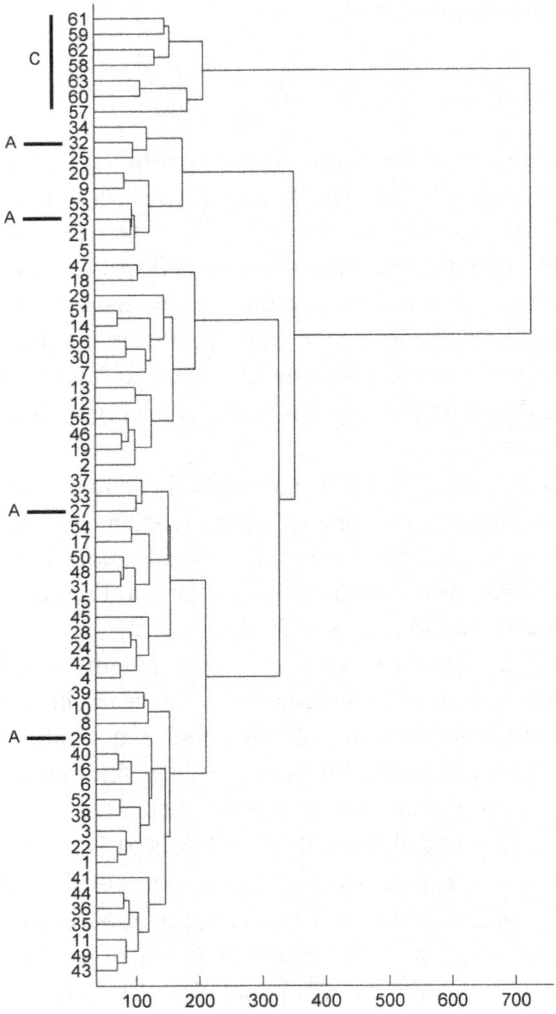

Figure 3.26: Cluster analysis of normalized MDECTE matrix

be accurate, and, conversely, the shorter the transcription the less likely this will be. It is consequently possible that a very short transcription will give very inaccurate probability estimates for the segment types. The normalization procedure will then accentuate this inaccuracy, and this will in turn affect the validity of the clustering. The obvious solution to the problem of poor population probability estimation by short documents or transcriptions is to determine which documents in the collection of interest are too short to provide reasonably good estimates and to eliminate the corresponding rows from the data matrix. But how short is too short? The answer lies in statistical sampling theory; for further details see Moisl (2011).

3.2.2.3 Dimensionality reduction

The dimensionality of data is the number of variables used to describe the data objects: data describing humans in terms of height and weight are two-dimensional, the weather in terms of temperature, atmospheric pressure, and wind speed are three-dimensional, and so on to any number of dimensions n. Reducing the dimensionality of data as much as possible with as little loss of information as possible is a major issue in data analysis across a wide range of research disciplines (Lee and Verleysen 2007; Verleysen 2003), and it is so for cluster analysis as well. One frequently cited and self-evidently important reason for this in the cluster analysis literature is that demand for computational resources is reduced by minimizing the number of variables included in the analysis (Kaufman and Rousseeuw 1990). Another is noise reduction: typically, not all variables included in an analysis are equally important in describing the objects in the research domain, and elimination of the less important ones removes random noise which can adversely affect the results (ibid.). There is, however, a third and rather deep reason for dimensionality reduction which has to do with the mathematical characteristics of high-dimensional spaces (Jimenez and Landgrebe 1998; Köppen 2000; Lee and Verleysen 2007; Verleysen 2003), and this is reviewed before going on to look at dimensionality reduction methods.

As will be seen, cluster analysis is based on measurement of proximity between and among data objects in n-dimensional space, and the discussion of data geometry has presented some proximity measures. For low-dimensional spaces, that is, for spaces where $n = 2$ and $n = 3$ which can be graphically represented, these measures are intuitively reasonable. In Figure 3.27, for example, the visual intuition is that the distance and angle between vector v and vector w are greater than the distance and angle between w and x, and quantitative measures like Euclidean distance and cosine confirm this intuition.

The manifold for MDECTEconsists of 63 vectors in a 156-dimensional vector space; it cannot be shown as in Figure 3.27 because dimensionalities

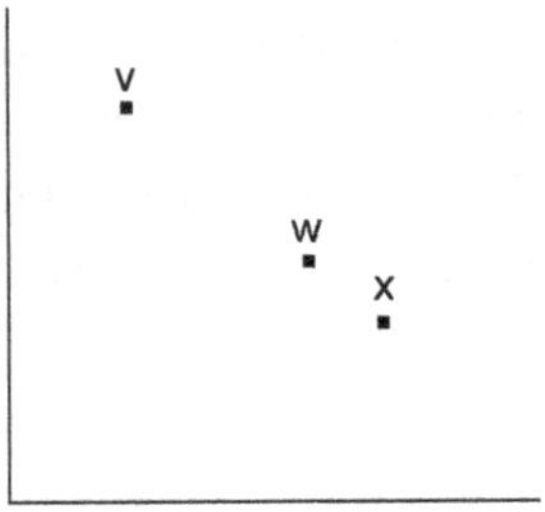

Figure 3.27: Relative proximities of vectors

greater than 3 cannot be directly graphically represented, but the principle for distance and angle measurement between vectors is the same. Or rather, it appears to be the same. Research into the properties of high-dimensional spaces has shown that geometrical intuitions based on conventional low-dimensional spaces are unreliable with respect to higher-dimensional ones, and that, as dimensionality grows, counterintuitive effects become ever more prominent (Bishop 2006; Jimenez and Landgrebe 1998; Köppen 2000; Lee and Verleysen 2007; Verleysen 2003). It was noted earlier that two and three dimensional spaces are a useful metaphor for conceptualizing higher dimensional spaces, but that they are no more than that and can easily mislead. The limitations of the metaphor quickly become apparent when one tries to apply intuitions about physical space based on human experience of the world to higher-dimensional data space. Consider the concept of vector space size (Donoho 2000; Köppen 2000; Verleysen 2003). Figure 3.27 shows what happens to the size of a cube as dimensionality increases from 3 to 100, where size is measured in terms both of volume and of the length of the diagonal from the origin to the opposite corner.

There are some highly counterintuitive effects here: in Figure 3.28a the length of the diagonal increases even though the volume remains constant; in Figure 3.28b the length of the diagonal increases even though the volume converges to 0; in Figure 3.28c and d the volume quickly starts to grow at a much greater rate than the length of the diagonal. The intuitive expectation based on experience of the three-dimensional physical world is that there should be proportionality between volume and diagonal length irrespective of the scaling of the data values, but that is not the case: rescaling of the data to axis lengths that are less than, equal to, or greater than 1 fundamentally alters their relationship. Volume is a human intuition based on experience

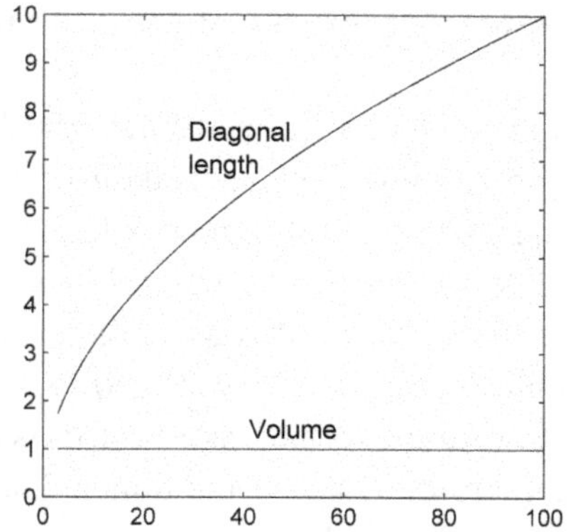

(a) Size of cube with side length 1 for dimensionality 3..100. Volume: Constant 1. Diagonal: 1.7321 .. 10.0000

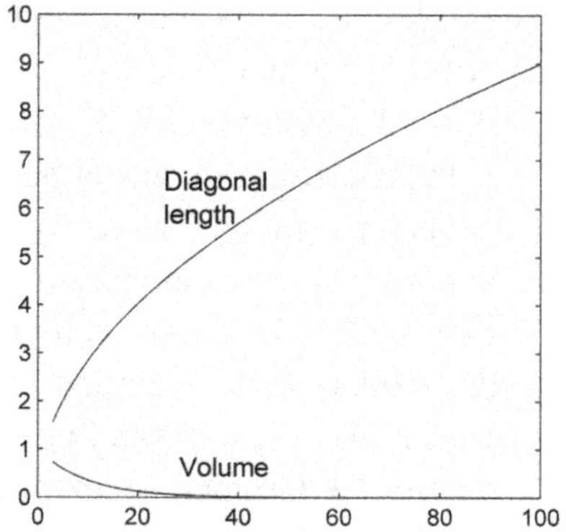

(b) Size of cube with side length 0.9 for dimensionality 3..100. Volume: 0.7290 .. 0.0000. Diagonal: 1.5588 .. 9.0000

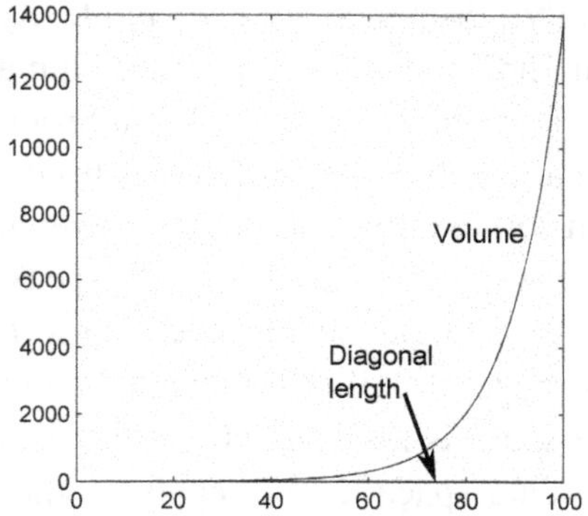

(c) Size of cube with side length 1.1 for dimensionality 3..100. Volume: 1.3310 .. 13780.6123. Diagonal: 1.9053 .. 11.0000

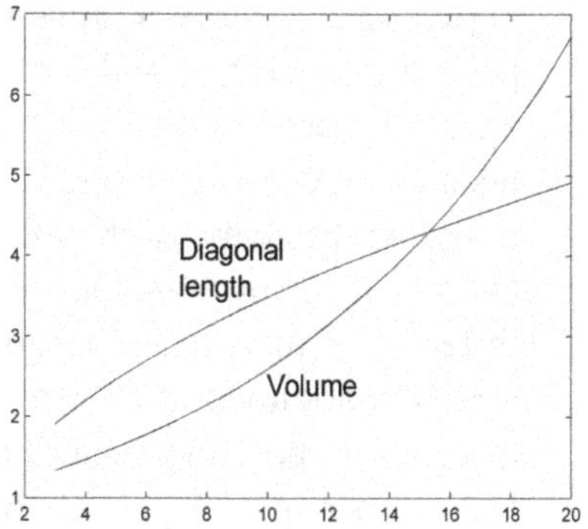

(d) Magnification of range 0..20 of (c)

Figure 3.28: Effect of dimensionality increase on the size of a cube

of the physical world, and the mathematical formulation of it captures the intuition for dimensionality 3. Beyond that dimensionality volume becomes intuitively meaningless, and the mathematical formulation of it reduces to the well known effects of multiplying values less than, equal to, or greater than 1 n times.

The following properties of high-dimensional spaces are of particular relevance for cluster analysis:

- For a fixed number of data vectors m and a uniform and fixed variable value scale, the manifold becomes increasingly sparse as their dimensionality n grows. To see this, assume some bivariate data in which both variables take values in the range 0..9: the number of possible vectors like $(0,9)$, $(3,4)$, and so on is $10 \times 10 = 100$. For trivariate data using the same range the number of possible vectors like $(0,9,2)$ and $(3,4,7)$ is $10 \times 10 \times 10 = 1000$. In general, the number of possible vectors is r^d, where r is the measurement range (here 0..9) and d the dimensionality. The r^d function generates an extremely rapid increase in data space size with dimensionality: even a modest $d = 8$ for a 0..9 range allows for $100,000,000$ vectors.

  This very rapid increase in data space size with dimensionality is widely known as the 'curse of dimensionality', discussed in, for example, Köppen (2000) and Lee and Verleysen (2007), and it is a problem in many areas of science and engineering. For cluster analysis it is a problem because, the higher the dimensionality, the more difficult it becomes to define the shape of the manifold sufficiently well to achieve reliable analytical results. Assume that we want to analyse, say, 24 speakers in terms of their usage frequency of 2 linguistic variables; these features are assumed to be rare, so a range of 0..9 is sufficient. The ratio of actual to possible vectors in the space is $24/100 = 0.24$, that is, the vectors occupy 24 percent of the data space. If one analyses the 24 speakers in terms of 3 phonetic segments, the ratio of actual to possible vectors is $24/1000 = 0.024$ or 2.4 percent of the data space. In the 8-dimensional case it is $24/100000000$, or 0.00000024 percent. A fixed number of vectors occupies proportionately less and less of the data space with increasing dimensionality. In other words, the data space becomes so sparsely inhabited by vectors that the shape of the manifold is increasingly poorly defined (Verleysen 2003).

  What about using more data? Let's say that 24 percent occupancy of the data space is judged to be adequate for manifold resolution. To achieve that for the 3-dimensional case one would need 240 vectors, 2400 for the 4-dimensional case, and 24,000,000 for the 8-dimensional one. This may or may not be possible. And what are the prospects for dimensionalities higher than 8?

– As dimensionality grows, the distances between pairs of vectors in the space become increasingly similar. In the relevant information retrieval and data mining literature, proximity between vectors in a space is articulated as the 'nearest neighbour' problem: given a set V of n-dimensional vectors and an n-dimensional vector w not in V, find the vector v in V that w is closest to in the vector space. This is an apparently straightforward problem easily soluble by, for example, calculating the Euclidean distance between w and each of the v in V, and selecting the shortest one.
  As dimensionality increases, however, this straightforward approach becomes increasingly unreliable because "under certain broad conditions ... as dimensionality increases, the distance to the nearest neighbour approaches the distance to the farthest neighbour. In other words, the contrast in differences to different data points becomes nonexistent" (Beyer et al. 1999); on this see further: Aggarwal, Hinneburg, and Keim (2001), François, Wertz, and Verleysen (2007), Hinneburg, Aggarwal, and Keim (2000), Korn, Pagel, and Faloutsos (2001), and Steinbach, Ertöz, and Kumar (2004). This effect can, moreover, appear for dimensionalities as low as 10–15 (Beyer et al. 1999).
  To demonstrate this, a sequence of 1000 matrices, each with 100 rows containing random values in the range $0 \ldots 1$, was generated such that the first matrix had dimensionality $k = 1$, the second had dimensionality $k = 2$, and so on to dimensionality $k = 1000$, as shown in Table 3.18.

Table 3.18: Fragments of matrices of increasing dimensionality

	1	1	2	1	2	...	1000
1	0.492	0.098	0.472	0.595	0.773	...	0.816
2	0.278	0.393	0.749	0.196	0.825	...	0.113
...	...	...	...	...	...	...	...
100	0.397	0.516	0.073	0.649	0.772	...	0.776

For each k a distance matrix containing the Euclidean distances between all possible pairings of the 100 row vectors was calculated; a fragment of the distance matrix for $k = 3$, for example, is shown in Table 3.19.

Table 3.19: Euclidean distance matrix for one of the 1000 matrices of Table 3.18

	1	2	...	1000
1	0	0.686	...	0.741
2	0.686	0	...	0.250
...	...	...	...	...
100	0.741	0.250	...	0

The Euclidean distance between row vector 1 and itself is 0, between row vector 1 and 2 is 0.686, and so on. For each of the 1000 distance matrices the maximum distance value max_k and the minimum distance value min_k in the matrix were found and the ratio min_k/max_k was calculated. The result is plotted in Figure 3.29.

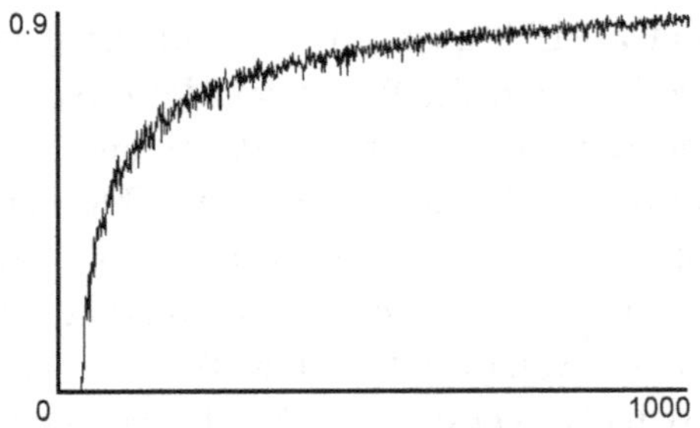

Figure 3.29: Min/max ratios for a sequence of 1000 distance matrices for dimensionality $1 \ldots 1000$

Figure 3.29 shows that, as dimensionality increases, (i) the ratio of minimum to maximum distance among vectors approaches 1, that is, they become increasingly similar, and (ii) the increase in similarity occurs very rapidly at relatively low dimensionality and then levels off. This means that it quickly becomes increasingly difficult to distinguish points from one another on the basis of distance. This phenomenon, where pairwise distances are the same for all points, is called 'concentration of distances' and makes some of the concepts we take for granted in low dimensional spaces meaningless, such as 'nearest neighbour'.

The implication for clustering is straightforward: because the most popular cluster analysis methods group vectors on the basis of their relative distances from one another in a vector space, as the distances between vectors in the space approach uniformity it becomes less and less possible to cluster them reliably.

One response to these characteristics of high-dimensional data is to use it as is and live with the consequent unreliability. The other is to attempt to mitigate their effects by reducing the data dimensionality. The remainder of this section addresses the latter alternative by presenting a range of dimensionality reduction methods.

The foregoing discussion of data creation noted that variables selected for a research project are essentially a first guess about how best to describe the domain of interest, and that the guess is not necessarily optimal. It may, therefore, be the case that the initial selection can be refined and, more specifically, that the number of variables can be reduced without losing too much relevant information. Given the importance of dimensionality reduction in data processing generally, there is an extensive literature on it and that literature proposes numerous reduction methods. The following account of these methods cannot be exhaustive. Instead, the aim is to provide an overview of the main approaches to the problem and a discussion of the methods that are most often used and / or seem to the author to be most intuitively accessible and effective for corpus linguists.

The methods for dimensionality reduction are of two broad types. One type selects a subset of the more important data variables and eliminates the remainder from the analysis, using some definition of importance. The other type abstracts a new and usually much smaller set of variables on the basis of the existing ones. In the relevant machine learning, artificial intelligence, and cognitive science literatures these approaches to dimensionality reduction are called 'feature selection' and 'feature extraction', but the present discussion has so far used the more generic term 'variable' for what these disciplines call features, and will continue to do so.

The dimensionality of data can be reduced by retaining variables which are important and eliminating those which are not, relative to some criterion of importance; for data in vector space format, this corresponds to eliminating the columns representing unimportant variables from the data matrix. The literature on variable selection is extensive – for summary accounts see Guyon and Elisseeff (2003), Kim, Street, and Menczer (2003), and Liu and Motada (2008) – but most of it is not directly relevant to cluster analysis. As subsequent discussion explains, there is a fundamental distinction between clustering, where the number and composition of clusters is inferred from data, and classification, where the number and characteristics of the clusters is prespecified and used to assign data items to the correct cluster. Almost all the variable selection literature is concerned with classification: given a set of variables, the various methods described in that literature work either by adding variables incrementally to a small seed selection or by deleting variables incrementally from the full set and observing the effect relative to an objective function which evaluates the 'goodness' of the current variable selection in predicting cluster membership. In other words, variable selection methods for classification require prespecification of clusters to work. The present discussion is concerned only with clustering, however, and as such these classification-oriented methods can be disregarded; this section deals only with variable selection for clustering, on which see Devaney and Ram (1997), Dy (2008), Dy and Bodley (2004), Jain, Murty, and Flynn (1999), Kim, Street, and Menczer (2003), Liu and Motada (2008), Tan, Steinbach, and Kumar (2006), and Xu and Wunsch (2009).

The point of cluster analysis is to group objects in a domain of interest in terms of their relative degrees of similarity based on the variables used to describe them. Intuitively, a variable is useful for this purpose if it has the following characteristics: frequency, variability, and nonrandomness. We will first briefly discuss these three characteristics in general before presenting ways of selecting variables for each.

- *Frequency*: In general, a variable should represent something which occurs often enough for it to make a significant contribution to understanding of the domain. For example, in the DECTE interviews the two most frequent phonetic segments occur 12454 and 8255 times respectively out of a total of 157116 segment tokens across all 63 speakers, but 13 segments occur only once. The frequent segments are prominent features which any attempt to understand the phonetics of Tyneside speech must take into account, whereas the infrequent ones tell one little about Tyneside speech and may well be just noise resulting from speaker mispronunciation or transcription error.
- *Variability*: The values which the variable takes should vary substantially. As the discussion of variable scaling has already noted, real-world objects can be distinguished from one another in proportion to the degree to which they differ: identical objects cannot be distinguished, objects that differ moderately from one another are moderately easy to distinguish, and so on. Data variables used to describe real-world objects are therefore useful for distinguishing the objects

they describe in proportion to the variability in their values: a variable with no variability says that the objects are identical with respect to the characteristic it describes and can therefore contribute nothing to distinction of objects, a variable with moderate variability days that the corresponding objects are moderately distinguishable with respect to the associated characteristic and therefore moderately useful for the purpose, and again so on.

- *Nonrandomness*: The variation in the values which the variable takes should be nonrandom. Random variation of the aspect of the domain which the variable describes means that there is no systematic variation among objects, and all one can say on this basis is that, in this respect, the objects differ, which is obvious from the outset. A variable is, therefore, useful for clustering to the extent that the values which it takes have a nonrandom distribution of variability among objects.

The remainder of this section presents ways of selecting variables for each of these criteria, then identifies associated problems, and finally proposes a way of resolving the problems.

Frequency

An $m \times n$ frequency matrix F is constructed in which the value at F_{ij} is the number of times variable j (for $j = 1 \ldots n$) occurs in document i (for $i = 1 \ldots m$). The frequency of occurrence of variable j across the whole corpus is given in Equation (3.19).

$$freq(F_j) = \sum_{i=1..m} F_{ij} \tag{3.19}$$

Frequencies for all the columns of F are calculated, sorted, and the less frequent variables are removed from F, thereby reducing the dimensionality of F. MDECTE is a frequency matrix, so the summation and sorting process can be applied directly; the result of doing so is shown in Figure 3.30.

Variability

The degree of variability in the values of a variable is described by its variance or, expressed in the original units of measurement, its standard deviation. Given a data matrix in which the rows are the objects of interest and the

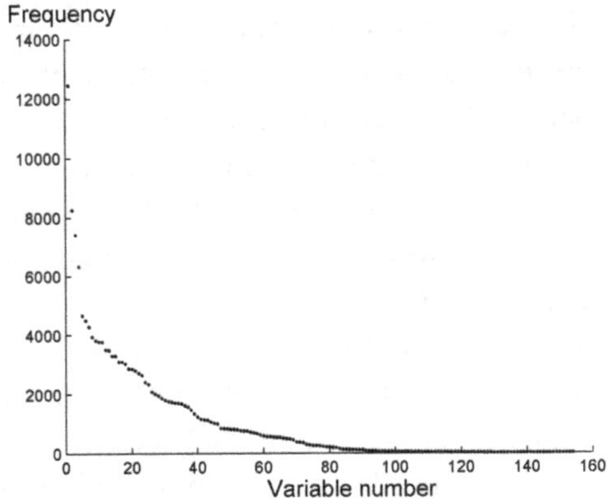

Figure 3.30: Sorted MDECTE variable frequencies

columns are variables describing them, the application of variance to dimensionality reduction of a data matrix is again straightforward: sort the column vectors in descending magnitude of variance and use a plot of the values to decide on a suitable threshold k below which all columns are eliminated. Figure 3.31 shows this for MDECTE.

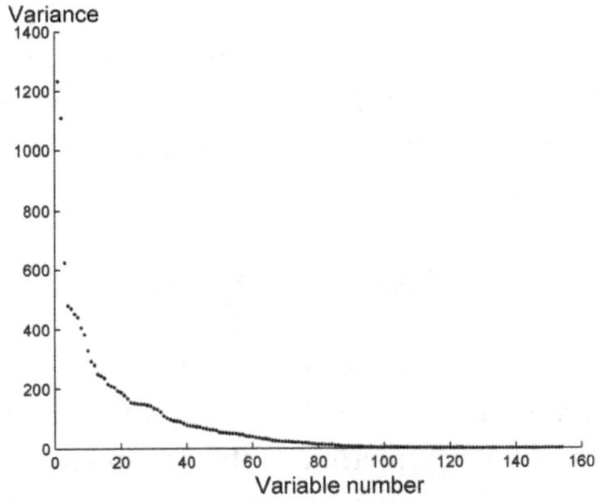

Figure 3.31: Sorted MDECTE variable variances

Note that, where variables are measured on different scales, conclusions about their relative variabilities based on the magnitudes of their variances can be misleading. The foregoing discussion of variable scaling made a distinction between absolute and intrinsic variability, where the first is the amount of variation in values expressed in terms of the scale on which those values are measured and the second is the amount of variation expressed independently of scale. Absolute magnitude of variability is measured by standard deviation, and comparison of the standard deviations of a set of variables therefore offers a scale-dependent assessment of their variabilities. The discussion of variable scaling also showed why scale dependence can be misleading – essentially, because the magnitude of a variable's standard deviation is strongly influenced by the magnitude of its values, so that, judged by its standard deviation, a variable with a relatively lower intrinsic variability but relatively larger values can appear to have greater variability than one with relatively higher intrinsic variability but relatively smaller values. For this reason, intrinsic variability as measured by the coefficient of variation, introduced earlier, should used as the criterion for variable selection where variables are measured on different scales. All the variables in MDECTE are measured on the same scale, segment frequency, and so this problem does not arise.

Nonrandomness

Two approaches to assessing nonrandomness in the distribution of variability are considered here: Poisson distribution and term frequency – inverse document frequency.

A widely used measure of nonrandomness is the ratio of the variance of a set of values to their mean. To understand this measure it is first necessary to understand the Poisson distribution, a statistical model of randomness. This part of the discussion briefly introduces the Poisson distribution, then shows how the variance-to-mean ratio relates to it, and finally describes the application of the ratio to dimensionality reduction.

The Poisson distribution models the number of times that a random and rare event occurs in some specified spatial or temporal interval; see for example (Walpole et al. 2007: 161ff.). More specifically, it models data generated by physical stochastic processes, where a physical stochastic process is one that generates events randomly over some interval of time or space in a domain of interest – the fluctuation in share values on the stock market over a given week, for example. A Poisson process is a stochastic process in which (i) the random events occur independently of one another, and (ii) the probability of occurrence of the random events over some designated interval i is described by the probability density function p shown in Equation (3.20).

$$p(x = r) = \frac{e^{-\lambda}\lambda^{r}}{r!} \tag{3.20}$$

where:

- p is a probability.
- x is the variable in question.
- r is the number of events that occur over an interval i, and $r!$ is r *factorial.*
- e is the base of the natural logarithm, that is, 2.71828.
- λ is the mean value of x over many intervals i.

For a Poisson process whose mean rate of occurrence of events λ over the designated interval i is known, therefore, this function gives the probability that some independently specified number r of events occurs over i. For example, assume that 7 cars pass through a rural intersection on Thursday, 3 on Friday, and 5 on Saturday; the mean number λ of cars passing through the intersection on a given day is 5. What is the probability that 4 cars will pass through on Sunday? The calculation is given in Equation (3.21)

$$p(x = 4) = \frac{2.72^{-5} \times 5^{4}}{4!} = 0.175 \tag{3.21}$$

The Poisson distribution can be used to test whether the values in a given data variable are random or not: if there is a close fit between the data and the theoretical distribution, it is probable that the data was generated by a random process.

How can degrees of adherence to the Poisson distribution be determined with respect to a set of variable values? A characteristic of the theoretical Poisson distribution is that its mean and variance are identical. Given a frequency matrix whose columns represent the variables of interest, therefore, the degree to which any column j diverges from Poisson can be determined by calculating the degree to which j's mean and variance differ. This ratio is known as the 'variance-to-mean ratio' (vmr) and is defined on a vector x by Equation (3.22).

$$vmr(x) = \frac{var(x)}{\mu(x)} \tag{3.22}$$

Vmr is also known as the 'index of dispersion', which indicates its use as a measure of dispersion of some set of values relative to a statistical distribution. Relative to a Poisson distribution it measures degree of divergence from randomness.

The *vmr* can be used for dimensionality reduction as follows; a document collection D containing m documents is assumed. The production of a natural language document, and more specifically the successive occurrence of tokens of variables which constitutes the document, is taken to be a Poisson process. For each of the n variables x_j describing the documents of $D(j = 1 \ldots n)$

- The intervals of interest are the m component documents of D.
- The mean rate of occurrence λ_j of x_j in the m documents is the total number of occurrences of x_j in D divided by m.
- The actual number of occurrences of x_j in document $d_i(i = 1 \ldots m)$ is r_{ij}.
- The question being asked with respect to x_j is: since the documents are taken to be generated by a Poisson process, and therefore that each document d_i is expected, on average, to contain λ_j tokens of x_j, how probable is the actual number of occurrences r_{ij} in each of the d_i?
- If the probability of x_j is high across all the d_i, then it fits the Poisson distribution, that is, the occurrence pattern of x_j is random and it can therefore be eliminated as a variable. If, however, the probability of x_j is low for one or more of the documents, then x_j diverges from the distribution –in other words, x_j occurs nonrandomly to a greater or lesser degree and should therefore be retained. In the ideal case the probability of x_j is low for a proper subset of the documents in D and high elsewhere, indicating that its occurrence pattern is nonrandom in some documents and random in the remainder and that it is therefore a good criterion for document classification.

The assumption that any natural language document or document collection is generated by a stochastic process is, of course, unjustified (Kilgarriff 2005). Lexical sequencing in natural language text, for example, is not generated by the proverbial monkeys sitting at keyboards but by writers who carefully choose the words that express what they want to express and sequence those words in accordance with the rules of syntax, and speakers do not utter phonetic segments at random but rather follow the phonological rules of the language being spoken. It is, therefore, not to be expected that the distribution of any feature or features of natural language, lexical or otherwise, will have a true Poisson distribution. Empirical results have, however, shown that several categories of lexical type are almost-Poisson: 'function' words, 'content' words which occur very frequently across all documents in a collection, and content words that are very infrequent across a collection. The relative degree of adherence to the Poisson distribution can, therefore, still be used as a dimensionality reduction criterion – some words are more random than others, and the aim is to identify the relatively non-random ones. Much the same applies to non-lexical variables such as the phonetic segments on which the MDECTE data matrix is based: degree of adherence to the Poisson distribution can be used to identify segments that are relatively more non-random than others, and should therefore be retained. Because the Poisson is not an ideal model for lexical type distribution in natural language text, exact correspondence cannot in general be expected, but the relative degree of divergence from mean-variance equivalence can nevertheless be used to distinguish variables on a continuum of randomness, ranging from those that are near-Poisson and can therefore be eliminated to those that are far from Poisson and should therefore be retained.

The *vmr* values for the column vectors of MDECTEwere calculated, sorted in descending order of magnitude, and plotted as shown in Figure 3.32. A threshold k is, again, selected and the variables which fall below the threshold are eliminated.

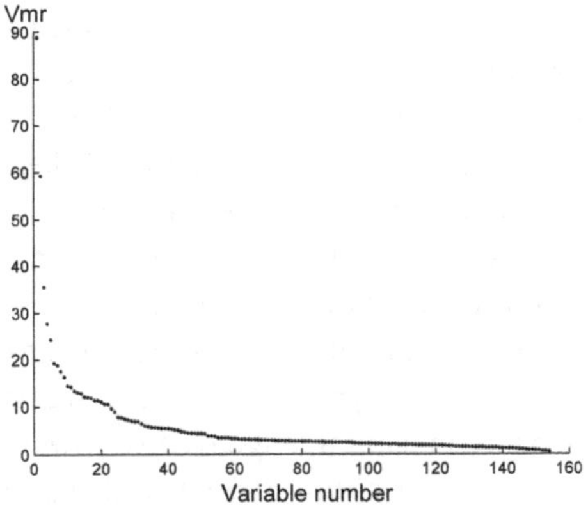

Figure 3.32: Sorted MDECTE variable *vmr* scores

The other approach to assessing nonrandomness is term frequency – inverse document frequency ($tf - idf$), which was developed by Spärck Jones (1972) and is extensively used by the Information Retrieval community – cf., for example, Manning, Raghavan, and Schütze (2008: Ch. 6). It focuses on the distribution of lexical items in document collections, but is readily adaptable to variables of other types.

Spärck Jones (1972) proposed what was to become a standard principle in Information Retrieval: that a lexical type's usefulness for differentiating documents is determined not by its absolute frequency across a collection, but by the pattern of variation in its frequency across the documents. To gain an intuition for this, assume a collection of documents related to the computer industry. At one end of the range are very low frequency words that, as expected, are of little or no use for document differentiation: a word like 'coffee' that occurs a few times in one or two documents that caution against spills into keyboards is insignificant in relation to the semantic content of the collection as a whole, and a word like 'bicycle' that occurs only once tells us only that the document in which it appears is unique on that criterion. At the other end of the range, a word like 'computer' and its morphological variants is likely to be both very frequent across the collection and to occur in most if not all the documents, and as such is also a poor criterion for differentiating documents despite its high absolute frequency: if all the documents are about computers, being about computers is not a useful distinguishing criterion.

In short, word frequency on its own is not a reliable clustering criterion. The most useful words are those whose occurrences are, on the one hand, relatively frequent, and on the other are not, like 'computer', more or less randomly spread across all collection documents but rather occur in clumps such that a relatively few documents contain most or all the occurrences and the rest of the collection few or none; the word 'debug', for example, can be expected to occur frequently in documents that are primarily about computer programming and compiler design, but only infrequently if at all in those about, say, word processing. On this criterion, the usefulness of lexical types is assessed in accordance with their 'clumpiness' of occurrence across documents in a collection.

When she proposed clumpiness of distribution as a criterion, Spärck Jones also provided a method for calculating it. That method, together with some emendments to it made by (Robertson 1972), became known as 'inverse document frequency' (idf). Relative to a lexical variable or 'term' t_j in a set T of n terms that occur across all the documents d_i in a collection D of m documents (for $i = 1 \dots m$ and $j = 1 \dots n$), idf is defined as in Equation (3.23).

$$idf(t_j) = log_2 \frac{m}{df_j} \tag{3.23}$$

where df_j is the document frequency, that is, the number of documents belonging to D in which t_j occurs. The inverse document frequency of a term, therefore, is the ratio of the total number of documents in a collection to the number of documents in which the term occurs; log_2 is not conceptually part of idf, but merely scales the m/df_j ratio to a convenient interval.

There is a problem with idf: it says that terms which occur only once in a corpus are the most important for document classification. Assuming a 1000-document collection, the idf of a term that occurs in one document is $log_2(1000/1) = 9.97$, for a term that occurs in two documents $log_2(1000/2) = 8.97$ and so on in a decreasing sequence. This is counterintuitive – does one really want to agree with idf that a lexical type which occurs once in a single document is a better criterion for document classification than one which occurs numerous times in a small subset of documents in the collection? It also contradicts the empirically-based and widely used principle (Luhn 1957) that medium-frequency words are the best discriminatory criteria, as well as standard practice in Information Retrieval of disregarding frequency-1 words. In short, idf cannot be right. It was rescued by building word frequency into the clumpiness measure, as in Equation (3.24).

$$tfidf(t_j) = tf(t_j) log_2 \frac{m}{df_j} \tag{3.24}$$

where $tf(t_j)$ is the frequency of term t_j across all documents in D. Using this formulation, the $tf - idf$ of some lexical type A that occurs once in a single document is $1 \times log_2(1000/1) = 9.97$, and the $tf - idf$ of a type B that occurs 400 times across 3 documents is $400 \times log_2(1000/3) = 3352.3$, that is, B is far more useful for document differentiation than A, which is more intuitively satisfying than the alternative (Robertson 2004; Spärck Jones 2004).

The notion of clumpiness in the distribution of lexical items across document collections extends naturally to other types of variables such as the MDECTE phonetic segments. Its application to dimensionality reduction is analogous to that of the methods already presented: the columns of the data matrix are sorted in descending order of $tf - idf$ magnitude, the $tf - idf$ values are plotted, the plot is used to select a suitable threshold k, and all the columns below that threshold are eliminated. The plot for MDECTE is given in Figure 3.33.

Though $tf - idf$ has been and is extensively and successfully used in Information Retrieval, it has a characteristic which compromises its utility for dimensionality reduction. Because clumpiness in $tf - idf$ relative to some

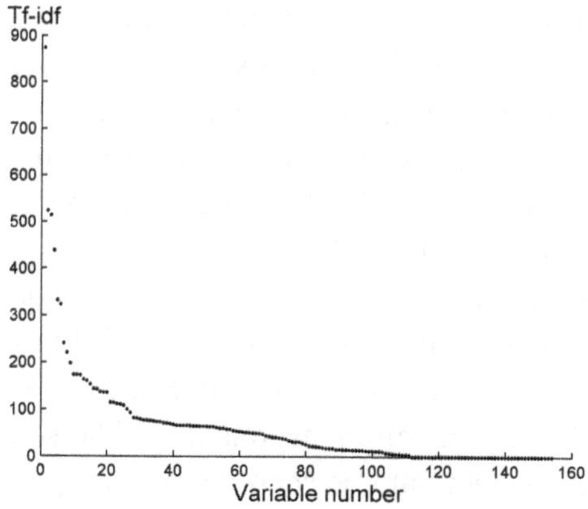

Figure 3.33: Sorted MDECTE variable $tf-idf$ scores

variable v is based on the ratio of the total number of documents m in a collection to the number of documents df in which v occurs, as df approaches m the ratio approaches 1 and the idf correspondingly approaches $log_2(1) = 0$. Where the idf is near 0 the $tf-idf$ of v is very small, and when it is at 0 – that is, where v occurs in every document – the $tf-idf$ remains 0 irrespective of v's frequency. It is, however, possible that v is nonrandomly distributed across the documents even where it occurs in every document – some documents might, for example, contain only one or two tokens of v, while others might contain scores or hundreds – and $tf-idf$ cannot identify such a distribution. Use of $tf-idf$ for dimensionality reduction therefore runs the risk of eliminating distributionally-important variables on account of the definition of clumpiness on which it is based. A small change to the formulation of $tf-idf$ prevents the idf term evaluating to zero and this allows relative frequency to remain a factor, as shown in Equation (3.25).

$$tfidf(t_j) = tf(t_j)log_2\frac{m+1}{df_j} \tag{3.25}$$

Since $(m+1)$ must always be greater than df the idf and consequently the $tf-idf$ are always greater than 0.

All the methods for dimensionality reduction presented so far, from frequency through to $tf-idf$, suffer two general problems. The first is that selection of a threshold k below which variables are discarded is problematic. Visual intuition based on plotting indicates that, the further to the left of the plot one goes, the more important the variable. But where, exactly, should the threshold be drawn? In Figure 3.33, for example, the variables to the right of the 110th or so can clearly be eliminated because they are at or near zero, but can the threshold be set further to the left without too much loss of information? Should it be set at, say, 80? Or 40? Or 25? And why? Unless one is able to give a principled reply, threshold selection is a subjective matter that runs the risk, on the one hand, of eliminating too few variables and thereby retaining excessive dimensionality, or on the other of eliminating too many and potentially compromising the analysis by disregarding relevant information.

The second problem is incommensurateness. The selection criteria focus on different aspects of data, and as such there is no guarantee that, relative to a given data matrix, they will select identical subsets of variables. Indeed, the expectation is that they will not: a variable can have high frequency but little or no variability, and even if it does have significant variability, that variability might or might not be distributed nonrandomly across the matrix rows. This expectation is fulfilled by MDECTE, as shown in Figure 3.34.

Each of the rows (a)–(d) of Figure 3.34 is based on a different sorting of the MDECTE matrix columns: 3.34a is based on the columns sorted in descending order of frequency, 3.34b of variance, 3.34c of *vmr*, and 3.34d of $tf-idf$. For each, vectors of column frequencies, variances, *vmr*s, and $tf-idf$s were calculated, *z*-standardized for comparability, and then co-plotted. Interpretation proceeds as follows. The frequency values of Figure 3.34a, shown as circles, decline smoothly and gradually from the high-frequency to the low-frequency columns of MDECTE because the columns were sorted by descending frequency, and the corresponding variance, *vmr*, and $tf-idf$ values scatter around the frequency curve with no obvious pattern except for a general diminution as the frequencies approach zero. When a threshold k based on the shape of the frequency curve is selected, all the variable columns to the left of k in the plot are retained and all those to the right are discarded; for $k = 60$, say, some high-variance, high-*vmr*, and high-$tf-idf$ variables are retained along with the high-frequency ones, but so are quite a few relatively low-variance, low-*vmr*, and low-$tf-idf$ ones, and quite a few relatively high-variance, high-*vmr*, high-$tf-idf$ ones are eliminated along with low-frequency ones. This pattern is repeated for Figures 3.34b–3.34d.

Given that the four variable selection criteria in the foregoing discussion can be expected to, and for MDECTE do, select different subsets of variables, which of them is to be preferred or, alternatively, how can their selections be reconciled?

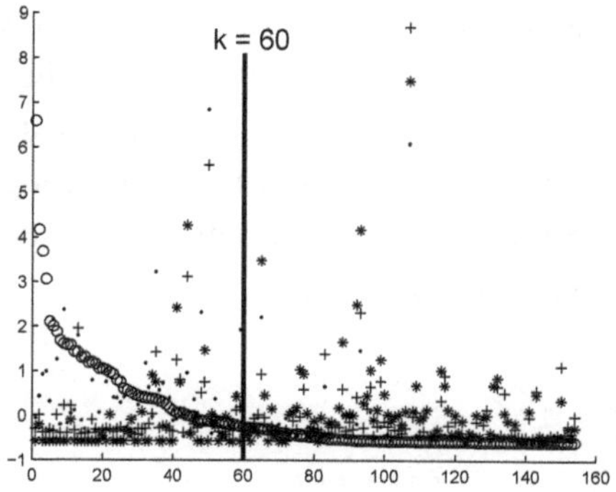

(a) Size of cube with side length 1 for dimensionality 3..100. Volume: Constant 1. Diagonal: 1.7321 .. 10.0000

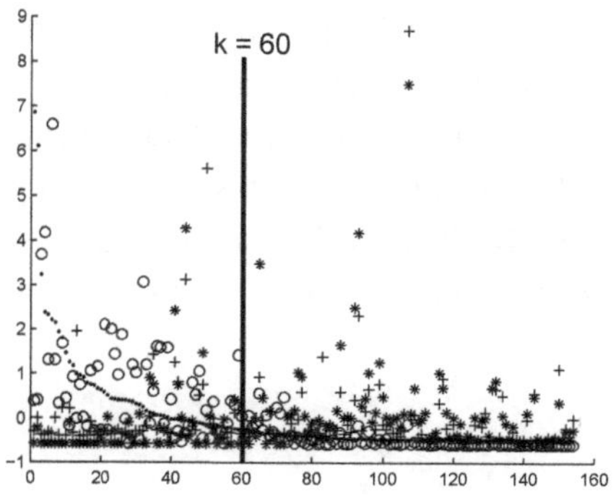

(b) Size of cube with side length 0.9 for dimensionality 3..100. Volume: 0.7290 .. 0.0000. Diagonal: 1.5588 .. 9.0000

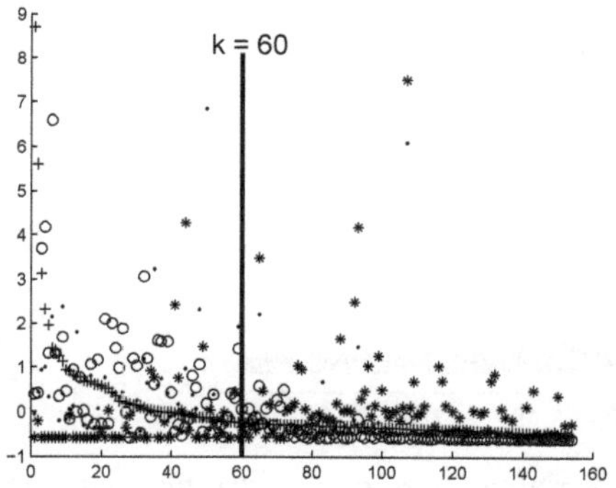

(c) Size of cube with side length 1.1 for dimensionality 3..100. Volume: 1.3310 .. 13780.6123. Diagonal: 1.9053 .. 11.0000

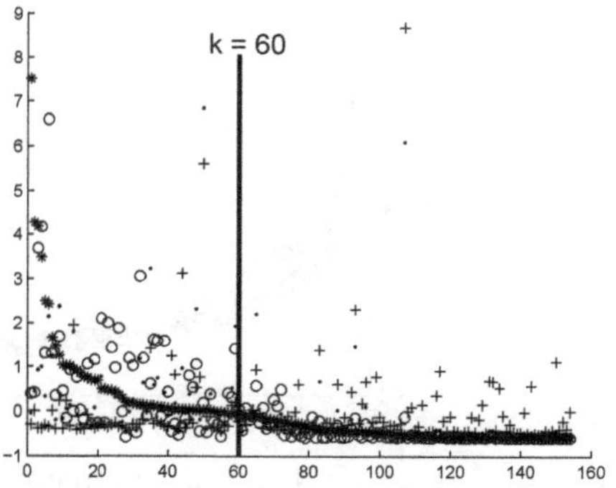

(d) Magnification of range 0..20 of (c)

Figure 3.34: Comparisons of sorted MDECTE variable *frequency*, *variance*, *vmr* and $tf-idf$ scores. Key: o = frequency, . = variance, + = *vmr*, * = $tf-idf$

With respect to threshold selection, the literature appears to contain no principled resolution, and none is offered here; selection of a suitable threshold remains at the discretion of the researcher. The remainder of this discussion deals with the problem of incommensurateness.

In any given research application there might be some project-specific reason to prefer one or another of the four variable selection criteria. Failing this, there is no obvious way of choosing among them. The alternative is to attempt to reconcile them. The reconciliation proposed here attempts to identify and eliminate the variables which fail to satisfy the principles of frequency, variability, and nonrandomness set out at the start of the discussion. This can be done by calculating frequency, variance, *vmr*, and $tf-idf$ values for each column of the data matrix, sorting each set of values in descending order of magnitude, *z*-standardizing for comparability, and then co-plotting. For MDECTE this amounts to co-plotting the frequency, variance, *vmr*, and $tf-idf$ curves from Figure 3.34, as shown in Figure 3.35.

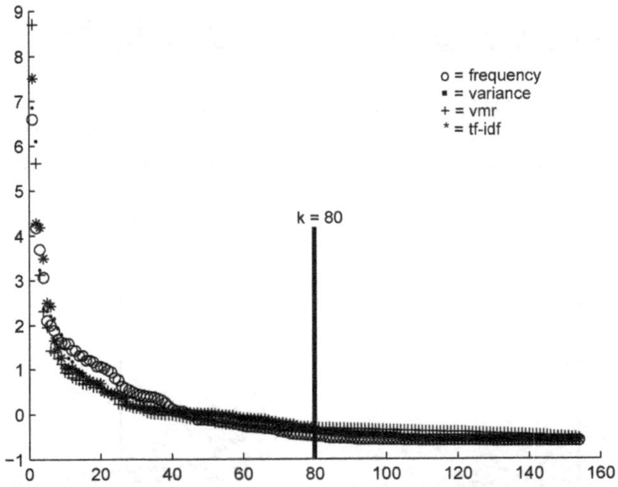

Figure 3.35: Co-plot of sorted MDECTE variable *frequency*, *variance*, *vmr* and $tf-idf$ scores

The variables with high-frequency, high-variance, and high-nonrandomness values are on the left of the plot, and these values diminish smoothly as one moves to the right. If a threshold is now selected, the variables to the right of it can be discarded and those to the left retained, yielding the required dimensionality reduction; as before, threshold selection is subjective, and in Figure 3.35 it could reasonably be anywhere between 20 and 80, though a conservative threshold of 80 has been selected. There is, of course, no guarantee that the frequency, variance, *vmr*, and $tf-idf$ distributions will agree as neatly as this in every application, but where they do this approach to reconciliation is effective.

It is, moreover, possible to further reduce dimensionality by refining the selection to the left of the threshold. This refinement is based on tabulation of the retained variables, as shown for MDECTE in Table 3.20, where the numbers in the columns of DECTE PDV codes, as described earlier.

Table 3.20: Categorization of variables retained in Figure 3.35

freq,var, vmr,tf-idf	freq,var, vmr	freq,var	vmr,tf-idf	freq	vmr	tf-idf
0016	0012	0002	0004	0246	0008	0030
0020	0014	0024	0028			0054
0060	0128	0034	0032			0056
0074	0130	0042	0036			0086
0080	0146	0050	0038			0136
0082	0160	0062	0046			0148
0104	0194	0084	0052			0152
0112	0202	0090	0064			0154
0116	0208	0214	0072			0186
0118	0210	0216	0076			0188
0120	0242	0218	0088			0200
0122	0244	0220	0092			0204
0134	0256	0222	0094			0206
0142	0260	0226	0106			0268
0144	0266	0228	0108			1260
0162	0270	0230	0114			1500
0164	0272	0232	0126			1520
0168	0276	0234	0132			2002
0176	0278	0236	0172			2880
0182	0282	0238	0180			5600
0196	0286	0240	0184			
0198	0288	0250	0248			
0212	0294	0252	0287			
0224		0254	1140			
0284		0258	1160			
1120		0262	1400			
1460		0264	1440			
2760		0274	1600			
		0280				
		1780				

Table 3.20 categorizes the retained variables according to how many of the methods selected them: the variables in column 1 were selected by all four, those in column 2 by all but $tf-idf$, and so on; other combinations of selection method are of course possible, but only those that actually occurred for MDECTE are listed. Refinement of the selection using this categorization proceeds as follows. Because the variables in column 1 were selected by all four methods, they are high-frequency, high-variance, and highly-nonrandom, and are thus kept because they satisfy all the retention criteria. Much the same can be said of the variables in column 2, though slightly less confidently because $tf-idf$ did not select them, and a decision on retention is therefore determined by whether or not one trusts *vmr* over $tf-idf$. The variables in the remaining columns can be discarded because they are not useful for clustering relative to the retention criteria.

- Column 3 contains high-frequency, high-variance variables, but the variance is near-randomly distributed.
- Column 4 contains variables whose values are nonrandomly distributed, but they are low-frequency and low-variance.
- Column 5 contains a high-frequency variable with little variance, and such variance as it has is near-random.
- Column 6 contains a low-frequency, low-variance variable whose values, on one measure, are nonrandomly distributed.
- Column 7 contains quite a large number of low-frequency, low-variance variables whose values on the $tf-idf$ measure are nonrandomly distributed.

The initial selection based on the threshold in Figure 3.35 comprised the 131 variables listed in Table 3.20. Retaining columns 1 and 2 of Table 3.20 reduces the selection to 51, which is a very substantial dimensionality reduction from the original 156. Other approaches to feature selection for clustering are in Devaney and Ram (1997), Dy (2008), Dy and Bodley (2004), Kim, Street, and Menczer (2003), and Roth and Lange (2004).

Finally, the various dimensionality reduction methods described thus far are general in the sense that they are applicable to any data matrix in which the rows represent objects to be clustered and the columns the variables describing those objects. Where the variables are lexical, however, there is additional scope for dimensionality reduction via stemming and elimination of so-called stop-words. This is a substantial topic in its own right, and is not discussed here; for further information see for example Frakes and Baeza-Yates (1992), Hull (1996), Xu and Croft (1998), and Ziviani and Ribeiro-Neto (1999).

As noted, the methods for dimensionality reduction are of two broad types. One type selects a subset of the more important data variables and eliminates the remainder from the analysis, using some definition of importance. The other type abstracts a new and usually much smaller set of variables on the basis of the existing ones. The preceding discussion has dealt with the first of these; the remainder of this section moves to the second, variable extraction.

The discussion of data creation noted that, because selection of variables is at the discretion of the researcher, it is possible that the selection in any given application will be suboptimal in the sense that there is redundancy among them, that is, that they overlap with one another to greater or lesser degrees in terms of what they represent in the research domain. Where there is such redundancy, dimensionality reduction can be achieved by eliminating the repetition of information which redundancy implies, and more specifically by replacing the researcher-selected variables with a smaller number of non-redundant variables that describe the domain as well as, or almost as well as, the originals. Slightly more formally, given an n-dimensional data matrix, dimensionality reduction by variable extraction assumes that the data can be described, with tolerable loss of information, by a manifold in a vector space whose dimensionality is lower than that of the data, and proposes ways of identifying that manifold.

For example, data for a study of student performance at university might include variables like personality type, degree of motivation, score on intelligence tests, scholastic record, family background, class, ethnicity, age, and health. For some of these there is self-evident redundancy: between personality type and motivation, say, or between scholastic record and family background, where support for learning at home is reflected in performance in school. For others the redundancy is less obvious or controversial, as between class, ethnicity, and score on intelligence tests. Variable extraction methods look for evidence of such redundancy between and among variables and use it to derive new variables which give a non-redundant, reduced-dimensionality representation of the domain. In the foregoing example, the researcher-defined variables personality type, motivation, scholastic record, and score on intelligence tests might be replaced by a 'general intelligence' variable based on similarity of variability among these variables in the data, and family background, class, ethnicity, age, and health with a 'social profile' one, thereby reducing data dimensionality from nine to two.

The following discussion of variable extraction first gives a precise definition of redundancy, then introduces the concept of intrinsic dimension, and finally presents some variable extraction methods.

If there is little or no redundancy in variables then there is little or no point to variable extraction. The first step must, therefore, be to determine the level of redundancy in the data of interest to see whether variable extraction is worth undertaking. The methods for doing this described below are all based on assessing the degrees of overlap between data variables in terms of the information about the domain that they represent, and they do this by measuring the similarity between and among the column vectors of the data matrix which represent the variables.

We have seen that the values in an n-dimensional vector are the coordinates of its location in n-dimensional space. The similarity of values in any two vectors in the space will consequently be reflected in the distance between them: vectors with very similar values will be close together, and to the extent that the differences in values increase they will be further apart. By calculating the distances between all unique pairings of column vectors in a data matrix, it is possible to identify degrees of similarity and therefore of redundancy between them. The Euclidean distances between all unique pairings of the 156 column vectors of MDECTE in 63-dimensional space were calculated, sorted in descending order of magnitude, and plotted in Figure 3.36; 'unique' means that all (v,w) pairs included in the calculation were different, (v,w) and (w,v) were regarded as identical because distance between two points is symmetrical, and (v,v), that is, distances of vectors to themselves, being always 0, were disregarded.

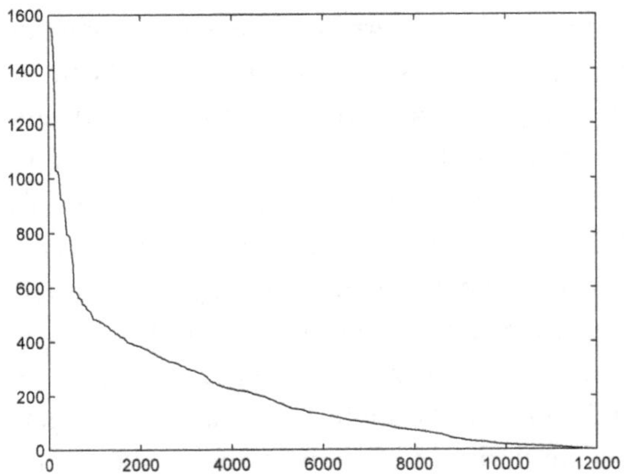

Figure 3.36: Sorted distances between pairs of column vectors in MDECTE

Figure 3.36 shows the shape of the redundancy in MDECTE. There is substantial variation in degree of redundancy between variables from relatively low on the left, where the column vectors are relatively far apart, to relatively high on the right where the vectors are relatively close together. The large number of relatively high-redundancy variable pairs from about 1000 rightwards indicates substantial scope for dimensionality reduction by variable extraction.

Another way of measuring redundancy is via the angle between them. The angle between a pair of vectors in a vector space reflects the distance between them, as discussed earlier, and degrees of similarity and therefore of redundancy between all unique pairings of column vectors of a data matrix can be found by calculating the cosines of the angles between them. The cosines between all unique pairings of the 156 column vectors of MDECTE were calculated, sorted in ascending order of magnitude, and plotted in Figure 3.37.

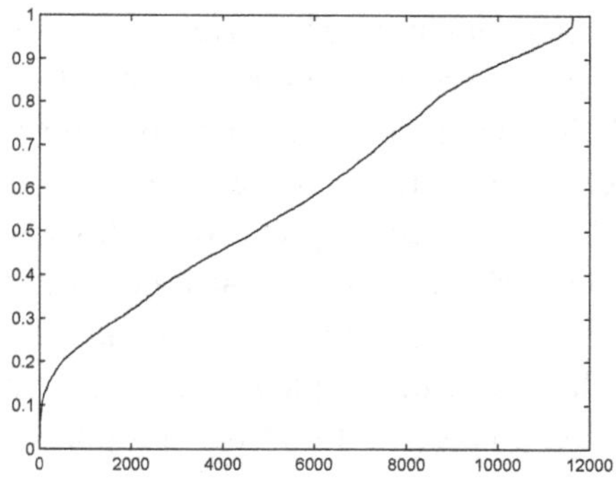

Figure 3.37: Sorted cosines between pairs of column vectors in MDECTE

The smaller the cosine the larger the distance between variables and thus the smaller the redundancy. There are relatively few non-redundant variable pairs at the left of the plot, but redundancy increases quite rapidly thereafter; as with distance, the indication is that there is substantial scope for dimensionality reduction by variable extraction.

Angle has an advantage over distance as an indicator of degree of redundancy. The magnitude of distances measured between vectors is determined by the scale on which the variables are measured, and as such it is difficult to know how to interpret a given distance in terms of degree of redundancy: does a distance of, say, 50 represent a lot of redundancy or only a little? A given distance is diagnostically useful only in relation to other distances, as in Figure 3.36. Angle, on the other hand, is independent of the scale on which variables are measured in that an angle between vectors does not change with vector length. The cosine measure has an absolute interval of $0 \dots 1$, and in relation to that interval it is possible to say that a given cosine value, say 0.2, is relatively small and indicates relatively low redundancy, and a value of 0.9 is relatively large and indicates relatively large redundancy.

A third and frequently used way of measuring redundancy is correlation. In probability theory two events A and B are said to be independent if the occurrence of A has no effect on the probability of B occurring, or vice versa, and dependent otherwise. Given two variables x and y and an ordered sequence of n observations at times $t_1, t_2 \dots t_n$ for each, if the measured value for x at time t_i (for $i = 1 \dots n$) has no predictive effect on what the measured value for y at t_i will be, then those variables are independent, or, failing this condition, dependent. In statistics, variables that are dependent are said to be associated, and the degree of association is the degree to which they depart from independence. Statistics provides various measures of association, the most often used of which, Pearson's product-moment correlation coefficient, or 'Pearson's correlation coefficient' for short, is described below.

To understand Pearson's correlation coefficient, one first has to understand the concept of covariance between any two variables x and y, which is a measure of the degree to which there is a linear relationship between the values taken at successive observations in the time sequence $t_1, t_2 \dots t_n$: as the observed values of x change in the sequence, do the values of y at each corresponding observation change in a constant proportion? Figure 3.38 gives some examples.

In Figure 3.38a there is a perfect linear relationship between x and y: for every increase of 2 in x in the t_i sequence there is a corresponding increase of 2 in y; the changes in x and y are in a constant proportion and their relationship is therefore perfectly linear, shown graphically by being able to draw a straight line through all the data points. The data in Figure 3.38b were generated by adding a different random positive value to each of the values in 3.38a, and the result is an imperfect linear relationship between x and y: the plot in 3.38b shows that the proportion of increase in x and y is no longer constant, but a straight line drawn through the data shows that there is still a discernible linear tendency inherited from 3.38a around which the randomized values of 3.38b fall. The data in Figure 3.38c, finally, were generated without reference to that in 3.38a or 3.38b, and consist entirely of random values: any line drawn through a plot of these values will fail to identify any systematic relationship between x and y, as one would expect.

	x	y
t1	2.00	3.00
t2	4.00	5.00
t3	6.00	7.00
t4	8.00	9.00
t5	10.00	11.00
t6	12.00	13.00
t7	14.00	15.00
t8	16.00	17.00
t9	18.00	19.00
t10	20.00	21.00

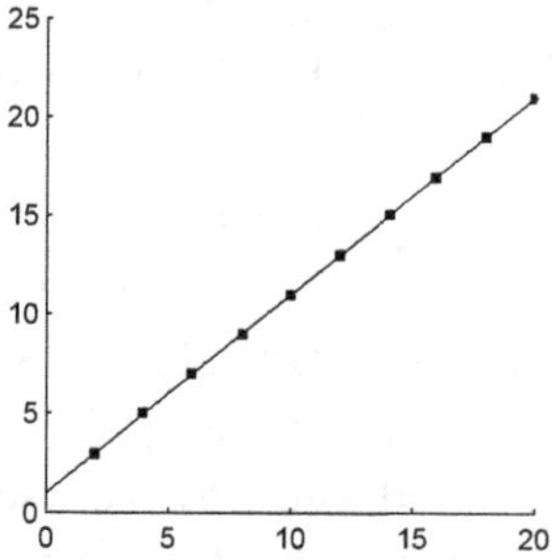

(a)

	x	y
t1	2.68	2.12
t2	8.47	7.00
t3	7.49	13.61
t4	9.42	13.69
t5	10.32	20.88
t6	14.91	17.24
t7	16.58	18.34
t8	18.17	19.26
t9	20.90	26.60
t10	22.65	27.41

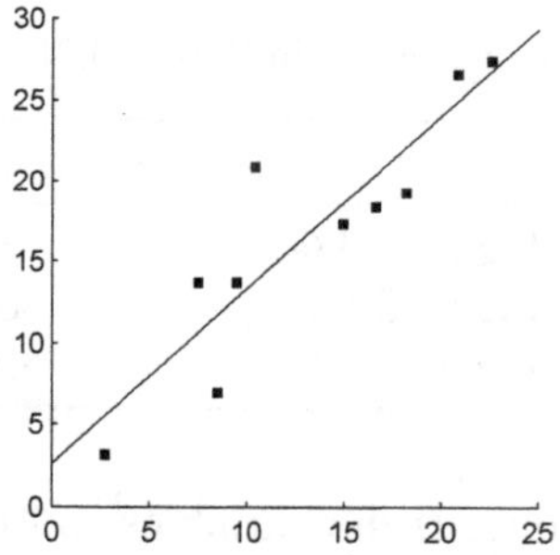

(b)

	x	y
t1	5.02	9.12
t2	18.80	16.34
t3	11.07	13.63
t4	19.06	1.42
t5	14.47	1.21
t6	9.97	7.32
t7	20.99	0.36
t8	18.43	23.30
t9	23.76	18.90
t10	10.53	11.90

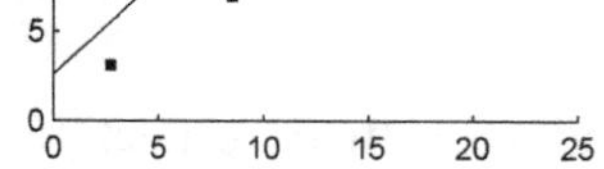

(c)

Figure 3.38: Examples of covariance between two variables

The covariance between two variables $cov(x,y)$ is a quantitative measure of the degree of linear relationship between them. It is defined as Equation (3.26)

$$cov(x,y) = \frac{\sum_{i=1..n}(x_i - \mu_x)(y_i - \mu_y)}{n-1} \tag{3.26}$$

where where μ_x and μ_y are the means of x and y respectively, n is the number of observations in x and y, and the $(x_i - \mu_x)(y_i - \mu_y)$ expression is the inner product of vectors x and y adjusted by subtracting their respective means. Using this formula, the covariances of the variables in Figure 3.38 are given in Table 3.21.

Table 3.21: Covariances of variables in Figure 3.38

	$cov(x,y)$
Figure 3.38a	33
Figure 3.38b	40.23
Figure 3.38c	5.87

This result is puzzling. If the variables in Figure 3.38a are most linearly related, those in 3.38b least, and those in 3.38c in between, one expects the numerical measure to reflect this, but it does not. The reason is that the magnitude of the covariance is dependent on the numerical magnitudes in the vectors, and the magnitudes in 3.38a are sufficiently larger than those in 3.38a to override the greater linearity in 3.38a. The solution is to standardize the vectors to a common scale, and this is what Pearson's correlation coefficient, given as *Pcorr* in Equation (3.27) does.

$$Pcorr(x,y) = \frac{cov(x,y)}{\sigma_x \sigma_y} \tag{3.27}$$

Pcorr divides x and y by their respective standard deviations, thereby transforming their values to a common scale and so eliminating the effect of scale. This is shown in Table 3.22.

The Pearson coefficient captures the intuition gained from examination of the plots in Figure 3.38 that 3.38a has the greatest linear relationship, 3.38b less, and 3.38c very little. In general, the Pearson coefficient states the correlation of any pair of variables as a value in the range $-1 \ldots 1$, where -1 is perfect negative and $+1$ perfect positive correlation, thereby allowing the correlations of pairs of variables to be compared irrespective of scale.

Table 3.22: Comparison of covariances and Pearson correlations of variables in 3.21

	$cov(x,y)$	Pearson
Figure 3.38a	33	1
Figure 3.38b	40.23	0.896
Figure 3.38c	5.87	0.139

Like angle, the Pearson coefficient has the advantage that it is independent of variable scale and can be interpreted in relation to a fixed $-1\ldots1$ interval. This interpretability can be enhanced by squaring the correlation value, which yields r^2, the coefficient of determination; r^2 represents the proportion of variance that two vectors share and, expressed as a percentage, gives an intuitively clearer impression than the correlation coefficient of the relationship between two vectors. For example, a correlation of 0.9 looks very strong, but an r^2 of 0.81 or 81 percent slightly less so; a correlation of 0.7 still looks reasonably strong but corresponds to an r^2 of only 0.49 or 49 percent of variance shared by the vectors; a correlation of 0.6, which one might regard as moderate, corresponds to only 36 percent shared variance. In other words, the coefficient of determination provides an intuitive correction to over-interpretation of the correlation coefficient.

Figure 3.39 co-plots the Pearson correlation coefficients and coefficients of determination of all unique pairings of column vectors of MDECTE; in the former case absolute values of negative correlations were used since, for present purposes, all that matters is degree of redundancy and not its direction. Like distance and angle, this criterion reveals a significant amount of redundancy and thus scope for dimensionality reduction by variable extraction.

Though they derive from different conceptual frameworks, that is, from vector space geometry and probability, cosine similarity and Pearson correlation are very closely related in that, with respect to determination of vector redundancy, both are based on the inner product of the vectors, both recognize that disparity of scale between the vectors is a problem, and both address this problem by normalization. The only substantive difference between them is that Pearson correlation subtracts their respective means from the vectors, with the result that the vectors are now mean-centred, that is, the sums of their values are 0; if vectors are mean-centred prior to cosine calculation, cosine similarity and Pearson correlation are equivalent.

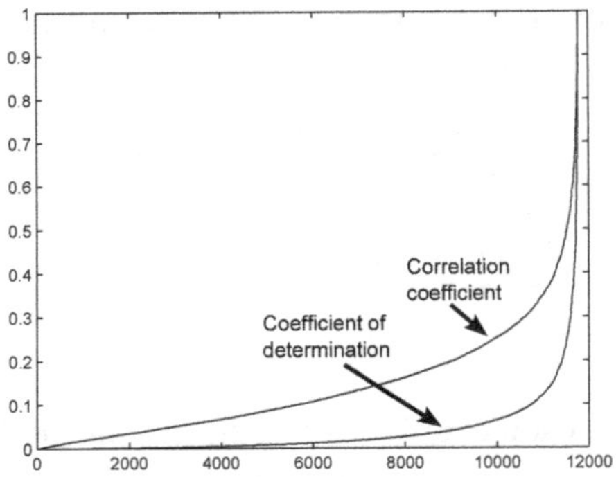

Figure 3.39: Correlation coefficients and coefficients of determination of the column vectors of MDECTE

The theoretical basis for variable extraction is the concept of intrinsic dimension. We have seen that an $m \times n$ matrix defines a manifold in n-dimensional space. In such a space, it is possible to have manifolds whose shape can be described in k dimensions, where $k < n$. Figure 3.40a and 3.40b show, for example, a three dimensional matrix and the corresponding plot in three dimensional space.

The data in Figure 3.40a describe a straight line in three dimensional space. That line can also be described in two dimensions, as in Figure 3.40b, and in fact can be described in only one dimension, its length 10.69, that is, by its distance from 0 on the real-number line, as in 3.40c. In general, a straight line can be described in one dimension, two dimensions, three dimensions, or any number of dimensions one likes. Essentially, though, it is a one-dimensional object; its intrinsic dimension is 1. In other words, the minimum number of dimensions required to describe a line is 1; higher-dimensional descriptions are possible but unnecessary.

Another example is a plane in three-dimensional space, shown in Figure 3.41a.

This plane can be redescribed in two-dimensional space, as in Figure 3.41b, and again in any number of dimensions one likes. Clearly, however, it cannot be described in one dimension; the intrinsic dimension of a plane is 2. Similarly, the intrinsic dimension of a cube is 3, in that the minimum-dimensionality data set that can describe it is the three x, y and z coordinates of the points that comprise it. A cube can of course, exist not only in 3 dimensional space but also in four, ten, twenty, and n-dimensional spaces, in which case it would be a k-dimensional manifold of intrinsic dimension $k = 3$ embedded in n-dimensional space, where $k < n$ and n is the embedding

v1	v2	v3
0	0	0
1	0.5	0.4
2	1	0.8
3	1.5	1.2
4	2	1.6
5	2.5	2
6	3	2.4
7	3.5	2.8
8	4	3.2
9	4.5	3.6

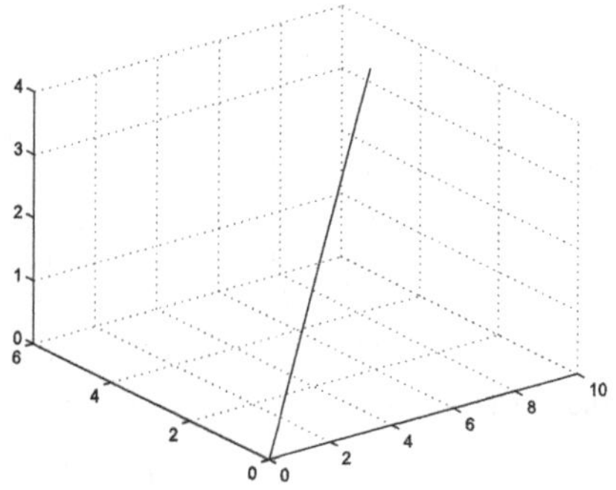

(a)

v1	v2
0	0
1	0.64
2	1.28
3	1.92
4	2.56
5	3.20
6	3.84
7	4.48
8	5.12
9	5.76

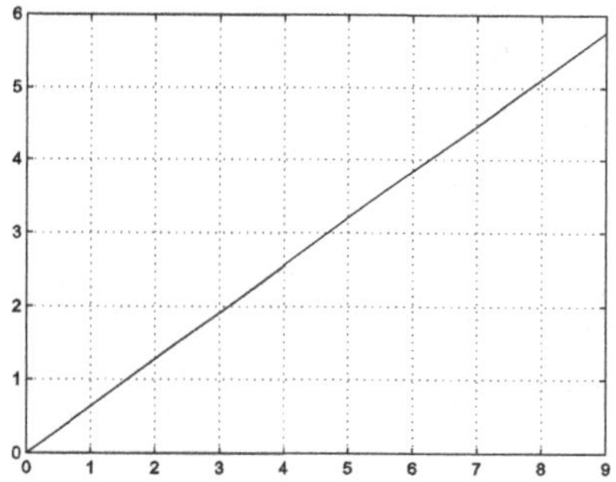

(b)

v1
10.69

(c)

Figure 3.40: A line in 3, 2, and 1 dimensional spaces

dimensionality.

The concept of intrinsic dimension is straightforwardly relevant to dimensionality reduction. The informational content of data is conceptualized as a k dimensional manifold in the n-dimensional space defined by the data variables. Where $k = n$, that is, where the intrinsic dimension of the data corresponds to the number of data variables, no dimensionality reduction is possible without significant loss of information. Where, however, there is redundancy between and among variables in a data matrix, it is possible to represent this information using a smaller number of variables, thus reducing the dimensionality of the data. In such a case, the aim of dimensionality reduction of data is to discover its intrinsic dimensionality k, for $k < n$, and to redescribe its informational content in terms of those k dimensions.

v1	v2	v3
0.6	0	0
0.6	0.6	0
...	...	...
0.6	5.4	0.12
...	...	...
0.6	11.4	0.24
...	...	...
0.6	5.4	0.72
...	...	...
0.6	8.4	1.08
...	...	...
0.6	11.4	1.14

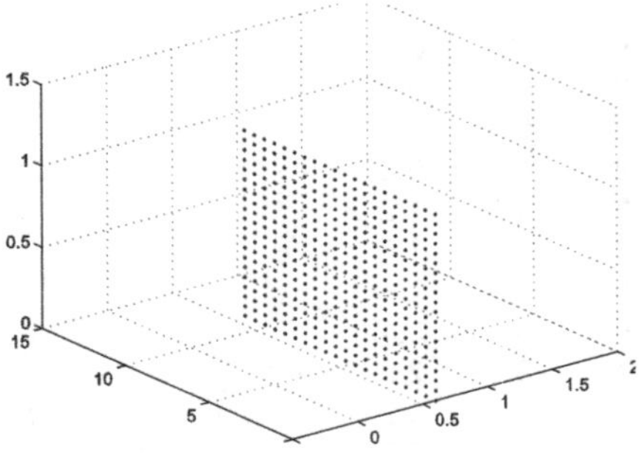

(a)

v1	v2
0	0
0.6	0.6
...	...
5.4	0.12
...	...
11.4	0.24
...	...
5.4	0.72
...	...
8.4	1.08
...	...
11.4	1.14

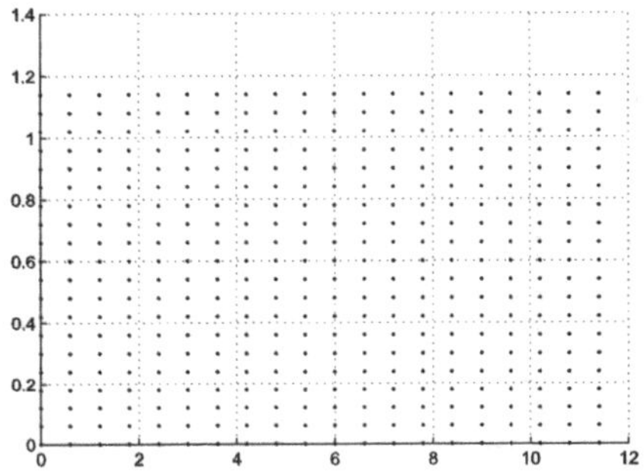

(b)

Figure 3.41: A plane in 3 and 2 dimensional spaces

Unfortunately, it is not usually obvious what the intrinsic dimension of given data is, and there is currently no known general solution for finding k; reviews of existing methods are found in Lee and Verleysen (2007) and Martinez, Martinez, and Solka (2011: Ch. 2.7). Existing dimensionality reduction methods require the researcher to approximate k using a variety of criteria; the remainder of this section describes the main linear and nonlinear methods. For discussion of intrinsic dimension see Jain and Dubes (1988: Ch. 2.6), Camastra (2003), Verleysen (2003), Lee and Verleysen (2007: Ch. 3).

This account of variable extraction methods first presents the standard method, linear principal component analysis (PCA), and then goes on to survey a range of other methods.

Because real-world objects can be distinguished from one another by the degree to which they differ, the data variables used to describe those objects are useful for clustering in proportion to how well they describe that variability, as already noted. In reducing dimensionality, therefore, a reasonable strategy is to attempt to preserve variability, and that means retaining as much of the variance of the original data in the reduced-dimensionality representation as possible. Redundancy, on the other hand, is just repeated variance, and it can be eliminated from data without loss of information. PCA reduces dimensionality by eliminating the covariance while preserving most of the variance in data. Because it is the standard dimensionality reduction method, PCA is described in greater or lesser degrees of detail and clarity by most publications in the field. The standard reference works are those of Jolliffe (2002) and Jackson (2003); selected briefer accounts are in Bishop (1995: 310ff.), Everitt and Dunn (2001: 48ff.), Tabachnick and Fidell (2007: Ch. 13), and Izenman (2008: Ch. 7).

Given an n-dimensional data matrix containing some degree of redundancy, linear PCA replaces the n variables with a smaller set of k uncorrelated variables called principal components which retain most of the variance in the original variables, thereby reducing the dimensionality of the data with only a relatively small loss of information. It does this by projecting the n-dimensional data reduction into a k-dimensional vector space, where $k < n$ and closer than n to the data's intrinsic dimensionality. This is a two-step process: the first step identifies the reduced-dimensionality space, and the second projects the original data into it.

Figure 3.42 shows a small two-dimensional data matrix and the corresponding manifold in the vector space, both in relation to two orthogonal basis vectors v_1 and v_2.

Vectors v_1 and v_2 both have a substantial degree of variance, as shown both by the standard deviation and the scatter plot, and the coefficient of determination shows that they are highly redundant in that they share 90 percent of their variance. The aim is to reduce the dimensionality of this data from 2 to 1 by eliminating the redundancy and retaining the total data variance, that is, the combined variance of v_1 and v_2.

The first step is to centre the data on 0 by subtracting their respective means from v_1 and v_2. This restates the data in terms of a different orthogonal basis but does not alter either the variable variances or their covariance. The mean-centred variables and corresponding plot are shown in Figure 3.43.

The basis vectors are now rotated about the origin, preserving their orthogonality, so that one or the other of them – in this case the horizontal one – becomes the line of best fit to the data distribution, as shown in Figure 3.44.

The line of best fit is the one that minimizes the sum of squared distances between itself and each of the data points. Two of the distances are shown by

	v1	v2
1	5.58	4.85
2	5.78	3.98
3	4.90	5.72
4	4.37	8.78
5	7.65	6.88
6	10.67	10.48
7	9.84	10.12
8	8.05	11.65
9	9.81	10.82
10	11.55	12.88
11	11.82	13.59
12	13.31	14.38
13	16.44	15.00
14	16.25	18.66
15	16.14	16.34
16	16.76	17.69
17	19.69	18.01
18	18.39	21.22
19	19.53	20.15
20	20.02	21.90
Mean:	12.33	13.15
Variance:	26.63	28.56
Standard deviation:	5.16	5.34

Covariance (v1, v2): 26.24
Correlation coefficient: 0.95
Coefficient of determination: 0.90

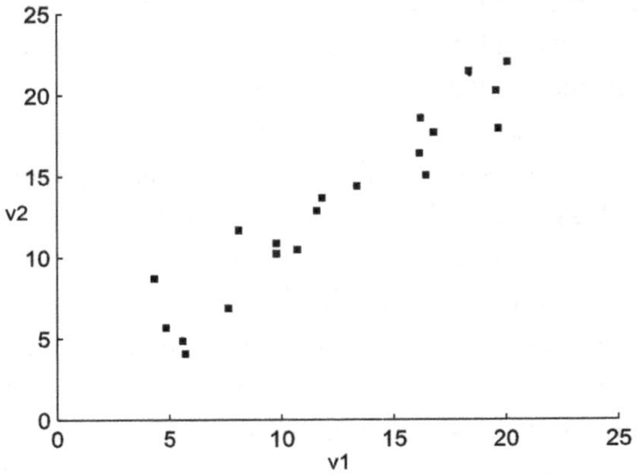

Figure 3.42: A two-dimensional matrix with high redundancy between variables

way of example in Figure 3.44. The distances are squared to prevent negative and positive values from cancelling one another; a fuller discussion of using minimization of squared distances to find a line of best fit to data is given later in the discussion. This again changes the basis, and the variable values are again adjusted relative to the new basis.

Note that:

- The total amount of variance in the data remains constant irrespective of basis: it is $(26.63 + 28.56) = 55.19$ in Figures 3.42 and 3.43, and $(53.85 + 1.34) = 55.19$ in Figure 3.44.
- The distribution of the variance between the variables has radically altered. Almost all the variance after rotation is in v_1 and very little in v_2, as shown in the data table in Figure 3.44; graphically, this corresponds to the basis vector for v_1 lying along the direction of greatest variability in the data, and the basis vector for v_2 lying along a direction with little variability.

	v1	v2
1	-6.75	-8.31
2	-6.55	-9.18
3	-7.43	-7.44
4	-7.96	-4.38
5	-4.68	-6.28
6	-1.66	-2.68
7	-2.49	-3.04
8	-4.27	-1.51
9	-2.51	-2.34
10	-0.78	-0.27
11	-0.51	0.44
12	0.98	1.23
13	4.11	1.85
14	3.92	5.51
15	3.81	3.19
16	4.43	4.53
17	7.36	4.86
18	6.06	8.07
19	7.20	7.00
20	7.69	8.74
Mean:	0	0
Variance:	26.63	28.56
Standard deviation:	5.16	5.34

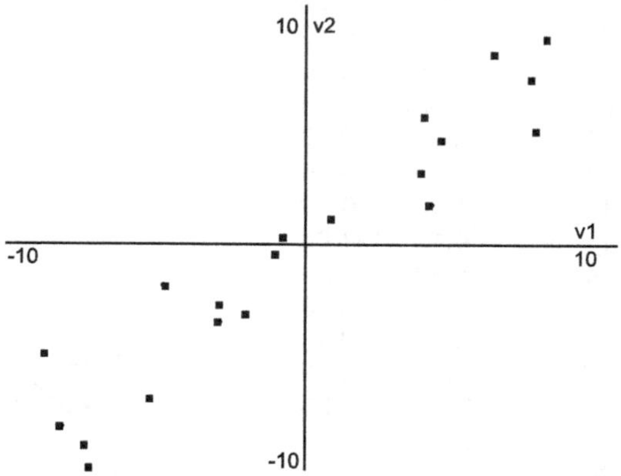

Covariance (V1, v2): 26.24
Correlation coefficient: 0.95
Coefficient of determination: 0.90

Figure 3.43: Mean-centred version of matrix in Figure 3.42

– The redundancy between v_1 and v_2 has been eliminated. The covariance, correlation coefficient, and coefficient of determination are all tiny and effectively zero.

This change of basis reveals the scope for dimensionality reduction. Elimination of v_2 would lose $1.34/55.19 = 2.4$ percent of the variance; if one were prepared to discard v_2, then data dimensionality would be reduced from $n = 2$ to $n = 1$, and that reduction would both retain most of the original variance while eliminating the original redundancy.

This idea extends to any dimensionality n and always proceeds in the same three steps:

– The data manifold is mean-centred.

	v1	v2
1	-10.66	-0.91
2	-11.15	-1.65
3	-10.51	0.19
4	-8.67	2.69
5	-7.76	-0.99
6	-3.08	-0.66
7	-3.91	-0.32
8	-4.05	2.03
9	-3.43	0.19
10	-0.74	0.37
11	-0.04	0.66
12	1.56	0.14
13	4.18	-1.68
14	6.69	0.99
15	4.94	-0.53
16	6.34	-0.40
17	8.61	-1.93
18	10.01	1.23
19	10.03	-0.33
20	11.63	0.53
Mean:	0	0
Variance:	53.85	1.34
Standard deviation:	7.34	1.16

Covariance (v1, v2): -2.98e-15
Correlation coefficient: -3.50e-16
Coefficient of determination: 1.22e-31

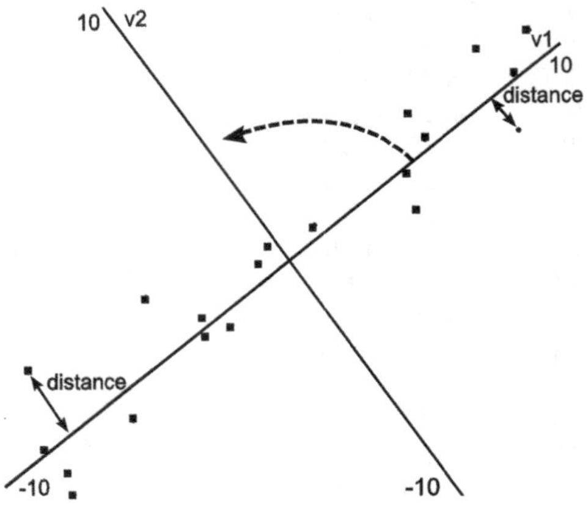

Figure 3.44: Basis vectors of Figure 3.43 rotated so that v_1 becomes the line of best fit to the data distribution

- A new orthogonal basis for the mean-centred data is found in which the basis vectors are aligned as well as possible along the main directions of variance in the manifold.
- Dimensionality is reduced by identifying and discarding the variables corresponding to the basis vectors with negligible variance.

The second step in this procedure is the key, and that is what PCA offers: it finds an orthogonal basis for any given n-dimensional data matrix such that the basis vectors lie along the main directions of variance in the data manifold. These basis vectors are the principal components of the data. Given a data matrix D whose m rows represent the m objects of interest and whose n columns represent the n variables describing those objects, PCA creates two matrices which we shall call $EVECT$ and $EVAL$:

- $EVECT$ is an $n \times n$ matrix whose column vectors are the principal components of D and constitute an orthonormal basis for D.
- $EVAL$ is an $n \times n$ diagonal matrix, that is, one in which the only nonzero values are in the diagonal from its upper left to its lower right corner. These diagonal values are the lengths of the basis vectors in $EVECT$, and represent the magnitudes of the directions of variance in the data manifold.

The diagonal values in $EVAL$ are sorted in descending order of magnitude and are synchronized with $EVECT$ such that, for $j = 1 \dots n$, $EVAL_j$ is the length of basis vector $EVECT_j$. Using $EVAL$, therefore, less important directions of variance can be identified and the corresponding basis vectors eliminated from $EVECT$, leaving a $k < n$ dimensional space into which the original data matrix D can be projected. Where such elimination is possible, the result is a dimensionality-reduced data matrix.

In the following example D is taken to be a fragment of length-normalized MDECTE small enough for convenient exposition, and is shown in Table 3.23a; a dimensionality reduction of the full MDECTE matrix using PCA is presented at the end of the discussion. D is first mean-centred, which involves calculating the mean of the values for each matrix column D_j (for $j = 1 \dots n$) and then subtracting that mean μ from each of the values in D_j: $DMC_j = D_j - \mu$.

Table 3.23: A fragment of MDECTE and its mean-centred version

(*a*) Matrix *D* with column means							(*b*) Mean-centred version *DMC* of *D*						
	ə$_1$	ə$_2$	o:	ə$_3$	Ī	eĪ		ə$_1$	ə$_2$	o:	ə$_3$	Ī	eĪ
g01	6.2	0	61.4	78.1	40.6	25.0	g01	-1.3	-0.8	20.8	0.8	2.7	-1.4
g02	8.9	0	11.9	83.3	31.7	44.6	g02	1.3	-0.8	-28.7	6.0	-6.0	18.2
g03	3.1	1.0	5.85	102.3	3.1	26.8	g03	-4.4	0.1	15.1	25.0	-3.7	0.4
g04	15.9	0.9	12.1	76.6	2.5	23.3	g04	8.3	0.0	-28.5	-0.6	-16.3	-3.0
g05	14.5	6.8	46.2	71.0	49.6	0	g05	6.9	5.9	5.5	-6.2	11.8	-26.4
g06	6.0	0	41.2	71.5	33.2	23.1	g06	-1.5	-0.8	0.6	-5.8	-4.6	-3.2
g07	5.8	0	11.6	59.1	45.5	31.9	g07	-1.7	-0.8	-29.0	-18.2	7.6	5.5
g08	3.1	0	44.5	64.2	32.1	44.5	g08	-4.4	-0.8	3.9	-13.0	-5.7	18.1
g09	5.4	0	30.7	104.1	69.0	10.8	g09	-2.1	-0.8	-9.9	26.8	31.2	-15.6
g10	6.6	0	90.6	62.3	20.7	34.0	g10	-0.9	-0.8	50.0	-14.9	-17.0	7.5
μ	7.6	0.9	40.7	77.3	37.9	26.5							

The covariance of each unique pair of columns DMC_i and DMC_j (for $i, j = 1 \ldots n$) is calculated as described in earlier in the discussion of redundancy and stored in C, where C is an $n \times n$ matrix in which both the rows i and columns j represent the variables of DMC, and the value at $C_{i,j}$ is the covariance of variable column i and variable column j in DMC. The values on the main diagonal are the 'covariances' of the variables with themselves, that is, their variances. Table 3.24 shows C derived from DMC.

Table 3.24: Covariance matrix C for DMC

	ə1	ə2	o:	ə3	ī	eī
ə1	19.20	5.67	-36.34	-10.96	-9.03	-27.16
ə2	5.67	4.57	3.05	-1.93	6.89	-20.40
o:	-36.34	3.05	643.77	-36.99	-78.14	-18.39
ə3	-10.96	-1.93	-36.99	242.91	96.09	-62.66
ī	-9.03	6.89	-78.14	96.09	205.14	-115.87
eī	-27.16	-20.40	-18.39	-62.66	-115.87	190.64

An n-dimensional orthonormal basis for C is constructed such that each basis vector is the least-squares best fit to one of the n directions of variation in C, as described above: the first basis vector b_1 is the line of best least-squares fit along the direction of greatest variance in C, the second basis vector b_2 is the line of best fit along the second-greatest direction of variance in C and orthogonal to b_1, the third basis vector b_3 is the line of best fit along the third-greatest direction of variance in C and orthogonal to both b_1 and b_2, and so on to b_n. Each of the b_i, for $i = 1 \ldots n$, is a principal component of C; the b_i are stored as the columns of the matrix $EVECT$ in descending order of the variance they represent.

This orthonormal basis is found by calculating the eigenvectors of C. Calculation of eigenvectors is a fairly complex matter and is not described here because the details are not particularly germane to the discussion. Most linear algebra textbooks provide accessible accounts; see for example (Lay 2010). The main thing is to realize that the eigenvectors of the covariance matrix constitute an orthogonal basis for it.

Table 3.25 shows the eigenvector matrix $EVECT$ for the covariance matrix of Table 3.24, in which the columns are the eigenvectors that constitute an orthonormal basis for the covariance matrix.

The orthonormal basis is n-dimensional, just like the original data matrix. To achieve dimensionality reduction, a way has to be found of eliminating any

Table 3.25: Eigenvector matrix $EVECT$ of the covariance matrix C

-0.05	0.01	-0.15	0.35	0.85	0.35
0.00	-0.04	-0.09	0.10	0.33	-0.93
0.97	-0.21	0.03	-0.08	0.09	0.03
-0.13	-0.57	0.78	0.22	0.05	-0.01
-0.20	-0.57	-0.26	-0.72	0.22	0.05
0.03	0.55	0.55	-0.54	0.33	-0.02

basis vectors that lie along relatively insignificant directions of variance. The criterion used for this is the relative magnitudes of the eigenvalues associated with the eigenvectors in $EVAL$.

The calculation of eigenvectors associates an eigenvalue with each eigenvector, as already noted, and the magnitude of the eigenvalue is an indication of the degree of variance represented by the corresponding eigenvector. Since the eigenvalues are sorted by magnitude, all the eigenvectors whose eigenvalues are below some specified threshold can be eliminated, yielding a k dimensional orthogonal basis for C, where $k < n$. This is shown in Table 3.26.

Table 3.26: Eigenvalue matrix $EVAL$ and column-reduced eigenvector matrix $EVECT$ of the covariance matrix C

666.21	0	0	0	0	0	-0.05	0.01	-0.15	0.35
0	384.69	0	0	0	0	0.00	-0.04	-0.09	0.10
0	0	167.41	0	0	0	0.97	-0.21	0.03	-0.08
0	0	0	82.12	0	0	-0.13	-0.57	0.78	0.22
0	0	0	0	4.31	0	-0.20	-0.57	-0.26	-0.72
0	0	0	0	0	1.51	0.03	0.55	0.55	-0.54

The values in the main diagonal of the matrix in the left-hand side of Table 3.26 are the eigenvalues for the corresponding columns of the eigenvector matrix in Table 3.25: 666.21 at location (1,1) of the eigenvalue matrix is the eigenvalue for column 1 of the eigenvector matrix, 384.69 at location (2,2) is the eigenvalue for column 2 of the eigenvector matrix, and so on. The final two eigenvalues are much smaller than the others, which indicates that the variance represented by the corresponding eigenvectors is small enough to be disregarded without much loss of information. Removal of the two rightmost columns of the eigenvector matrix yields the 4-dimensional orthonormal basis for the covariance matrix shown on the right of Table 3.26; deciding how many basis vectors to retain is not always obvious, and is further discussed below.

Once the reduced-dimensionality space has been found, the mean-centred version DMC of the original n-dimensional data matrix D is projected into the reduced k-dimensional space, yielding a new $n \times k$ data matrix $D_{reduced}$ that still contains most of the variability in D. This is done by multiplying DMC^T on the left by the reduced-dimensionality eigenvector matrix $EVECT^T_{reduced}$, where the superscript T denotes matrix transposition, that is, re-shaping of the matrix whereby the rows of the original matrix become columns and the columns rows. This multiplication is shown in Equation (3.28).

$$D^T_{reduced} = EVECT^T_{reduced} \times DMC^T \qquad (3.28)$$

$D^T_{reduced}$ can now be transposed again to show the result in the original format, with rows representing the data objects and columns the variables, as in Table 3.27.

Table 3.27: Projection of original 6-dimensional data matrix of Table 3.23 into four-dimensional space

	v_1	v_2	v_3	v_4
g01	19.54	-7.28	-0.00	-3.26
g02	-26.96	16.13	15.34	-1.48
g03	12.36	-15.30	21.74	5.24
g04	-24.81	14.16	0.06	18.49
g05	2.72	-19.00	-23.82	7.10
g06	2.28	4.04	-4.75	3.12
g07	-26.97	15.27	-13.57	-10.96
g08	7.47	19.80	2.15	-10.65
g09	-19.77	-39.65	4.24	-7.95
g10	54.15	11.82	-1.40	0.34

The original six-dimensional data has been reduced to four dimensions at the cost of some information loss; more is said about this cost below. The full MDECTE matrix has already been shown to contain a substantial degree of redundancy, and can therefore usefully be dimensionality-reduced. The matrices $EVECT$ and $EVAL$ were calculated for it, and the distribution of values in the diagonal of $EVAL$ are shown in Figure 3.45.

Most of the variance in MDECTE is captured by the first few basis vectors, and virtually all of it is captured in the first 20 or so. MDECTE can, therefore, be reduced from 156-dimensional to 20-dimensional with a small loss of information about the phonetic usage of the speakers that it represents.

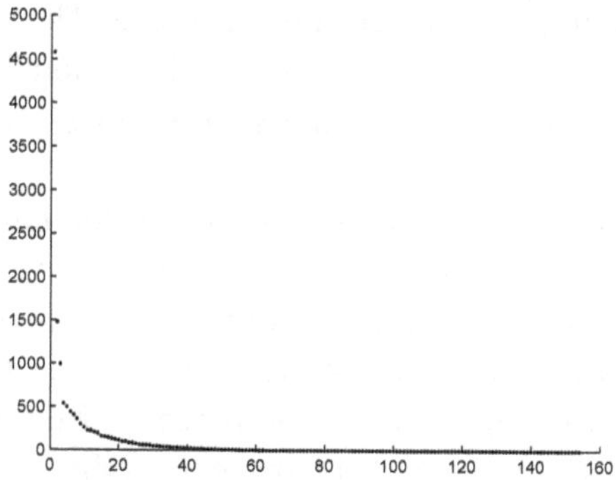

Figure 3.45: Eigenvalues of the full MDECTE matrix

Or, seen another way, 20 or so variables are sufficient to describe the phonetic usage that the original 156 variables described redundantly.

Several issues arise with respect to PCA:

- selection: Probably the most important issue is selection of a dimensionality threshold below which components are eliminated. There is no known general and optimal way of determining such a threshold, and selection of one is therefore subjective. There are, however, criteria to guide the subjectivity.
  - *A priori criterion:* The number k of dimensions to be selected is known in advance, so that the eigenvectors with the k largest eigenvalues are chosen. If, for example, one wants to represent the data graphically, only the first two or three dimensions are usable. The obvious danger here is that too few dimensions will be selected to retain sufficient informational content from the original matrix, with potentially misleading results, but this is a matter of judgement in particular applications.
  - *Eigenvalue criterion:* Only eigenvectors having an eigenvalue ≥ 1 are considered significant and retained on the grounds that significant dimensions should represent the variance of at least a single variable, and an eigenvalue < 1 drops below that threshold.
  - *Scree test criterion:* The scree test is so called by analogy with the erosion debris or scree that collects at the foot of a mountain. The eigenvalues are sorted in descending order of magnitude and plotted; the 'scree' descends from the 'mountain' at the left of the plot to the 'flat' on the right, and the further to the right one goes the

less important the eigenvalues become. The scree shown in Figure 3.45, was used to reduce the dimensionality of MDECTE to 20. Selection of a scree threshold is a matter of researcher judgement.

- *Percentage of variance criterion:* On the basis of the eigenvectors, it is possible to calculate the cumulative percentage of variance captured by successive dimensions, and most PCA software provides this information. Thus, the dimension with the largest eigenvector might capture 68 percent of the total data variance, and the second-largest 17 percent, giving a cumulative 85 percent, the third-largest 8 percent giving a cumulative 93 percent, and so on. The question, of course, is what percentage is enough; this is a matter for the researcher to decide relative to a particular application. The cumulative percentages in the case of MDECTE, where the first 20 eigenvectors were selected on the basis of the scree test, are given in Table 3.28.

Table 3.28: Cumulative percentages of variance captured by the first $n \leq 20$ eigenvectors

Eigenvector	Cumulative percent	Eigenvector	Cumulative percent
1	35.5	11	78.4
2	47.1	12	80.1
3	54.8	13	81.7
4	58.9	14	83.3
5	62.8	15	84.5
6	66.2	16	85.7
7	69.4	17	86.9
8	72.2	18	87.9
9	74.5	19	88.9
10	76.6	20	89.8

The first 20 eigenvectors capture 90 percent of the variance in the original data; the first 30 capture 95 percent, and the first 61 100 percent. Even the safest option of keeping the first 61 eigenvectors would result in a very substantial reduction in dimensionality, but one might take the view that the small gain in terms of data loss over, say, 20 or 30 is not worth the added dimensionality.

- *Variable interpretation:* In any data matrix the variables typically have labels that are semantically significant to the researcher in the sense that they denote aspects of the research domain considered to be relevant. Because PCA defines a new set of variables, these labels are no longer applicable to the columns of the dimensionality-reduced matrix. This is why the variables in the PCA-reduced matrices in the foregoing discussion were given the semantically-neutral labels $v_1 \ldots v_4$; the values for these variables are self-evidently not interpretable as the frequencies of the original data since some of them are negative. In applications where the aim is simply dimensionality reduction and semantic interpretation of the new variables is not an issue, this doesn't matter. There are, however, applications where understanding the meaning of the new variables relative to the original ones might be useful or essential, and there is some scope for this. Jolliffe (2002: 63) notes that "there is no reason, a priori, why a mathematically derived linear function of the original variables, which is what the PCs are, should have a simple interpretation", but that "it is remarkable how often it seems to be possible to interpret the first few PCs", though "it is probable that some interpretations owe a lot to the analyst's ingenuity and imagination". In other words, there is no guarantee that an intuitively satisfying interpretation of the principal components is there to be found, and, if one is proposed, there is no obvious way of assessing its validity.
- *Linearity:*PCA is an effective and very widely used dimensionality reduction technique, but because it takes no account of manifold shape it can yield misleading results when applied to nonlinear manifolds. To deal with the latter, nonlinear versions of PCA have been developed. One of these infers a reduced-dimensionality representation from data using the learning capability of artificial neural networks (ANN) – see Diamantaras and Kung (1996), Carreira-Perpinan (1997), and Haykin (1999: Ch. 8). Another, the Generative Topographic Mapping (GTM) (Bishop, Svensen, and Williams 1998), is a latent variable model which assumes that a relatively small number of low-dimensional probability distributions underlie high-dimensional observed data, and whose aim is to model the observed data in terms of those distributions. A third and increasingly widely used nonlinear variant of PCA is Kernel PCA, which projects the nonlinear manifold into a higher-dimensional space in which it becomes linear and then applies the eigendecomposition procedure described above to that linear representation, thereby

in effect taking account of the original data nonlinearity in the dimensionality reduction: Schölkopf, Smola, and Müller (1998), Schölkopf, Smola, and Müller (1999), Ham et al. (2004), Shawe-Taylor and Cristianini (2004: Ch. 6), Lee and Verleysen (2007: Ch. 4.4.1).
- *Other issues:* Other PCA issues are the use of the correlation rather than covariance matrix and the effect of outliers, for which see: Jolliffe (2002) and Jackson (2003); selected briefer accounts are in Bishop (1995: 310ff.), Everitt and Dunn (2001: 48ff.), and Tabachnick and Fidell (2007: Ch. 13).

The remainder of this part of the discussion describes a range of other variable extraction methods, categorized by whether or not they take data nonlinearity into account.

Singular value decomposition (SVD)

SVD(Jackson 2003; Jolliffe 2002; Wall, Rechtsteiner, and Rocha 2003) is a theorem in linear algebra whereby any matrix D with m rows and n columns can be represented as the product of three matrices, as in Equation (3.29).

$$D_{m,n} = U_{m,m} S_{m,n} V^T_{n,n} \tag{3.29}$$

where
- U, S, and V are the matrices whose product gives D.
- The column vectors of U are an orthonormal basis for the column vectors of D.
- The column vectors of V are an orthonormal basis for the row vectors of D; the T superscript denotes transposition, that is, V is rearranged so that its rows become columns and its columns rows.
- S is a diagonal matrix, that is, a matrix having nonzero values only on the diagonal from $S_1, 1$ to S_m, n, and those values in the present case are the singular values of D in descending order of magnitude. These singular values are the square roots of the eigenvectors of U and V.

Because the column vectors of V are an orthonormal basis for D and the values in S are ranked by magnitude, SVD can be used for dimensionality reduction in exactly the same way as PCA. Indeed, when D is a covariance or correlation matrix, SVD and PCA are identical. SVD is more general than PCA because it can be applied to matrices of arbitrary dimensions with unrestricted numerical values whereas PCA is restricted to square matrices containing covariances or correlations, but in practice it is a straightforward matter to calculate a covariance or correlation matrix for whatever data matrix one wants to analyze, so the choice between SVD and PCA is a matter of preference.

3.2.2.4 Factor Analysis (FA)

FA is very similar to PCA, and the two are often conflated in the literature. Relative to a redundant n-dimensional matrix D, both use eigenvector decomposition to derive a set of basis vectors from the variance / covariance matrix for D, both use the relative magnitudes of the eigenvalues associated with the eigenvectors to select a reduced number of basis vectors $k < n$, and the k vectors are taken to constitute a reduced-dimensionality basis into which D can be projected in order to reduce its dimensionality from n to k. They differ, however, both conceptually and, as a consequence, in how variability in data is analyzed.

PCA is a formal mathematical exercize that uses patterns of covariance in redundant data to find the main directions and magnitudes of variance, and these directions are expressed as a set of non-redundant synthetic variables in terms of which the original variables can be re-stated. These synthetic variables may or may not have a meaningful interpretation relative to the research domain that the original variables describe, but there is no explicit or implicit claim that they necessarily do; PCA is simply a means to a dimensionality reduction end. FA differs in that it does make a substantive claim about the meaningfulness of the variables it derives. Specifically, the claim is that observed data represent significant aspects of the natural process which generated them in a way that is obscured by various kinds of noise and by suboptimal selection of redundant variables, that these significant aspects are latent in the observed data, and that the factors which FA derives identifies these aspects. As such, FA offers a scientific hypothesis about the natural process to which it relates.

The basis for this claim is what distinguishes FA mathematically from PCA: in deriving components, PCA uses all the variance in a data matrix, but FA uses only a portion of it. To understand the significance of this, it is first necessary to be aware of the distinction between different kinds of variance that FA makes. Relative to a given observed data variable v_i in an n-dimensional matrix D, where i is in the range $1 \ldots n$:

- The common variance of v_i is the the variance that v_i shares will all the other variables in D; this is referred to as its communality.
- Specific variance of v_i is the variance unique to v_i.
- Error variance of v_i is the variance due to the noise factors associated with data.
- Total variance of v_i is the sum of its common, specific, and error variances.

PCA analyzes total variance, but FA analyzes common variance only, on the grounds that common variance reflects the essentials of the natural process which the data describes, and that analysis of the patterns of covariance in the data calculated only on communality allows scientifically meaningful factors to be extracted.

FA has two main disadvantages relative to PCA: (i) the common variance on which FA is based is complicated to isolate whereas the total variance on which PCA is based is straightforward, and (ii) the factors which FA generates are not unique, whereas PCA generates a unique and optimal summary of the variance in a matrix. On the other hand, FA has the advantage when meaningful interpretation of the derived variables is required by the researcher. Where, therefore, the aim is simply dimensionality reduction and interpretation of the extracted variables is not crucial, PCA is the choice for its simplicity and optimality, but where meaningful interpretation of the extracted variables is important, FA is preferred.

For discussions of FA see Jolliffe (2002), Jackson (2003), Tabachnick and Fidell (2007: Ch. 13), Hair et al. (2010: Ch. 3). Closely related to FA is Independent Component Analysis, whose aim is to model data as a linear mixture of underlying factors (Hyvärinen, Karhunen, and Oja 2001; Stone 2004).

Multidimensional Scaling (MDS)

PCA uses variance preservation as its criterion for retaining as much of the informational content of data as possible in dimensionality reduction. MDS uses a different criterion, preservation of proximities among data objects, on the grounds that proximity is an indicator of the relative similarities of the real-world objects which the data represents, and therefore of informational content; if a low-dimensional representation of the proximities can be constructed, then the representation preserves the informational content of the original data. Given an $m \times m$ proximity matrix P derived from an $m \times n$ data matrix M, MDS finds an $m \times k$ reduced-dimensionality representation of M, where k is a user-specified parameter. MDS is not a single method but a family of variants. The present section describes the original method on which the variants are ultimately based, classical metric MDS, and a variant, metric least squares MDS.

Classical MDS requires that the proximity measure on which it is to operate be Euclidean distance. Given an $m \times n$ data matrix M, therefore, the first step is to calculate the $m \times m$ Euclidean distance matrix E for M. Thereafter, the algorithm is:

- Mean-centre E by calculating the mean value for each row E_i (for $i = 1 \ldots m$) and subtracting the mean from each value in E_i.
- Calculate an $m \times m$ matrix S each of whose values $S_{i,j}$ is the inner product of rows E_i and E_j, where the inner product is the sum of the product of the corresponding elements as described earlier and the T superscript denotes transposition: $S_{i,j} = \sum_{k=1 \ldots m}(E_{i,k} \times E^T_{j,k})$.
- Calculate the eigenvectors and eigenvalues $EVECT$ and $EVAL$ of S, as already described.
- Use the eigenvalues, as in PCA, to find the number of eigenvectors k worth retaining.
- Project the original data matrix M into the k-dimensional space, again as in PCA: $M^T_{reduced} = EVECT^T_{reduced} x M^T$.

This algorithm is very reminiscent of PCA, and it can in fact be shown that classical MDS and PCA are equivalent and give identical results – cf. Borg and Groenen (2005: Ch. 24), Lee and Verleysen (2007: Ch. 4) –, and are therefore simply alternative solutions to a problem. The variant of classical MDS about to be described, however, extends the utility of MDS beyond what PCA is capable of, and provides the basis for additional dimensionality techniques more powerful than either of them.

Classical MDS and PCA both give exact algebraic mappings of data into a lower-dimensional representation. The implicit assumption is that the original data is-free. This is, however, not always and perhaps not even usually the case with data derived from real-world observation, and where noise is present classical MDS and PCA both include it in calculating their lower-dimensional projections. Metric least squares MDS recognizes this as a problem, and to compensate for it relaxes the definition of the mapping from higher-dimensional to lower-dimensional data as algebraically exact to approximate: it generates an $m \times k$ representation matrix M' of an $m \times n$ numerical-valued matrix M by finding an M' for which the distances between all distinct pairings of row vectors i, j in M' are as close as possible to the proximities p_{ij} between corresponding row vectors of M, for $i, j = 1 \ldots m$. The reasoning is that when the distance relations in M and M' are sufficiently similar, M' is a good reduced-dimensionality representation of M. Metric least squares MDS operates on distance measurement of proximity. This can be any variety of distance measure, but for simplicity of exposition it will here be assumed to be Euclidean.

The mapping f from M to M' could in principle be explicitly defined but is in practice approximated by an iterative procedure using the following algorithm:

1. Calculate the Euclidean distance matrix $D(M)$ for all distinct pairs (i, j) of the m rows of M, so that $\delta_{i,j} \in D(M)$ is the distance from row i to row j of M, for $i, j = 1 \ldots m$.
2. Select a dimensionality k and construct an $m \times k$ matrix M' in which m k-dimensional vectors are randomly located in the k-space.
3. Calculate the Euclidean distance matrix $D(M')$ for all distinct pairs i, j of the m rows of M', so that $d_{i,j} \in D(M')$ is the distance from row i to row j of M', for $i, j = 1 \ldots m$.
4. Compare the distance matrices $D(M)$ and $D(M')$ to determine how close they are, where closeness is quantified in terms of an objective function called a stress function. If the stress function has reached a predetermined threshold of acceptable closeness between $D(M)$ and $D(M')$, stop. Otherwise, adjust the values in the m row vectors of M' so that the distances between their new locations in the k-space more closely approximate the corresponding ones in $D(M)$, and return to step (3).

Finding M' is, in short, a matter of moving its row vectors around in the k-space until the distance relations between them are acceptably close to those of the corresponding vectors in M.

The stress function is based on the statistical concept of squared error. The difference or squared error e^2 between a proximity $\delta_{i,j}$ in M and a distance $d_{i,j}$ in M' is given in Equation (3.30).

$$e^2_{i,j} = (\delta_{i,j} - d_{i,j})^2 \tag{3.30}$$

The total difference between all δ_{ij} and d_{ij} is therefore as in Equation (3.31).

$$e^2 = \sum_{i,j=1..m} (\delta_{i,j} - d_{i,j})^2 \tag{3.31}$$

This measure is not as useful as it could be, for two reasons. The first is that the total error is a squared quantity and not easily interpretable in terms of the original numerical scale of the proximities, and the solution is to un-square it; the reasoning here is the same as that for taking the square root of variance to obtain a more comprehensible measure, the standard deviation. The second is that the magnitude of the error is scale-dependent, so that a small difference between proximities and distances measured on a large scale can appear greater than a large difference measured on a small scale, and the solution in this case is to make the error scale-independent; the reasoning in this case is that of the discussion of variable scaling earlier on. The reformulation of the squared error expression incorporating these changes is called stress, which is given in Equation (3.32).

$$stress = \sqrt{\frac{\sum_{i,j=1..m}(\delta_{i,j} - d_{i,j})^2}{\sum_{i,j=1..m} d_{i,j}^2}} \tag{3.32}$$

This is the stress function used to measure the similarity between $D(M)$ and $D(M')$. By iterating steps (3) and (4) in the above MDS algorithm, the value of this stress function is gradually minimized until it reaches the defined threshold and the iteration stops. Minimization of the stress function in MDS is a particular case of what has become an important and extensive topic across a range of science and engineering disciplines, function optimization. Various optimization methods such as gradient descent are available but all are complex and presentation would serve little purpose for present concerns, so nothing further is said about them here; for details of their application in MDS see Borg and Groenen (2005: Ch. 8).

As with other dimensionality reduction methods, a threshold dimensionality k must be determined for MDS. The indicator that k is too small is non-zero stress. If $k = n$, that is, the selected dimensionality is the same as the original data dimensionality, the stress will be zero. Any k less than the (unknown) intrinsic dimension will involve some increase in stress; the question is what the dimensionality should be to give an acceptable level. The only obvious answer is empirical. Starting with $k = 1$, MDS is applied for monotonically-increasing values of k, and the behaviour of the stress is observed: when it stops decreasing significantly with increasing k, an approximation to the intrinsic dimension of the data, and thus of optimal k, has been reached (ibid.: 4f.). Figure 3.46 shows this by applying MDS to MDECTE for each increment in k in the range $1 \ldots 156$ and plotting stress in each case.

The indication is that an appropriate dimensionality for MDECTE is in the range 20...30.

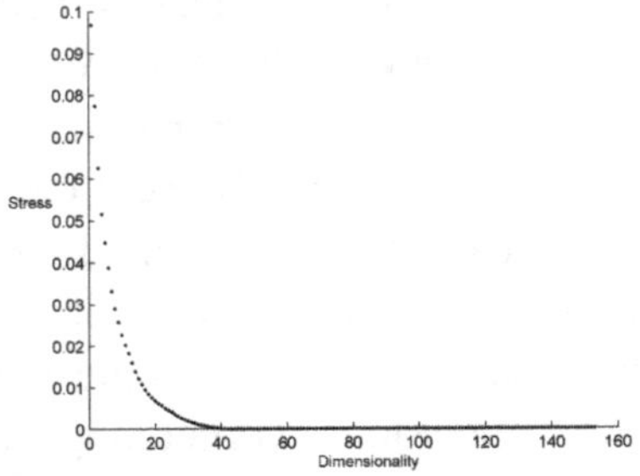

Figure 3.46: Sorted MDS stress values for $k = 1\ldots156$ with respect to MDECTE

Having generated a reduced-dimensionality representation, it is natural to ask how good a representation of the original data it is. The degree of stress is an obvious indicator, though an ambiguous one because there is no principled criterion for what stress value constitutes an acceptable threshold of closeness (Borg and Groenen 2005: 47ff.); essentially, a stress value is significant only in relation to its location in the profile of stress values for the given dataset, as shown for MDECTE in Figure 3.46. Another is the degree of correlation between the distances of the original data and the distances of the reduced-dimensionality representation, which can be stated as a correlation coefficient or visualized as a scatter plot, or both. Figure 3.47 shows this for a reduction of MDECTE to dimensionality 20 which Figure 3.46 indicates is a reasonable choice.

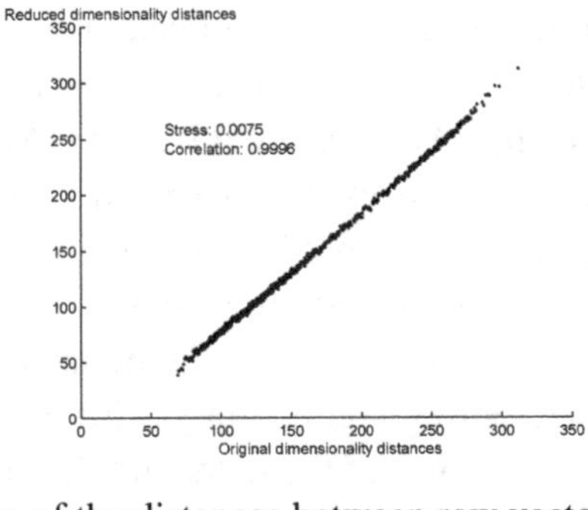

Figure 3.47: Correlation of the distances between row vectors in MDECTE and in a representation of MDECTE reduced to dimensionality 20 by MDS

The indicators support the conclusion that the 20-dimensional representation of MDECTE is a good one: the stress value of 0.0075 is low relative to the stress profile in Figure 3.46, the correlation is near-perfect, and the plot of proximities on the horizontal axis against distances on the vertical is almost perfectly linear.

For MDS see Kruskal and Wish (1978), Jolliffe (2002), Jackson (2003), and Borg and Groenen (2005). Good brief accounts are in Jain and Dubes (1988: Ch. 2.7), Groenen and Velden (2005), Izenman (2008: Ch. 13), Lee and Verleysen (2007: Ch. 4.4.2), Hair et al. (2010: Ch. 10), and Martinez, Martinez, and Solka (2011: Ch. 3).

Sammon's Mapping

Sammon's mapping is a nonlinear variant of metric least squares MDS. It differs from MDS in a single modification to the stress function shown in Table 3.29.

Table 3.29: Comparison of MDS and Sammon's Mapping stress functions

$stress = \sqrt{\frac{\sum_{i,j=1..m}(\delta_{i,j}-d_{i,j})^2}{\sum_{i,j=1..m} d_{i,j}^2}}$	$stress = \sqrt{\frac{\sum_{i,j=1..m}(\delta_{i,j}-d_{i,j})^2}{\sum_{i,j=1..m} \delta_{i,j}^2}}$
a: Linear metric MDS stress function	b: Sammon mapping stress function

The difference is in the normalization term: linear MDS normalizes by the distances d in the reduced-dimensionality space, and the Sammon mapping by the distances δ in the full-dimensionality one. This difference allows the Sammon version to incorporate nonlinearity in higher-dimensional data into the lower-dimensional representation which it generates. To see why so apparently small a change can have so fundamental an effect, consider the manifold in Figure 3.48.

The manifold is nonlinear, but linear measurement does not capture the geodesic distances between the points on it equally well. The distance from f to e, for example, is relatively short and the linear measure is a good approximation to the geodesic distance; from f to c and f to a it is longer and the linear measure is a less good approximation; from a to k it is less good still. The best way to capture the shape of the manifold is to add the distances $a \rightarrow b$, $b \rightarrow c$, and so on, and simply to disregard the remaining distances.

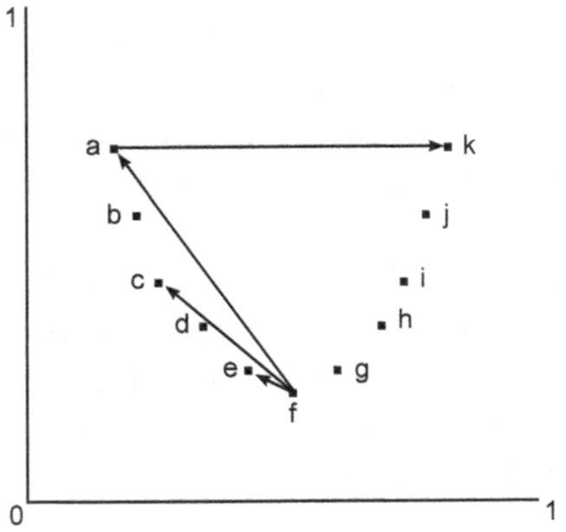

Figure 3.48: The effect of the normalization term in Sammon's Mapping

Sammon's mapping is based on this idea. When the normalization term, that is, the distance $\delta_{i,j}$, has a relatively large value, the value of the stress function is relatively small, but if $\delta_{i,j}$ is relatively small the stress function value is relatively large; because the iterative procedure underlying metric least squares MDS and Sammon minimizes the stress function, this means that the minimization is based much more on the smaller than on the larger linear distances in the data: the larger the value of the stress function the greater the adjustment to the output matrix in the MDS algorithm. In other words, in generating the reduced-dimensionality matrix Sammon's mapping concentrates on the smaller linear distances in the input matrix because these are better approximations to the shape of whatever nonlinearity there is in the data, and incrementally ignores the distances as they grow larger.

As with metric least squares MDS, the reduced dimensionality k is user specified and can be estimated by applying Sammon's mapping to the data for incrementally increasing values of k, recording the stress for each k, and then plotting the stress values to see where they stop decreasing significantly. Figure 3.49 shows this for the range $k = 1 \ldots 156$.

Also as with metric least squares MDS, the goodness of the reduced-dimensionality representation can be assessed by calculating the degree of correlation between the distances of the original data and the distances of the reduced-dimensionality representation, which can be stated as a correlation coefficient or visualized as a scatter plot, or both.

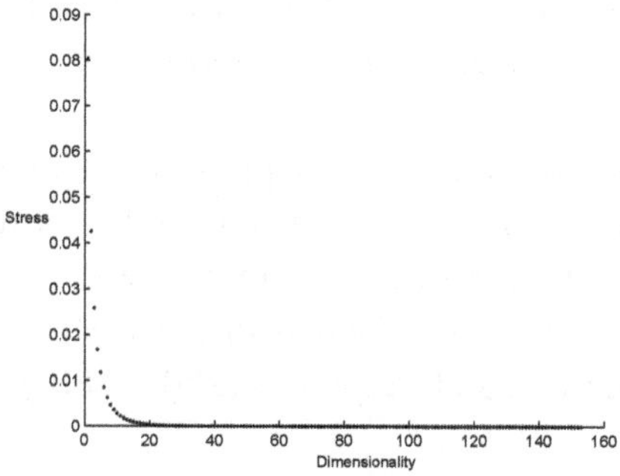

Figure 3.49: Sorted Sammon's mapping stress values for $k = 1 \ldots 156$ with respect to MDECTE

Figure 3.50 shows this for a reduction of MDECTE to dimensionality 12, which 3.49 indicates is a reasonable choice.

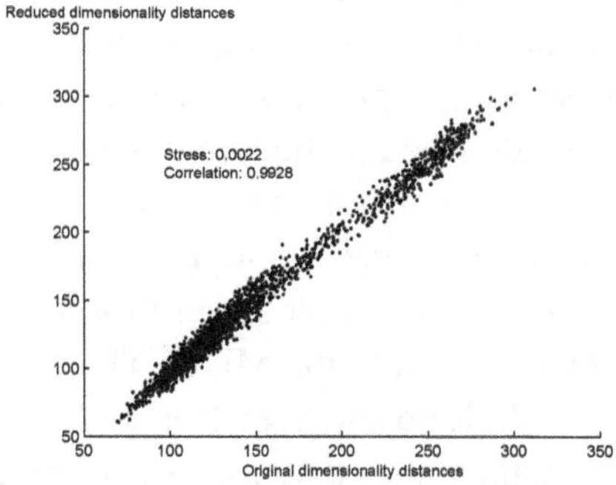

Figure 3.50: Correlation of the distances between row vectors in MDECTE and in a representation of MDECTE reduced to dimensionality 12 by Sammon's mapping

Though still close, the correlation between distances in the original-dimensionality and reduced-dimensionality spaces is here slightly less good than for MDS relative to the same data. Formally, therefore, the MDS result is better than the Sammon one, even if only marginally. This does not, however, warrant the conclusion that the MDS dimensionality reduction is better than the Sammon. Such a conclusion assumes that the original linear Euclidean distances accurately represent the shape of the MDECTE data manifold. But, as we have seen, MDECTE contains substantial nonlinearity. The Sammon dimensionality-reduced matrix represents that nonlinearity and is for that reason to be preferred; the slightly less good formal indicators arise because, in taking account of the nonlinearity, Sammon loses some linear distance information.

The original paper for Sammon's mapping is Sammon (1969); Lee and Verleysen (2007: Ch 4.2.3) provide a good recent discussion of it. For variants see Lee and Verleysen (ibid.: 86f.); probably the most important among these is curvilinear component analysis (Demartines and Hérault 1997), for further discussion of which see Lee and Verleysen (2007: 88ff.).

Isomap

Isomap is a variant of MDS which reduces dimensionality by operating on a nonlinear rather than on a linear distance matrix. Given a linear distance matrix D_L derived from a data matrix M, Isomap derives a graph-distance approximation to a geodesic distance matrix D_G from D_L, and D_G is then the basis for dimensionality reduction using either the classical or the metric least squares MDS procedure; graph distance approximation to geodesic distance has already been described in the foregoing discussion of data geometry. The Isomap approximation differs somewhat from the procedure already described, however, in that it uses the topological concept of neighbourhood. The present discussion gives a brief account of this concept before going on to describe Isomap and applying it to the MDECTE data.

Topology (Munkres 2000; Reid and Szendroi 2005; Sutherland 2009) is an aspect of contemporary mathematics that grew out of metric space geometry. Its objects of study are manifolds, but these are studied as spaces in their own right, topological spaces, without reference to any embedding metric space and associated coordinate system. Topology would, for example, describe a manifold embedded in the metric space of Figure 3.51a independently both of the metric defined on the space and of the coordinates relative to which the distances among points are calculated, as in Figure 3.51b.

Topology replaces the concept of metric and associated coordinate system with relative nearness of points to one another in the manifold as the mathematical structure defined on the underlying set; relative nearness of points is determined by a function which, for any given point p in the manifold, returns the set of all points within some specified proximity to p. But how, in the absence of a metric and a coordinate system, is the proximity characterized?

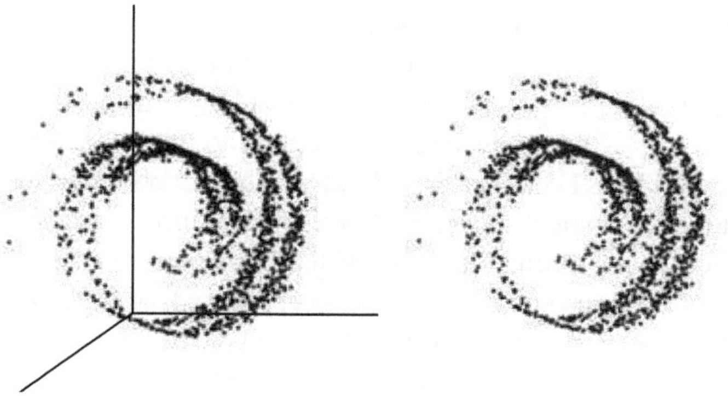

(a) *(b)*

Figure 3.51: A manifold embedded in metric space (a) and as a topological manifold (b)

The answer is that topological spaces are derived from metric ones and inherit from the latter the concept of neighbourhoods. In a metric space, a subset of points which from a topological point of view constitutes a manifold can itself be partitioned into subsets of a fixed size called neighbourhoods, where the neighbourhood of a point p in the manifold can be defined either as the set of all points within some fixed radius ε from p or as the k nearest neighbours of p using the existing metric and coordinates; in Figure 3.52 a small region of the manifold from Figure 3.51 is magnified to exemplify these two types of neighbourhood.

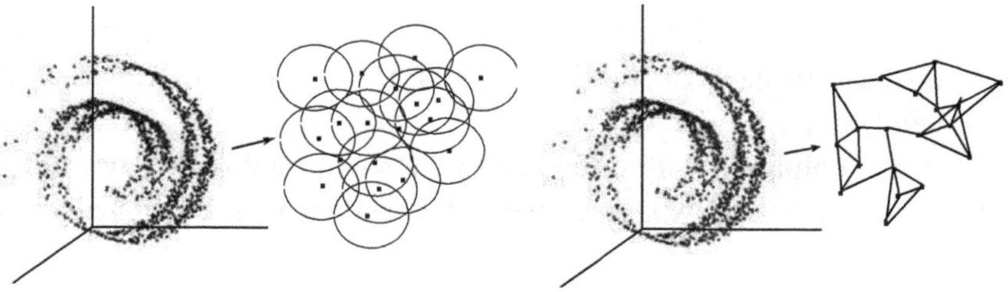

(a) Neighbourhoods of diameter ε *(b)* Neighbourhoods for k = 3 nearest neighbours

Figure 3.52: Neighbourhoods in a magnified fragment of a geometric object in metric space

In Figure 3.52a the neighbourhood of every point is the other points within a radius of ε, shown as circles within the magnification rectangle; in 3.52b a neighbourhood of any point is the k nearest points irrespective of distance, shown for $k = 3$ as lines connecting each point to the three nearest to itself. Once a manifold has been partitioned into neighbourhoods and thereby transformed into a topological space, the frame of reference is discarded and only the neighbourhoods defined in terms of the metric are retained. In this way, manifolds of arbitrary shape can be conceptualized as being composed of metric subspaces; if the original metric is Euclidean, for example, the manifold in Figure 3.52 can be understood as a patchwork of locally-Euclidean subspaces. Intuitively, this corresponds to regarding the curved surface of the Earth as a patchwork of flat neighbourhoods, which is how most people see it.

Topological spaces are supersets of metric spaces, so that every metric space is also a topological one. This observation is made for convenience of reference to geometrical objects in subsequent discussion: these are referred to as manifolds irrespective of whether they are embedded in a metric space or constitute a topological space without reference to a coordinate system.

Assume now the existence of an $m \times n$ data manifold M embedded in a metric space and a specification of neighbourhood size as a radius ε or as k nearest neighbours; in what follows, only the k nearest neighbour specification is used to avoid repetition. Isomap first transforms M into a topological manifold by constructing a set of k-neighbourhoods. This is done in two steps:

1. A matrix of linear distances between the data objects, that is, the rows of M, is calculated; assume that the measure is Euclidean and call the matrix D.
2. A neighbourhood matrix N based on D is calculated which shows the distance of each of the data objects M_i ($i = 1 \dots m$) to its k nearest neighbours.

This is exemplified with reference to the small randomly generated two-dimensional matrix M whose scatterplot is shown with row labels in Figure 3.53.

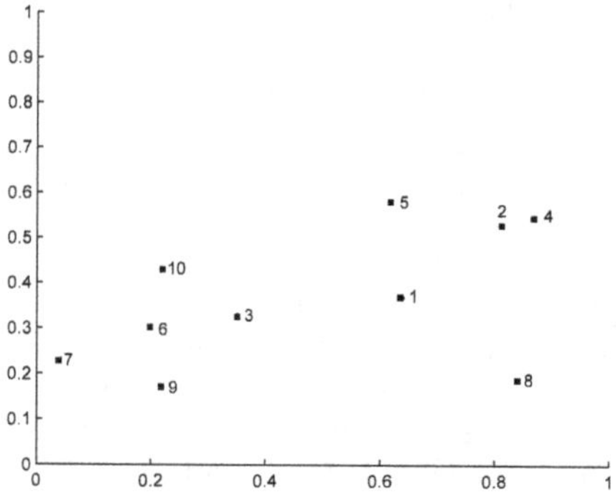

Figure 3.53: Scatter plot of a randomly generated two-dimensional matrix M

Table 3.30 shows the data matrix M underlying Figure 3.53, Table 3.31 the Euclidean distance matrix D for M, and Table 3.32 the corresponding neighbourhood matrix N for $k = 4$.

Table 3.30: Data matrix M underlying Figure 3.53

	v_1	v_2		v_1	v_2
1	0.64	0.37	6	0.20	0.30
2	0.81	0.53	7	0.04	0.23
3	0.35	0.33	8	0.84	0.19
4	0.87	0.55	9	0.22	0.17
5	0.62	0.58	10	0.22	0.43

Table 3.31: Euclidean distance matrix D for data in Table 3.30

	1	2	3	4	5	6	7	8	9	10
1	0	0.228	0.296	0.288	0.210	0.443	0.622	0.272	0.467	0.421
2	0.228	0	0.500	0.067	0.197	0.647	0.829	0.340	0.689	0.592
3	0.296	0.500	0	0.566	0.368	0.148	0.329	0.514	0.210	0.157
4	0.288	0.067	0.566	0	0.256	0.713	0.895	0.357	0.753	0.658
5	0.210	0.197	0.368	0.256	0	0.504	0.683	0.451	0.575	0.423
6	0.443	0.647	0.148	0.713	0.504	0	0.182	0.645	0.132	0.136
7	0.622	0.829	0.329	0.895	0.683	0.182	0	0.805	0.195	0.278
8	0.272	0.340	0.514	0.357	0.451	0.645	0.805	0	0.619	0.662
9	0.467	0.689	0.210	0.753	0.575	0.132	0.195	0.619	0	0.265
10	0.421	0.592	0.157	0.658	0.423	0.136	0.278	0.662	0.265	0

Table 3.32: Neighbourhood matrix N corresponding to Euclidean distance matrix in Table 3.31

	1	2	3	4	5	6	7	8	9	10
1	0	0.228	inf	0.288	0.210	inf	inf	0.272	inf	inf
2	0.228	0	inf	0.067	0.197	inf	inf	0.340	inf	inf
3	0.296	inf	0	inf	inf	0.148	inf	inf	0.210	0.157
4	0.288	0.067	inf	0	0.256	inf	inf	0.357	inf	inf
5	0.210	0.197	0.368	0.256	0	inf	inf	inf	inf	inf
6	inf	inf	0.148	inf	inf	0	0.182	inf	0.132	0.136
7	inf	inf	0.329	inf	inf	0.182	0	inf	0.195	0.278
8	0.272	0.340	inf	0.357	0.451	inf	inf	0	inf	inf
9	inf	inf	0.210	inf	inf	0.132	0.195	inf	0	0.265
10	inf	inf	0.157	inf	inf	0.136	0.278	inf	0.265	0

M and D are self-explanatory in the light of the discussion so far. N is less so. Note that, apart from 0 in the main diagonal, each row of N has exactly 4 numerical values, corresponding to $k = 4$. The numerical value at $N_{i,j}$ indicates both that j is in the k-neighbourhood of i and the distance between i and j; the k-neighbourhood of N_1, for example, includes N_2, N_4, N_5, and N_8, which can be visually confirmed by Figure 3.53. The zeros indicate that a data object is at a *nil* distance from itself, and the inf values (for 'infinity') that j is not in the neighbourhood of i.

Isomap now interprets N as a graph in which data objects are nodes, the numerical values are arcs labelled with distances between pairs of nodes, and the inf values indicate no arc. In graph representation, the N of Table 3.32 looks like Figure 3.54.

Using the graph, the shortest node-to-node distance between any two points in the data manifold can be calculated using one of the standard graph traversal algorithms (Gross and Yellen 2006). The shortest distance between node 8 and node 7, for example, follows the path $8 \rightarrow 5 \rightarrow 3 \rightarrow 6 \rightarrow 7$, and is $0.451 + 0.368 + 0.148 + 0.182 = 1.149$. Reference to Table 3.31 shows that this is greater than the Euclidean distance of 0.805, which goes directly from 8 to 7. Isomap calculates the graph distances between all points in the data manifold and stores them in a matrix G. The one derived from Figure 3.54 is shown in Table 3.33; note that the values shown in that table may not correspond exactly to those derivable from Figure 3.54 on account of round-off discrepancies. Where there is only a single arc traversal the graph and Euclidean distances are identical but for multi-arc traversals the graph distances

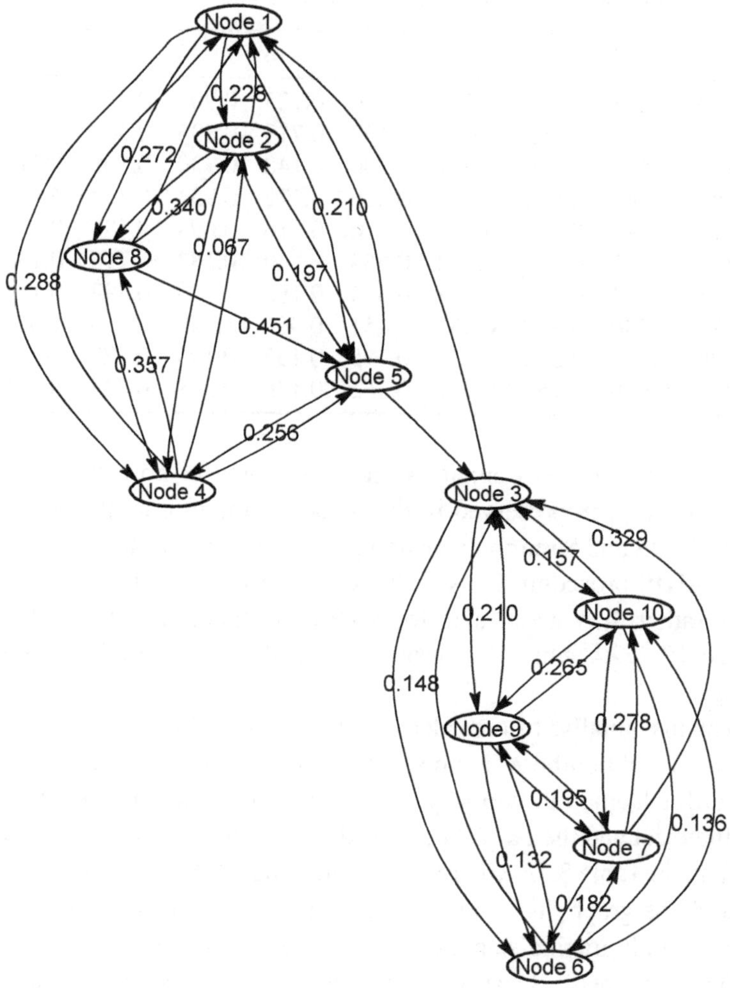

Figure 3.54: Graph interpretation of the neighbourhood matrix in Table 3.32

are larger where the path between data objects is not linear. Isomap applies the classical or metric least squares MDS procedure to such graph distance matrices G to reduce their dimensionality, as already described.

Table 3.33: Shortest-path graph distance table for Table 3.32 / Figure 3.54

	1	2	3	4	5	6	7	8	9	10
1	0	0.228	0.578	0.288	0.210	0.725	0.907	0.272	0.787	0.734
2	0.228	0	0.565	0.067	0.197	0.713	0.895	0.340	0.774	0.721
3	0.296	0.524	0	0.584	0.506	0.148	0.330	0.568	0.210	0.157
4	0.288	0.067	0.624	0	0.256	0.772	0.954	0.357	0.834	0.781
5	0.210	0.197	0.368	0.256	0	0.516	0.698	0.481	0.577	0.524
6	0.444	0.672	0.148	0.732	0.654	0	0.182	0.716	0.132	0.136
7	0.625	0.853	0.329	0.914	0.835	0.182	0	0.897	0.195	0.278
8	0.272	0.340	0.819	0.357	0.451	0.966	1.149	0	1.028	0.975
9	0.506	0.733	0.210	0.794	0.715	0.132	0.195	0.777	0	0.265
10	0.453	0.680	0.157	0.741	0.662	0.136	0.278	0.724	0.265	0

Selection of a dimensionality k and assessment of how well the original distances have been preserved in the reduced-dimensionality representation are as for MDS and Sammon's mapping, and are not repeated here; where the classical MDS procedure is used, the criterion for selection of k is residual variance rather than stress, and, for MDECTE, this indicates a value of k in the range $10 \ldots 20$, which is consistent with those for MDS and Sammon's mapping.

It remains, finally, to consider an important problem with Isomap. The size of the neighbourhood is prespecified by the user, and this can be problematical for Isomap in two ways. If k or ε is too small the neighbourhoods do not intersect and the graph becomes disconnected; Figure 3.55a shows this for the data in Table 3.32, where $k = 3$ rather than 4. Because the graph is disconnected, the graph distances between data objects cannot all be calculated; the matrix of all graph distances G for $k = 3$ is shown in Figure 3.55b, where inf indicates no path. Because the distance matrix is incomplete, the Isomap dimensionality reduction fails.

The way to deal with this problem is incrementally to increase k until the matrix of all distances G no longer contains inf entries. This potentially creates the second of the above-mentioned problems: too large a neighbourhood leads to so-called short-circuiting, where the connectivity of the neighbourhoods fails correctly to represent the manifold shape. To show what this involves, a simplified version of the Swiss roll manifold of Figure 3.56 in two-dimensional space is used.

The geodesic distance from P to R in Figure 3.56 follows the spiral, and the neighbourhood must be chosen to follow it. It is, however, visually clear

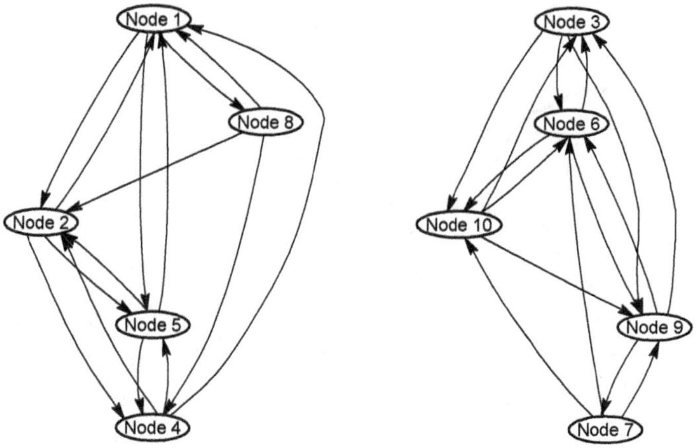

(a)

	1	2	3	4	5	6	7	8	9	10
1	0	0.228	inf	0.294	0.210	inf	inf	0.272	inf	inf
2	0.228	0	inf	0.067	0.197	inf	inf	0.499	inf	inf
3	inf	inf	0	inf	inf	0.148	0.405	inf	0.210	0.157
4	0.288	0.067	inf	0	0.256	inf	inf	0.560	inf	inf
5	0.210	0.197	inf	0.256	0	inf	inf	0.481	inf	inf
6	inf	inf	0.148	inf	inf	0	0.327	inf	0.132	0.136
7	inf	inf	0.329	inf	inf	0.182	0	inf	0.195	0.278
8	0.272	0.340	inf	0.357	0.481	inf	inf	0	inf	inf
9	inf	inf	0.210	inf	inf	0.132	0.195	inf	0	0.268
10	inf	inf	0.157	inf	inf	0.136	0.460	inf	0.265	0

(b)

Figure 3.55: Shortest-path graph distance table for the data in Table 3.32, for $k = 3$

in 3.56 that, for $k = 3$ or greater, the neighbourhood of a point can include other points geodesically distant from it: taking point P, $k = 2$ correctly includes S and T, but $k = 3$ includes Q rather than U because distance $d2$ is less than $d1$. This cuts all points U to Q out of the graph approximation and seriously misrepresents the shape of the manifold. Figure 3.56b shows this for the radius definition of neighbourhoods, where $\varepsilon = r1$ correctly includes S and T, but $\varepsilon = r2$ includes Q and points adjacent to it rather than U.

One might argue that the Swiss roll is a contrived example and that data manifolds in real-world applications are unlikely to present this problem. MDECTE seems to, however. The foregoing discussion leads one to expect

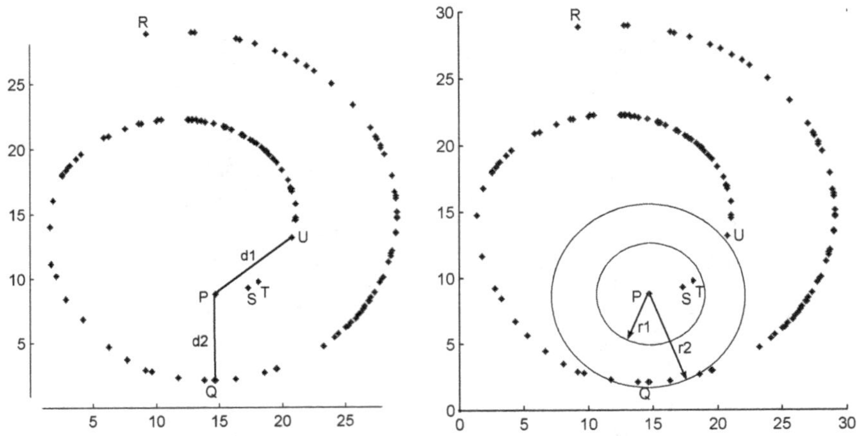

(a) k　　　　　　*(b)* ε

Figure 3.56: Short-circuiting for too-large values of k / ε

that the likelihood of short-circuiting grows with increasing neighbourhood size. As well shall see, the data objects in MDECTE form two well separated clusters, and to achieve full connectivity in the graph approximation of geodesic distance for it, a neighbourhood of $k = 30$, almost half the number of data objects, is required; in such high-dimensional data there is no possibility of checking for short-circuiting directly by inspection of data plots, as in Figure 3.55, but with so large a neighbourhood it seems intuitively likely. This suspicion will, moreover, apply to any data in which there are well-separated clusters – precisely the kind of data with which this book is concerned, and with respect to which dimensionality reduction is supposed to serve as a means to identification of the clusters. In other words, Isomap seems not to be well suited to dimensionality reduction of strongly clustered data.

The short-circuiting problem was identified by (Balasumbramanian and Schwartz 2002) and currently remains unresolved. Until it is, an alternative is simply to dispense with the neighbourhoods and to derive the matrix of graph distances directly from the linear distance matrix, as described in the discussion of nonlinearity detection below. This has the drawback that any noise present in the data is included in the graph distance calculation, which can be a problem for data with a large noise element, but has the major advantage that the possibility of short-circuiting is eliminated because the graph distances all represent minimum spanning tree traversals.

Isomap was proposed by Tenenbaum, deSilva, and Langford (2000), and modified to deal with a greater range of nonlinear manifold types in DeSilva and Tenenbaum (2003). Other useful accounts are in Lee and Verleysen (2007: 102ff.), Izenman (2008: Ch. 16), and Xu and Wunsch (2009: Ch. 9.3.3). Numerous other nonlinear dimensionality reduction methods exist. Nonlinear versions of PCA have already been mentioned; others are Locally Linear Embedding (Roweis and Saul 2000), Principal Curves (Hastie and Stuetzle 1989), Principal Manifolds (Gorban et al. 2007), Curvilinear Component Analysis (Demartines and Hérault 1997), Curvilinear Distance Analysis (Lee, Lendasse, and Verleysen 2004), and Laplacian Eigenmaps (Belkin and Niyogi 2003). For discussions of these and others see: Carreira-Perpinan (1997), Carreira-Perpinan (2011), Fodor (2002), Lee and Verleysen (2007), Izenman (2008: Ch. 16), and Maaten, Postma, and Herik (2009).

Identification of nonlinearity

The foregoing discussion has made a distinction between dimensionality reduction methods appropriate to linear data manifolds and methods appropriate to nonlinear ones. How does one know if a manifold is linear or nonlinear, however, and therefore which class of reduction methods to apply? Where the data are low-dimensional the question can be resolved by plotting, but this is impossible for higher-dimensional data; this section describes various ways of identifying nonlinearity in the latter case.

Data abstracted from a natural process known to be linear are themselves guaranteed to be linear, but data abstracted from a known nonlinear process are not necessarily nonlinear. To see why, consider the sigmoid function used to model a range of processes such as population growth in the natural world, shown in Figure 3.57.

The linearity or otherwise of data based on empirical observation of a natural process is determined by what part of the process is observed. If observation comes from the linear part (A) of the theoretical distribution in Figure 3.57 then the data will be linear, but if from the nonlinear part (B) then the data will be nonlinear. Ideally, of course, empirical observation should be representative in the statistical sense, but the representativeness of data is not usually known. The only way to find out if data is nonlinear is to test for it.

In practice, data abstracted from observation are likely to contain at least some noise, and it is consequently unlikely that strictly linear relationships

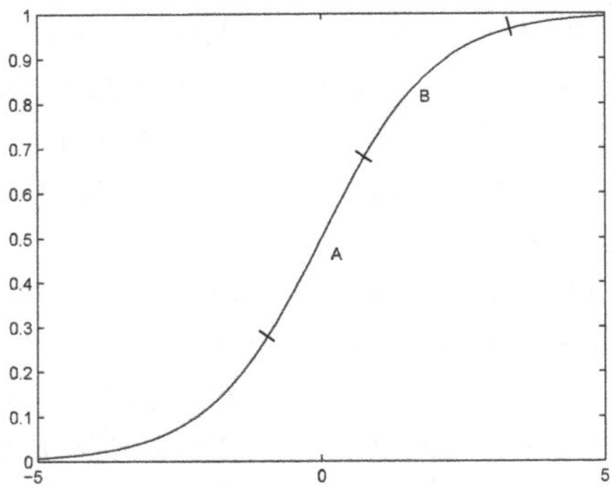

Figure 3.57: Data sampling from a natural process described by the function $y = 1/(1+e^{-x})$

between variables will be found. Instead, one is looking for degrees of deviation from linearity. Three ways of doing this are presented.

The graphical method is based on pairwise scatter-plotting of variables and subsequent visual identification of deviation from linearity. In Figure 3.58a, for example, the essentially linear relationship of variables v_1 and v_2 is visually clear despite the scatter, and the nonlinear relationship in 3.58b equally so.

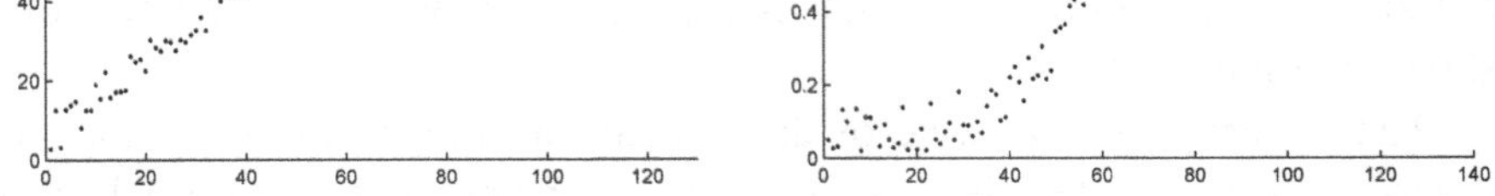

(a) Linear relationship

(b) Nonlinear relationship

Figure 3.58: Scatter plots of essentially linear and essentially nonlinear bivariate data (variables v_1 and v_2)

Looking for nonlinearity in this way involves plotting of all possible distinct pairings of data variables, that is, of columns in the data matrix, and visual identification of any nonlinearity. Such plotting provides an intuition for the shape of the data, but the number of plots *np* required to visualize all distinct pairs of variables as in Equation (3.33)

$$np = \frac{n(n-1)}{2} \tag{3.33}$$

where n is the number of variables; for $n = 100$, there would be 4950 different variable pairs to consider. This can be reduced by examining only a tractable subset of the more important variables in any given application, and so is not typically an insuperable problem; for what is meant by important variables and how to select them, see Huan and Motada (1998) and Manning, Raghavan, and Schütze (2008: 251f.). Visual interpretation of scatter plots is subjective, moreover, and where the shape of the relationship between variables is not as unambiguous as those in Figure 3.58 different observers are likely to draw different conclusions. For example, is the relationship in Figure 3.59 linear with substantial noise, or nonlinear?

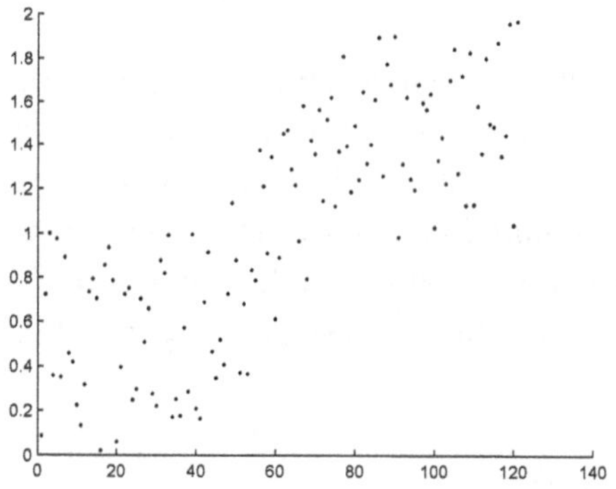

Figure 3.59: Possibly noisy linear, possibly nonlinear bivariate data

To be fully useful, graphically-based identification of nonlinearity needs to be supplemented by quantitative measures of the degree of nonlinearity. Regression analysis provides this; for what follows see Draper and Smith (1998), Izenman (2008), Miles and Shevlin (2001), Motulsky and Christopoulos (2004), and Seber and Wild (2003). Regression attempts to model the relationship between one or more independent variables whose values can value freely, and a dependent variable whose values are hypothesized to be causally determined by the independent one(s), by finding a curve and the corresponding mathematical function which best fits the data distribution. The outline of regression in this section simplifies in two main respects, neither of which compromises the generality of subsequent discussion based on it:

– Because the aim is simply to decide whether given data are linear or nonlinear rather than to find the optimal mathematical fit, the discussion confines itself to parametric regression, where a specific mathematical model for the relationship between independent and dependent variables is proposed a priori, and does not address nonparametric regression, where the model is inferred from the data.
– The discussion is based on the simplest case of one independent variable; the principles of regression extend straightforwardly to multiple independent variables.

The first step in parametric regression is to select a mathematical model that relates the values of the dependent variable y to those of the independent variable x. A linear model proposes a linear relationship between x and y, that is, a straight line of the general form given in Equation (3.34).

$$y = ax + b \tag{3.34}$$

where a and b are scalar constants representing the slope of the line and the intercept of the line with the y-axis respectively. In regression a and b are unknown and are to be determined. This is done by finding values for a and b such that the sum of squared residuals, that is, distances from the line of best fit to the dependent-variable values on the y-axis, shown as the vertical lines from the data points to the line of best fit in Figure 3.60, is minimized.

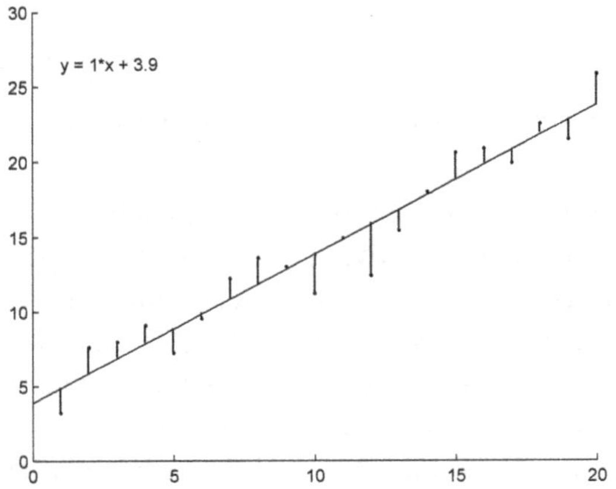

Figure 3.60: Linear model with residuals

A nonlinear model proposes a nonlinear relationship between x and y. Numerous nonlinear models are available. Frequently used ones in regression are polynomials with the general form given in Equation (3.35).

$$y = a_n x^n + a_{n-1} x^{n-1} + \cdots + a_2 x^2 + a_1 x^1 + a_0 \tag{3.35}$$

where the $a_n \ldots a_0$ are constants and n is the order of the polynomial; where $n = 1$ the polynomial is first-order, where $n = 2$ it as second-order and so on, though traditionally orders 1, 2, and 3 are called 'linear', 'quadratic', and 'cubic' respectively. As with linear regression, nonlinear regression finds the line best fit by calculating the coefficients $a_n \ldots a_0$ which minimize the sum of squared residuals between the line and the y values; Figure 3.61 shows this for quadratic and cubic polynomials. How the coefficents $a_n \ldots a_0$ are found is described in relevant textbooks, including those cited above.

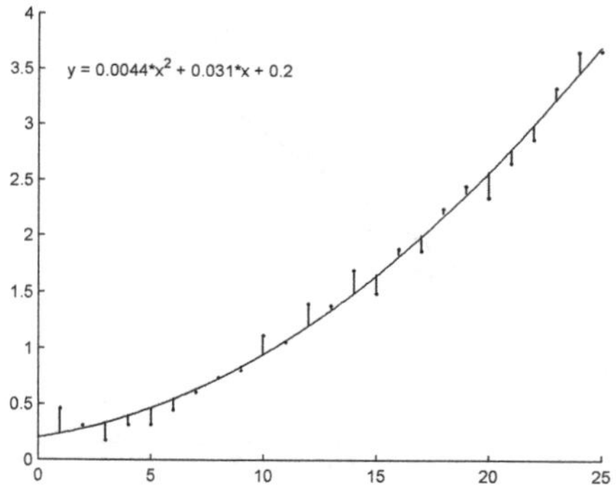

(a) Quadratic

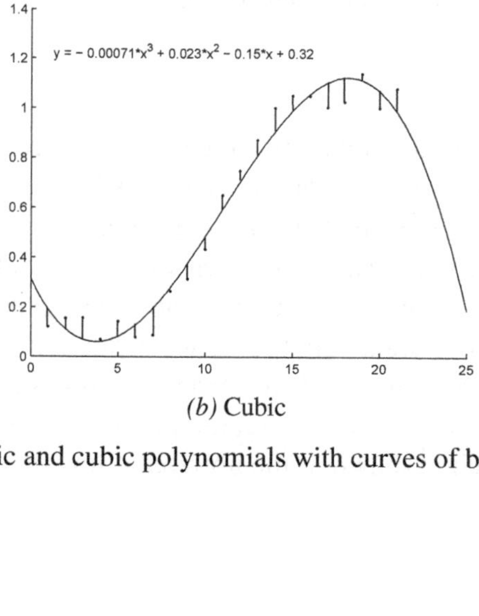

(b) Cubic

Figure 3.61: Quadratic and cubic polynomials with curves of best fit

Using regression to identify nonlinearity in data would appear simply to be a matter of comparing the goodness of fit of the linear model with that of whatever nonlinear model has been chosen: the data are linear if a straight line provides as good a fit (using the least-squares criterion) as any other mathematical function, and nonlinear if the nonlinear model is a significantly better fit than the linear one (Mark and Workman 2005a). In Figure 3.62, for example, the cubic model looks like it fits the data best, the quadratic less well, and the linear least well; based on visual inspection, one would say that this data is nonlinear.

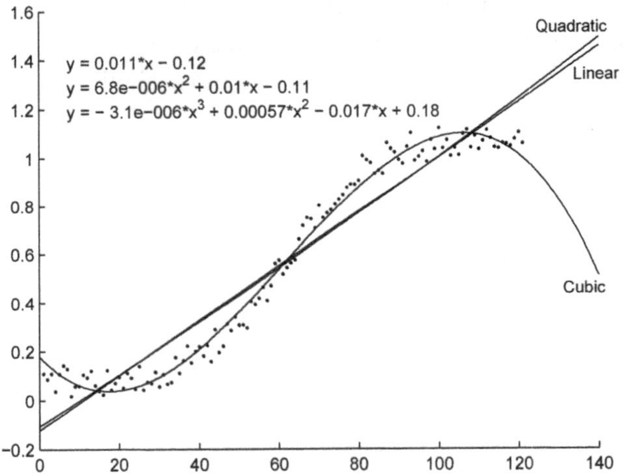

Figure 3.62: Linear, quadratic and cubic polynomials with curves of best fit

Such direct visual interpretation can be corroborated in several ways.

A statistical model differs from a deterministic one in that the functional relationship between independent and dependent variables which it posits does not necessarily hold exactly for each value of the independent variable, as for a deterministic model, but only on average. One therefore expects a random scatter of dependent variable values about a line of best fit.

The upper plot in Figure 3.63, for example, shows a linear model for bivariate data, and the lower shows a scatter plot of the residuals relative to the line of best fit, which is shown as the horizontal line at 0 on the *y*-axis.

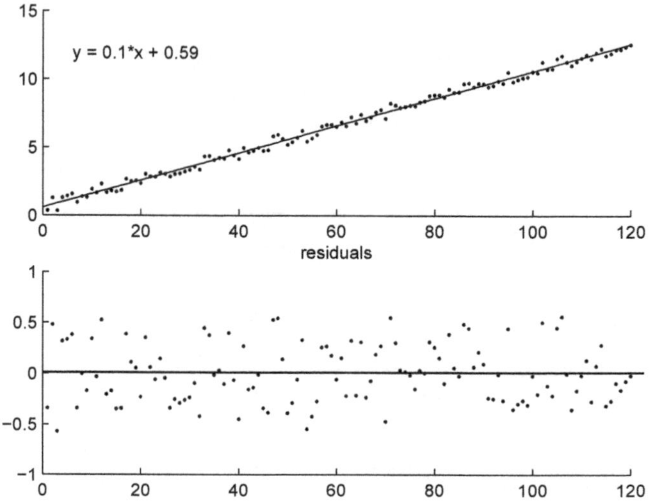

Figure 3.63: Linear model with corresponding residual plot

The residual plot emphasizes the deviation from the line of best fit and in this case shows no systematic pattern, indicating a good fit. In Figure 3.64, on the other hand, the residual plot corresponding to the linear model shows substantial deviation from randomness, and the one for the cubic model in Figure 3.65, though better, is still not optimal.

Various goodness-of-fit statistics can be used to corroborate the above graphical methods. Some often-used ones are briefly outlined below; others are discussed in Mark and Workman (2005c), Mark and Workman (2005b), Mark and Workman (2006).

- Runs test: The runs test is a quantification of the intuitions underlying residual plots. A run is a series of consecutive data points that are either all above or all below the regression line, that is, whose residuals are all positive or all negative. If the residuals are randomly distributed above and below the regression line, then one can calculate the expected number of runs: where N_a of the number of points above the line and N_b the number of points below, one expects to see the number of runs given by Equation (3.36).

$$N = \left(\frac{2N_aN_b}{N_a + N_b}\right) + 1 \tag{3.36}$$

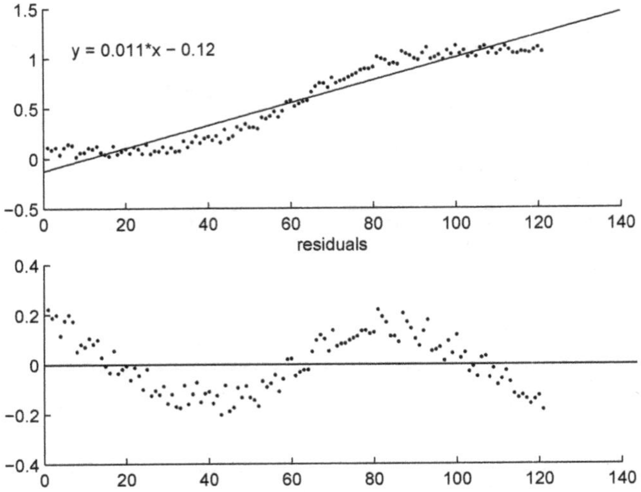

Figure 3.64: Linear model with corresponding residual plot

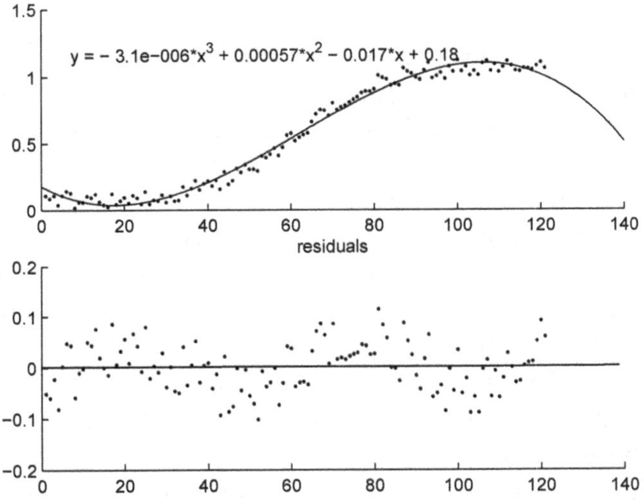

Figure 3.65: Cubic model with corresponding residual plot

Using this formula 51 runs are expected for 100 data points. The scatter of residuals around the line of best fit for Figure 3.63 looks random, and the number of runs, 52, confirms this. By contrast, for the data in Figures 3.64 and 3.65, the linear regression has only 8 runs and the cubic one 28; the result of the runs test for the cubic regression, though still far from optimal, is much closer to what is expected than the linear one, and confirms the visual impression from the lines of best fit and the residual plots that the cubic fit is the better.

- Summed square of errors (*SSE*): This measure is also called 'summed square of residuals', which clarifies what it is: the sum of deviations of the dependent variable values from the line of best fit, squared to prevent the positive and negative residuals from cancelling one another. This is shown in Equation (3.37)

$$SSE = \sum_{i=1..n} (y_i - \hat{y}_i)^2 , \tag{3.37}$$

where n is the number of data points, the y_i are the actual values of the dependent variable, and the $\hat{y}_i$ the corresponding values on the regression line. The smaller the *SSE* the less deviation there is in dependent variable values from the corresponding ones on the regression line, and consequently the better the fit. In Figure 3.62 the *SSE* for the linear regression line is 348.6, for the quadratic one 346.2, and for the cubic 54.31, which supports the graphical evidence that the cubic is the preferred model.

- Root mean squared error (*RMSE*): This is also known as the standard error of regression, and is the standard deviation of the residuals from the regression line, shown in Equation (3.38).

$$RMSE = \sqrt{\frac{SSE}{rdf}} \tag{3.38}$$

where rdf is the residual degrees of freedom, defined as the number n of data points minus the number of fitted coefficients c in the regression: $rdf = n - c$. For the $n = 120$ data points in Figure 3.62, the rdf for the linear polynomial model would be $120 - 2$, for the quadratic model $120 - 3$, and for the cubic model $120 - 4$. As with *SSE*, the smaller the *RMSE* the less deviation there is from the regression line, and consequently the better the fit.

- R^2, also known as the coefficient of determination, is a measure of how much of the variability of the dependent variable is captured by the regression model, and is defined by Equation (3.39).

$$R^2 = 1 - \frac{SSE}{SST} \tag{3.39}$$

SSE is defined as above, and *SST* is the sum of squares of *y*-value deviations from their mean, as in Equation (3.40).

$$SST = \sum_{i=1..n} (y_i - \bar{y}_i)^2 \tag{3.40}$$

where $\bar{y}$ is the mean of the dependent variable values and y_i is the *i*th of those values. The SSE/SST term is therefore the ratio of the variability of the dependent variable relative to the regression model and its total variability relative to its mean. The numerically smaller the ratio, and hence the larger the R^2, the better the fit: if the model fits perfectly then there is no residual variability, $SSE = 0$, and $R^2 = 1$, but if not *SSE* approaches *SST* as the fit becomes less and less good, and R^2 approaches 0.

R^2 is a widely used measure of goodness of fit, but discussions of it in the literature generally advise that it be so used only in combination with visual inspection of data plots. This caution derives from (Anscombe 1973), who presented four data sets, both linear and nonlinear, whose plots showed them to be very differently distributed but whose statistics, including R^2, were identical; on its own, R^2 is not a reliable indicator of nonlinearity (Mark and Workman 2005c).

With this caveat in mind, the R^2 scores for different regression models of a given data set can be compared and the best model selected. For Figure 3.62, R^2 for the linear regression is 0.803, for the quadratic regression it is 0.804, and for the cubic 0.969; the cubic model has the best score and is corroborated by the scatter plot, and the conclusion on the basis of this criterion is again that the data are nonlinear.

These statistics all look reasonable, but they have an underlying conceptual problem. In general, for a given family of models such as polynomials, the model with more parameters typically fits the data better than one with fewer parameters; Figure 3.66 shows, for example, regressions using polynomials of order one, three, six, and nine together with the relevant statistics. The greater the number of parameters the more convoluted the line of best fit can be and thus the closer it can get to the data values, thereby affecting all the statistics.

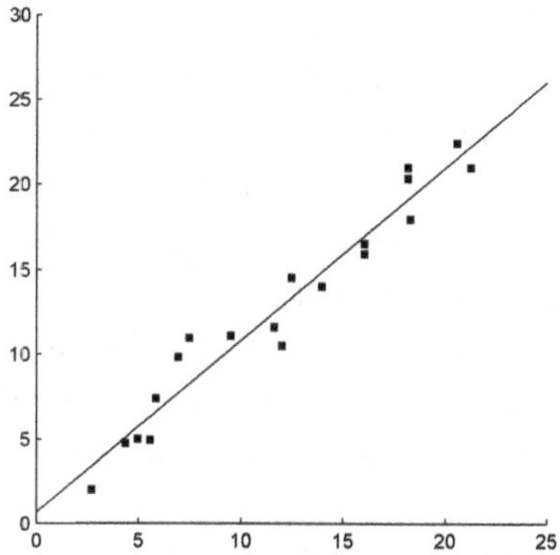

(a) First-order polynomial (linear): SSE: 32.24; RMSE: 1.338; R^2: 0.955

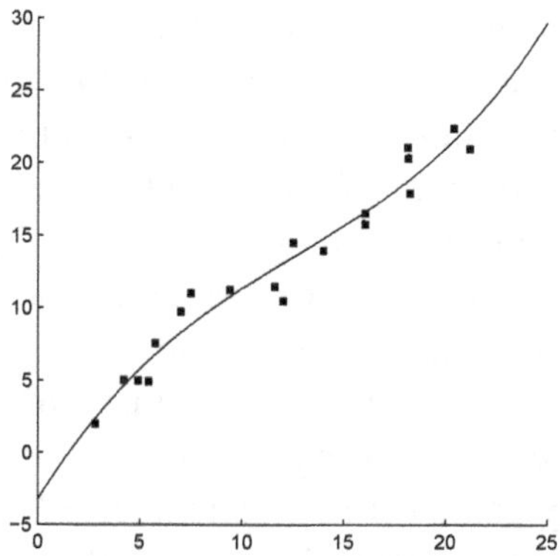

(b) Third-order polynomial (cubic): SSE: 28.17; RMSE: 1.327; R^2: 0.961

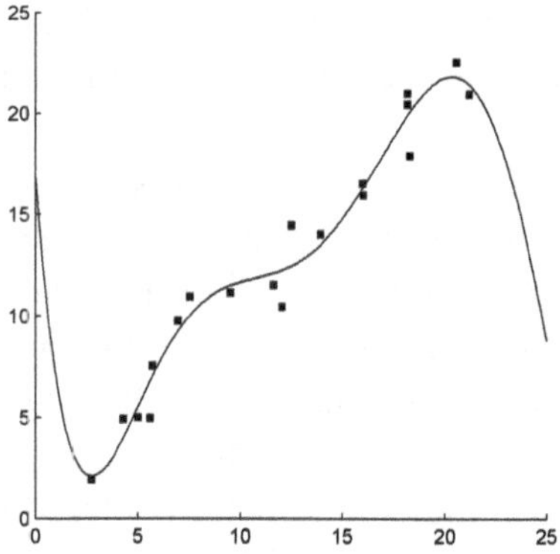

(c) Sixth-order polynomial: SSE: 16.75; RMSE: 1.135; R^2: 0.977

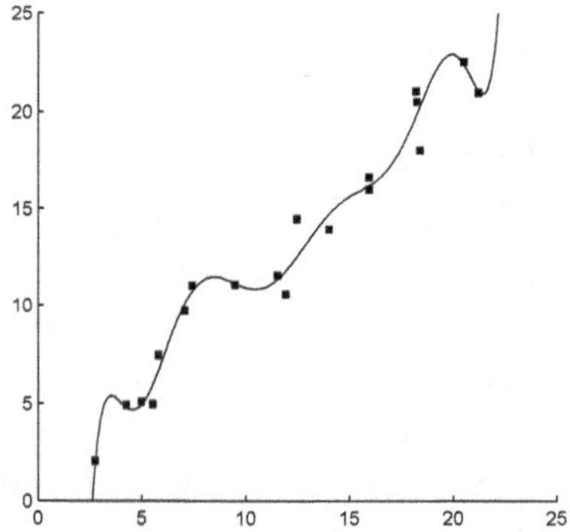

(d) Ninth-order polynomial: SSE: 14.9; RMSE: 1.22; R^2: 0.979

Figure 3.66: Polynomial models of increasing order fitted to the same data

Use of the foregoing statistics for identification of a nonlinear relationship between variables implies that the best model is always the one which comes closest to the data points. Where the relationship between variables is perfectly linear this is not a problem because increasing the number of parameters will not affect the statistics: the linear model is optimal. But, as already noted, empirical data typically contains noise, and that is where the problem lies. Given data that is not perfectly linear and a model for it with n parameters, $n > 2$, there are two possible interpretations. On the one hand, it may be that the model is fitting noise and thereby obscuring a relationship between the variables which is better captured by a model with fewer than *n* parameters; this is known as overfitting in the literature – cf., for example, Izenman (2008: 13f.). On the other, it may be that the nonlinearity is not noise but a genuine reflection of the nonlinear relationship between those aspects of the domain of interest which the data describes, and that the model with *n* parameters is the preferred one. Intuition based on visual examination of Figure 3.66a tells one that the relationship between variables *x* and *y* is essentially linear with a little noise which is best captured by a linear model with two parameters, and that Figures 3.66b–3.66b obscure this by fitting the noise increasingly accurately. But is this intuition correct, or are the data really nonlinear as 3.66d indicates?

Knowledge of the likelihood and scale of noise in the domain from which the data were abstracted can help in deciding, but this is supplemented by literature offering an extensive range of model selection methods (ibid.: Ch. 5). Two of the more frequently used methods are outlined and exemplified here, one based on statistical hypothesis testing and the other on information theory.

- Extra sum-of-squares *F*-test – cf. Motulsky and Christopoulos (2004: Ch. 22): Given two models which belong to the same family such as polynomials, one of which has more parameters than the other, this criterion tests the null hypothesis that the simpler model, the one with fewer parameters, is the correct one. It calculates a probability *p* from (i) the difference between the *SSE* values of each model, and (ii) the complexity of each model, that is, the number of degrees of freedom. If the value of *p* is less than a selected significance level, conventionally 0.05, then the null hypothesis is taken to be falsified and the more complex model is taken to fit the data significantly better than the simpler one; otherwise the conclusion is that there is no compelling evidence to support the more complex model, so the simpler model is accepted. As Motulsky and Christopoulos (ibid.: 138) note, the result of this test is not a definitive criterion for model correctness. What the test provides is an indication of whether there is sufficient evidence to reject the simpler null hypothesis model and to adopt the more complex one. For the data in Figure 3.67 (p. 146), for example, *p* for the linear model as null hypothesis and the cubic one as alternative hypothesis is 0.37. Relative to significance level 0.05, the indication is that the null hypothesis should be accepted, or, in other words, that on the available evidence the data should be interpreted as essentially linear with added noise.

– Akaike's Information Criterion (AIC) – cf. Burnham and Anderson (2002), Motulsky and Christopoulos (2004: Ch. 22): Given two candidate models, AIC works not by hypothesis testing like the extra sum-of-squares F-test, but rather uses information theory to calculate the models' probabilities of correctness. The probability that a given model is correct is given by Equation (3.41).

$$p = \frac{e^{-0.5\Delta}}{1 + e^{-0.5\Delta}} \tag{3.41}$$

where e is the entropy of the model and Δ is the difference in AIC scores between the two models. When the two AIC scores are identical, both models have a 0.5 probability of being correct. Otherwise the difference in probabilities can serve as the basis for model selection. In the case of the data for Figure 3.66, for example, the AIC probability is 0.98 that the linear model is correct and 0.02 that the cubic one is.

In both cases the data sample from Figure 3.66 is too small for the criteria to be meaningfully applied, and the examples are given for expository purposes only.

An alternative to regression is to make the ratio of mean nonlinear to mean linear distances among points on the data manifold the basis for nonlinearity identification. This is motivated by the observation that the shape of a manifold represents the real-world interrelationship of objects described by variables, and curvature in the manifold represents the nonlinear aspect of that interrelationship. Linear metrics ignore the nonlinearity and will therefore always be smaller than nonlinear ones; a disparity between nonlinear and linear measures consequently indicates nonlinearity, and their ratio indicates the degree of disparity.

The ratio of mean geodesic to mean Euclidean distance between all pairs of nodes in a graph gives a measure of the amount of nonlinearity in a data manifold. If the manifold is linear then the two means are identical and the ratio is 1; any nonlinearity makes the mean of geodesic distances greater than the Euclidean mean, and the ratio is greater than 1 in proportion of the degree of nonlinearity. Figure 3.67 gives examples for two-dimensional data using the graph-based approximation of geodesic distance described in the foregoing discussion of data geometry, though the principle extends straightforwardly to higher dimensions.

The linear manifold in Figure 3.67a has a means ratio of 1, 3.67b shows moderate nonlinearity with a ratio slightly more than 1, 3.67c shows sub-

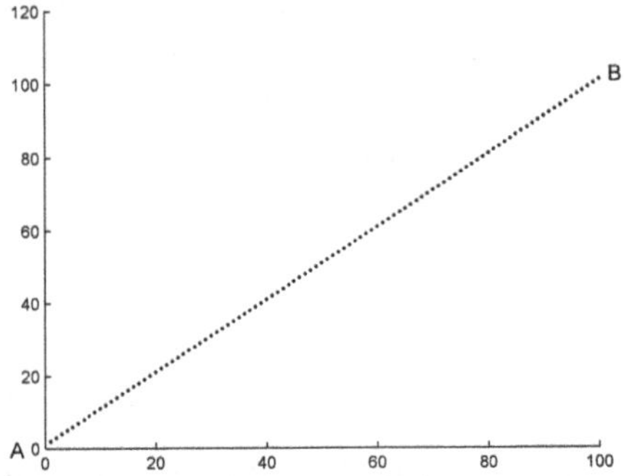

(a) Mean Euclidean distance: 47.10
Mean geodesic distance: 47.10
Ratio geodesic to Euclidean: 1
Euclidean distance A to B: 140.00
Geodesic distance A to B: 140.00
Ratio geodesic to Euclidean A to B: 1

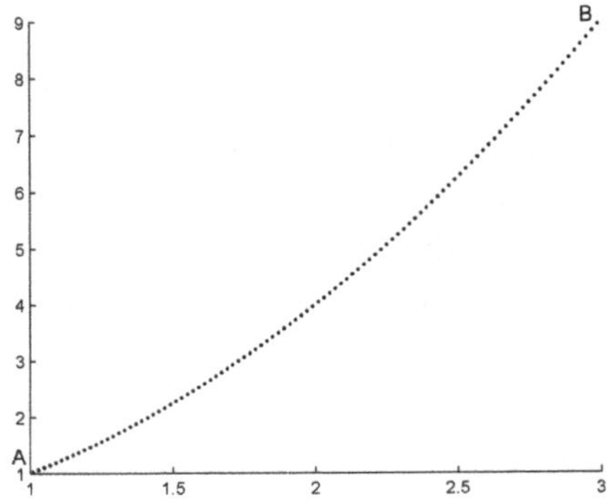

(b) Mean Euclidean distance: 0.48
Mean geodesic distance: 0.49
Ratio geodesic to Euclidean: 1.02
Euclidean distance A to B: 1.40
Geodesic distance A to B: 1.47
Ratio geodesic to Euclidean A to B: 1.05

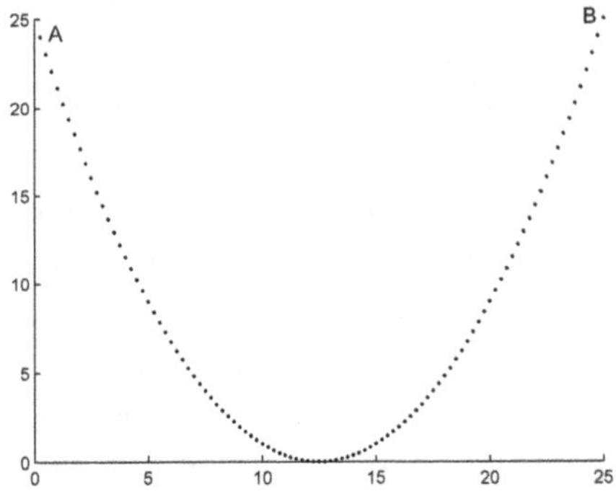

(c) Mean Euclidean distance: 12.71
Mean geodesic distance: 15.96
Ratio geodesic to Euclidean: 1.26
Euclidean distance A to B: 24.77
Geodesic distance A to B: 57.06
Ratio geodesic to Euclidean A to B: 2.30

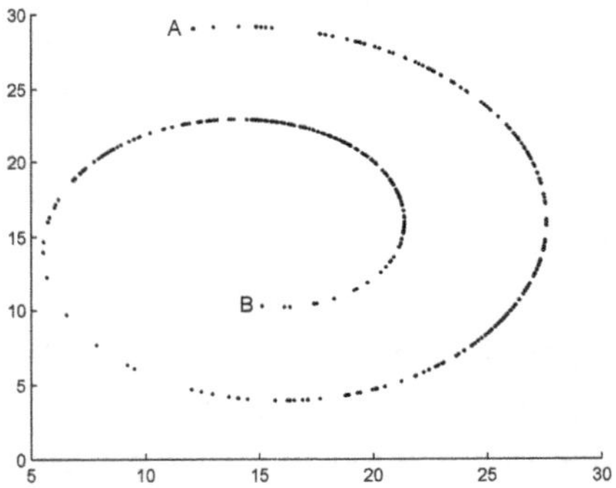

(d) Mean Euclidean distance: 11.30
Mean geodesic distance: 30.34
Ratio geodesic to Euclidean: 2.69
Euclidean distance A to B: 19.02
Geodesic distance A to B: 92.16
Ratio geodesic to Euclidean A to B: 4.85

Figure 3.67: Comparison of linear and geodesic distance measures for 1-dimensional manifolds in two-dimensional space

stantial nonlinearity and a correspondingly increased ratio, and 3.67d shows extreme nonlinearity with a ratio much larger than those of Figures 3.67a–3.67c. Figure 3.67 also includes distances and ratios for the end-points on each of the manifolds, labelled *A* and *B*, for which there is an analogous but enhanced progression.

Figure 3.68 shows an example which is less tidy than those in 3.67 and closer to what one might expect from empirical data. It is a version of 3.67d with random noise added, and shows the path of the shortest graph distance from *A* to *B*.

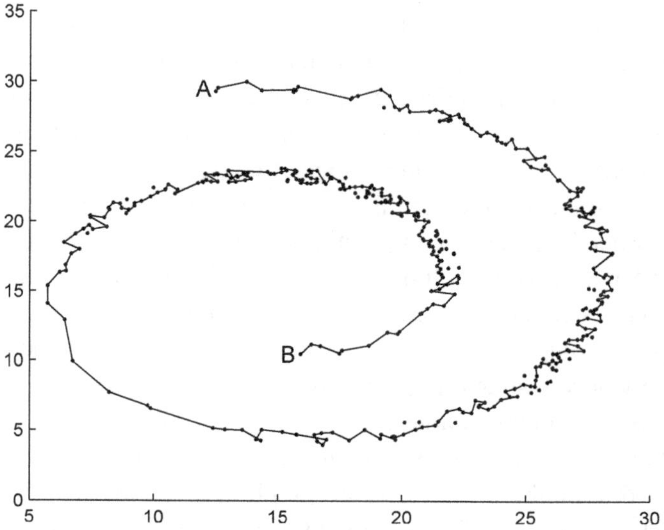

Figure 3.68: Randomized version of Figure 3.67d with Euclidean vs. graph distance statistics
Mean Euclidean distance: 11.38
Mean geodesic distance: 41.91
Ratio geodesic to Euclidean: 3.68
Euclidean distance *A* to *B*: 18.66
Geodesic distance *A* to *B*: 191.73
Ratio geodesic to Euclidean *A* to *B*: 6.60

An advantage of the graph-based approach to nonlinearity identification is that it gives a global measure of data manifold nonlinearity which is independent of dimensionality. The regression-based approach requires analysis of all unique pairings of columns in the data matrix, which can be onerous for high-dimensional data, whereas calculation of the Euclidean distance matrix and subsequent transformation into a geodesic distance matrix is done in a single operation.

An apparent problem with the graph-based approach is computational tractability. Central in theoretical computer science is the study of the intrinsic complexity of computing mathematical functions, and in particular the classification of mathematical functions according to the time and memory space resources which computation of them requires (Arora and Barak 2009; Goldreich 2008, 2010). An important criterion for this classification is the rate of increase of resource requirements relative to growth in the size of the problem that a function models. For some functions the relationship between resource requirements and problem size is linear, and computation of them is tractable even for very large problems, that is, feasible within reasonable time and space. For others the relationship is nonlinear, so that resource requirements grow at a greater rate than problem size, and computation becomes disproportionately more time and / or space consuming with increasing problem size. In some cases the nonlinearity is such that the computation rapidly becomes intractable; classic computations of this last type are the well known Towers of Hanoi and Travelling Salesman problems. The complexity of a computation is standardly given using the so-called big-O notation whereby, for example, a computation whose complexity is described as being $O(n^2)$ is read as one whose resource requirements increase on the order of the square of the size of a problem parameter n.

Calculation of a minimum spanning tree scales in time as $O(elog(v))$, where e is the number of arcs in the graph and v is the number of nodes (Cormen et al. 2009: Ch. 23); Figure 3.69 shows the scaling behaviour for a sequence of 200 Euclidean distance matrices derived from data matrices containing random values, where the first contains 10 rows / columns and each subsequent one is incremented by 10 to give the sequence 10, 20, 30. . . 2000. Each Euclidean matrix was interpreted as a complete graph and, for each graph, the minimum spanning tree was first calculated using Kruskal's algorithm (ibid.: 631ff.). The geodesic distance matrix between all pairs of nodes was then generated by tree traversal, thereby simulating the method for geodesic distance measurement proposed above.

The horizontal axis represents the dimensionality of the Euclidean distance matrix and the vertical axis the number of seconds required calculate the minimum spanning tree and pairwise graph distances for each 10-increment graph. Using a conventional desktop computer running Matlab at 2.6 GHz, for the 2000-row / column matrix 10.11 seconds were required and, extrapolating, 26.43 seconds are required for 3000, 63.37 seconds for 5000, 246.22 for 10000, and 1038.81 for 20000. These requirements do not seem excessive, and they could easily be reduced by using a faster computer. Eventually, of course, the shape of the curve guarantees that execution of the algorithm will become prohibitively time consuming for very large matrices. But 'large'

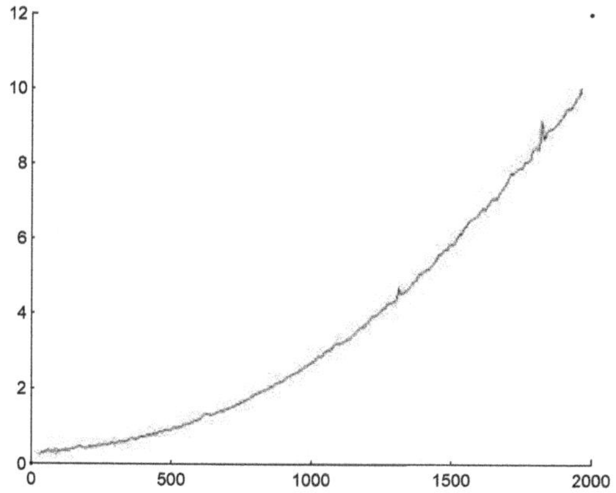

Figure 3.69: Scaling behaviour of minimum spanning tree calculation and traversal

is a relative concept: in research applications involving matrices up to, say, dimensionality 20000 computational!complexity is not a significant factor in using the proposed approach to nonlinearity identification and graph distance calculation.

A potential genuine disadvantage of the graph distance-based approach is that it does not make the distinction between model and noise that the regression-based approach makes, and treats the data matrix as a faithful representation of the domain from which the data was abstracted. Unless the data is noiseless, therefore, the graph distance-based approach includes noise, whether random or systematic, in its calculations, which may or may not be a problem in relation to the application in question.

Using the graphical and regression-based methods outlined above, no strictly or even approximately linear relationships between pairs of variables were found in MDECTE. In a relatively few cases the relationships looked random or near-random, but most showed a discernible pattern; the pair ɔ: (corresponding to the dialectal pronunciation of the vowel in 'goat') and ɑ: (corresponding to the dialectal pronunciation of the vowel in 'cold') is representative and is used as the basis for discussion in what follows.

A scatter plot of ɔ: on the horizontal axis and ɑ: on the vertical in Figure 3.70 shows a visually clear nonlinear relationship.

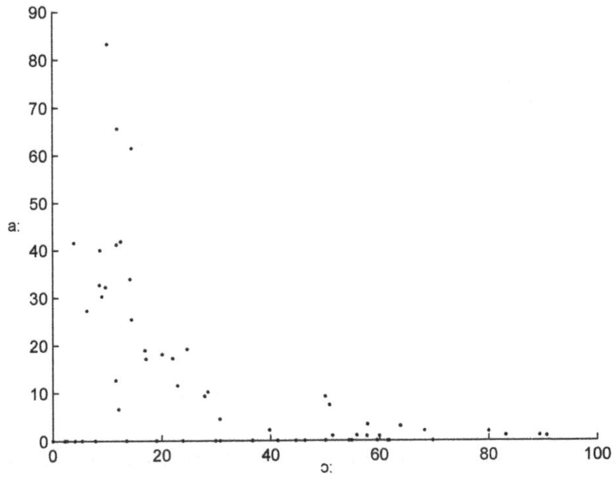

Figure 3.70: Scatter plot of column values in MDECTE representing the phonetic segments ɔ: and ɑ:

Using ɔ: as the independent variable and ɑ: as the dependent, a selection of polynomials was used to model the nonlinear relationship. These are shown in Figure 3.71.

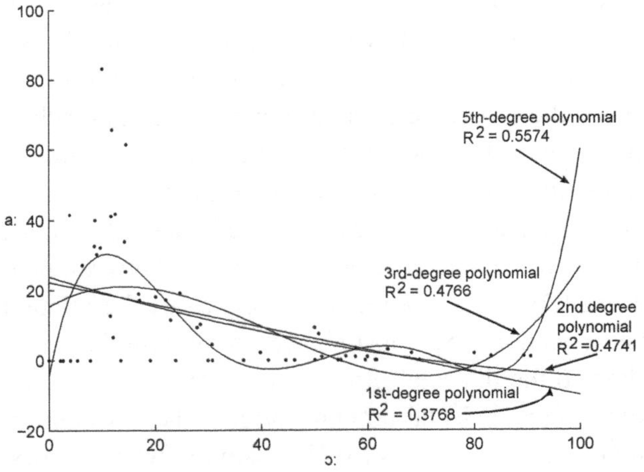

Figure 3.71: Polynomial regression models of the ɔ: / ɑ: relationship

Visually, the linear model appears to fit least well and the 5th-degree polynomial best, as expected, and this is confirmed by runs tests, residual plots, and the goodness of fit statistics in Table 3.34.

Table 3.34: Goodness of fit statistics for Figure 3.71

	SSE	*RMSE*	R^2
Degree 1	12420	15.03	0.3768
Degree 2	10480	13.93	0.4741
Degree 3	10390	14.00	0.4786
Degree 5	8821	13.15	0.5574

Table 3.35 contains the results obtained for the extra sum-of-squares F-test, and Table 3.36 those the AIC test.

Table 3.35: Extra sum-of-squares F-test for Figure 3.71

Null hypothesis	Alternative hypothesis	p-value	Conclusion
First order	Second order	0.0028	Reject null
First order	Third order	0.0097	Reject null
First order	Fifth order	0.0014	Reject null
Second order	Third order	0.5086	Accept null
Second order	Fifth order	0.0309	Reject null
Third order	Fifth order	0.0153	Reject null

Table 3.36: AIC test for Figure 3.71

Percent chance model 1 correct	Percent chance model 2 correct	Preferred model
First order 2.46	Second order 97.50	Second order
First order 6.17	Third order 98.83	Third order
First order 0.79	Fifth order 99.21	Fifth order
Second order 72.24	Third order 27.76	Second order
Second order 23.90	Fifth order 76.10	Fifth order
Third order 10.77	Fifth order 89.23	Fifth order

These results further support the indications so far: that the first-order model is worst, that second-order is better than the third, but that the fifth-order model is preferred.

The Euclidean 63×63 distance matrix E was calculated for MDECTE, the minimum spanning tree for E was found, and the graph distance matrix G was derived by tree traversal, all as described in the foregoing discussion. The distances were then linearized into vectors, sorted, and co-plotted to get a graphical representation of the relationship between linear and graph distances in the two matrices. This is shown in Figure 3.72.

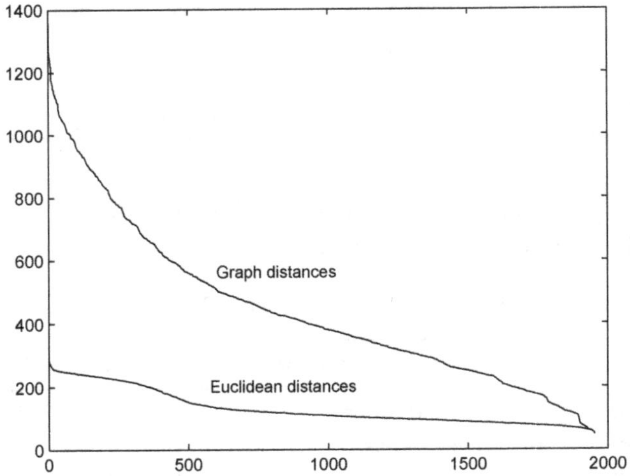

Figure 3.72: Comparison of Euclidean and geodesic distances for MDECTE

The graph distances between and among the speakers in MDECTE are consistently larger than the Euclidean ones over the entire range. This is summarized in the ratio $mean(G)/mean(E)$ of mean distances, which is 3.89. On these indicators, MDECTE can be said to contain a substantial amount of nonlinearity.

4 Cluster

The Introduction described cluster analysis as a family of mathematically-based computational methods for identification and graphical display of structure in data when the data are too large either in terms of the number of variables or of the number of objects described, or both, for them to be readily interpretable by direct inspection. Chapter 2 sketched how it could be used as a tool for generation of hypotheses about the natural process from which the data were abstracted, and Chapter 3 used it to exemplify various data issues, in both cases without going into detail on how it actually works. The present chapter now provides that detail.

The discussion is in three main parts: the first part attempts to define what a cluster is, the second presents a range of clustering methods, and the third discusses cluster validation, that is, the assessment of how well a clustering result has captured the intrinsic structure of the data. As in Chapter 3, the MDECTE matrix provides the basis for exemplification, but in a dimensionality-reduced form. Specifically, dimensionality is reduced to 51 using the variable selection method which combines the frequency, variance, *vmr* and $tf-idf$ selection criteria; variable selection rather than extraction was used for dimensionality reduction because the original variables will be required for hypothesis generation later in the discussion.

4.1 Cluster definition

In cluster analytical terms, identification of structure in data is identification of clusters. To undertake such identification it is necessary to have a clear idea of what a cluster is, and this is provided by an innate human cognitive capability. Human perception is optimized to detect patterning in the environment (Köppen 2000; Peissig and Tarr 2007), and clusters are a kind of pattern. Contemplation of a rural scene, for example, reveals clusters of trees, of farm buildings, of sheep. Looking up at the night sky reveals clusters of stars. And, closer to present concerns, anyone looking at the data plot in Figure 4.1 immediately sees the clusters.

A casual observer looking at the scatterplots in Figure 4.1 would say that 4.1a shows a few small concentrations of points but is essentially random,

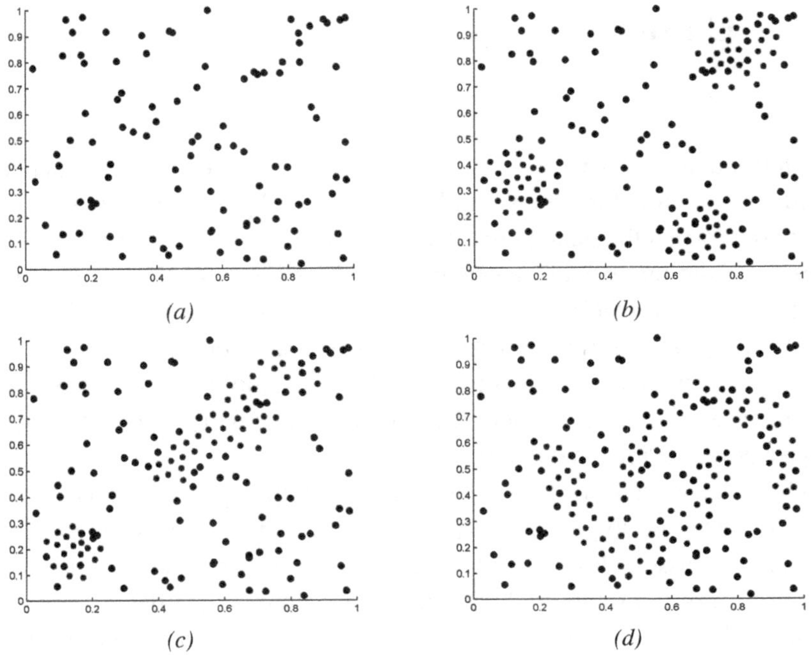

Figure 4.1: A selection of clusters in two-dimensional space

that 4.1b has three clearly identifiable clusters of roughly equal size, that 4.1c has two clusters of unequal size the smaller of which is in the lower-left corner of the plot and the larger elongated one in the upper right, and that 4.1d has two intertwined, roughly semi-circular clusters, all embedded in a random scatter of points. That casual observer would, moreover, have been able to make these identifications solely on the basis of innate pattern recognition capability and without recourse to any explicit definition of the concept 'cluster'.

Direct perception of pattern is the intuitive basis for understanding what a cluster is and is fundamental in identifying the cluster structure of data, but it has two main limitations. One limitation is subjectivity and consequent unreliability. Apart from the obvious effect of perceptual malfunction in the observer, this subjectivity stems from the cognitive context in which a given data distribution is interpreted: the casual observer brings nothing to the observation but innate capability, whereas the researcher who compiled the data and knows what the distribution represents brings prior knowledge which potentially and perhaps inevitably affects interpretation. In Figure 4.1c, for example, does the larger cluster on the upper right contain two subclusters or not? What would the answer be if it were known that the points represent cats in the upper part of the cluster and dogs in the lower? The other limitation is that reliance on innate perceptual capability for cluster identification is confined to what can be perceived, and in the case of data this means a maximum dimensionality of 3 or less for graphical representation; there is no way of perceiving clusters in data with dimensionality higher than that directly.

The obvious way to address these limitations is by formal and unambiguous definition of what a cluster is, relative to which criteria for cluster membership can be stated and used to test perceptually-based intuition on the one hand and to identify non-visualizable clusters in higher-dimensional data spaces on the other. Textbook and tutorial discussions of cluster analysis uniformly agree, however, that it is difficult and perhaps impossible to give such a definition, and, if it is possible, that no one has thus far succeeded in formulating it. In principle, this lack deprives cluster analysis of a secure theoretical foundation. In practice, the consensus is that there are intuitions which, when implemented in clustering methods, give conceptually useful results, and it is on these intuitions and implementations that contemporary cluster analysis is built.

The fundamental intuition underlying cluster analysis is that data distributions contain clusters when the data objects can be partitioned into groups on the basis of their relative similarity such that the objects in any group are more similar to one another than they are to objects in other groups, given some definition of similarity. In terms of the geometric view of data on which the present discussion is based, the literature conceptualizes similarity in two ways: as distance among objects in the data space, and as variation in the density of objects in the space. Two quotations from a standard textbook (Jain and Dubes 1988: 1) capture these:

- "A cluster is an aggregation of points in the test space such that the distance between any two points in the cluster is less than the distance between any point in the cluster and any point not in it".
- "Clusters may be described as connected regions of multi-dimensional space containing a relatively high density of points, separated from other such regions by a region containing a relatively low density of points".

These distance and density views of similarity may at first sight appear to be a distinction without a difference: data points spatially close to one another are dense, and dense regions of a space contain points spatially close to one another. There is, however, a substantive difference, and it corresponds to that between the metric space and topological geometries introduced in the discussion of data in the preceding chapter. The distance conceptualization of similarity uses a metric to measure the proximities of data points relative to a set of basis vectors which define the embedding space of the data manifold, whereas the density one uses relative proximity of points on the data manifold without reference to an embedding space. As we shall see, the difference is important because clustering methods based on density are able to identify a greater range of cluster shapes than at least some of the ones based on distance in metric space.

4.2 Clustering methods

The Introduction noted that a large number clustering methods is available and that only a selection of them can be included in the present discussion. It also gave some general selection criteria: intuitive accessibility, theoretically and empirically demonstrated effectiveness, and availability of software implementations for practical application. The discussion has now reached the stage where these criteria need to be applied.

Clustering methods assign the set of given data objects to disjoint groups, and the literature standardly divides them into two categories in accordance with the kind of output they generate: hierarchical and nonhierarchical. Given an $m \times n$ dimensional data matrix D, hierarchical methods regard the m row vectors of D as a single cluster C and recursively divide each cluster into two subclusters each of whose members are more similar to one another than they are to members of the other on the basis of some definition of similarity, until no further subdivision is possible: at the first step C is divided into subclusters c_1 and c_2, at the second step c_1 is divided into two subclusters $c_{1.1}$, $c_{1.2}$, and c_2 into $c_{2.1}, c_{2.2}$, at the third step each of $c_{1.1}, c_{1.2}, c_{2.1}, c_{2.2}$ is again subdivided, and so on. The succession of subdivisions can be and typically is represented as a binary tree, and this gives the hierarchical methods their name. Nonhierarchical methods partition the m row vectors of D into a set of clusters $C = c_1, c_2..c_k$ such that the members of cluster $c_i (i = 1 \ldots k)$ are more similar to one another than they are to any member of any other cluster, again on the basis of some definition of similarity. Both hierarchical and nonhierarchical methods partition the data; the difference is that the nonhierarchical ones give only a single partition into k clusters, where k is either pre-specified by the user or inferred from the data by the method, whereas the hierarchical ones offer a succession of possible partitions and leave it to the user to select one of them. Because these two categories offer complementary information about the cluster structure of data, examples of both are included in the discussion to follow, starting with non-hierarchical ones.

As might be expected from the foregoing comments, the literature on clustering is vast (Xu and Wunsch 2009: 9ff.), and no explicit or implicit pretence that all of it has been consulted is made. With a very few exceptions, the literature prior to 1988 has not been cited. On the face of it this looks arbitrary, but there is method to it. It is the year in which the excellent, still fundamental, and still extensively referenced book *Algorithms for Clustering Data* (Jain and Dubes 1988) appeared, and it is here taken to be an amalgamation of earlier work sufficient for present purposes; for a summary of earlier work on clustering and the state of research to 1996, with extensive bibliography, see Arabie and Hubert (1996). Since 1988 several textbooks have appeared, of which the following have been used: Everitt's *Cluster Analysis*, the first edition of which was published in 1974 and which has since then appeared in successive editions the most recent of which is Everitt et al. (2011), Kaufman and Rousseeuw (1990), Gordon (1999), Gan, Ma, and Wu (2007), Xu and Wunsch (2009), Mirkin (2013); briefer surveys are Jain, Murty, and Flynn (1999), Xu and Wunsch (2005), Manning, Raghavan, and Schütze (2008: Chs. 16-18), Izenman (2008: Ch. 12), Berkhin and Dhillon (2009). Because, moreover, cluster analysis is part of the more general field of multivariate data analysis and is also an important component in information retrieval and data mining, textbooks in these subjects typically contain accounts of it. There was no attempt at comprehensiveness here; Berkhin (2006), Tan, Steinbach, and Kumar (2006: Ch. 9), and Hair et al. (2010: Ch. 9f.) were found useful.

Two preliminary notes:

1. The literature subcategorizes hierarchical and non-hierarchical methods in accordance with the data representation relative to which they are defined: graph-based methods treat data as a graph structure and use concepts from graph theory to define and identify clusters, distributional methods treat data as a mixture of different probability distributions and use concepts from probability theory to identify clusters by decomposing the mixture, and vector space methods treat data as manifolds in a geometric space and use concepts from linear algebra and topology. The discussion of data creation and transformation in the preceding chapter was based on vector space representation, and

to provide continuity with that discussion only vector space clustering methods are included in this chapter. For information about graph-based methods see Schaeffer (2007), Gan, Ma, and Wu (2007: Ch. 11), Xu and Wunsch (2009: Ch. 4), and about distributional methods Fraley and Raferty (2002), Gan, Ma, and Wu (2007: Ch. 14); there are also clustering methods based on specific classes of algorithm such as artificial neural networks and genetic algorithms, for which see Berkhin (2006), Gan, Ma, and Wu (2007: Ch. 10). For current trends in clustering see Jain (2010).

2. A distinction between two types of data categorization needs to be made explicit. One type assigns data objects to a set of categories known a priori, and is variously called 'classification', 'discriminant analysis', and 'pattern recognition' in the literature. The other type does not use a priori categories but rather infers them from the data, and is referred to in the literature as 'clustering' or 'cluster analysis'. Unfortunately there is some inconsistency of terminological usage, particularly with respect to 'classification' and 'clustering / cluster analysis'. The focus of this book is on hypothesis generation based on discovery of structure in data. As such it is interested in the second of the above types of categorization and uses the terms 'clustering' and 'cluster analysis' with respect to it throughout the discussion, avoiding 'classification' altogether to forestall confusion.

4.2.1 Nonhierarchical clustering methods

As noted, given m n-dimensional row vectors of a data matrix D, nonhierarchical methods partition the vectors into a clustering C consisting of a set of k clusters $C = c_1 \ldots c_k$ such that the members of cluster $c_i (i = 1 \ldots k)$ are more similar to one another than they are to any member of any other cluster. The theoretical solution to finding such a partition is to define an objective function f, also called an error function or a criterion function, which measures the goodness of a partition relative to some criterion in order to evaluate each possible partition of the m vectors into k clusters relative to f, and, having done this, to select the partition for which the value of f is optimal. In practice, such exhaustive search of all possible clusterings for the optimal one rapidly becomes intractable. The rather complex combinatorial mathematics of this intractability are discussed in Jain and Dubes (1988: Ch. 3.3) and summarized in Equation (4.1):

$$S(n,k) = \frac{\sum_{i=1...k}(-1)^{k-i}\frac{k!}{i!(k-i)!}i^n}{k!} \quad (4.1)$$

where $S(n,k)$ is the number of possible partitions of n objects into k clusters. According to this formula, to partition a set of objects into, say, 3 clusters, 10 objects can be clustered in a relatively moderate 9330 different ways, 20 objects a much more demanding 580606446 different ways, 30 objects a huge $3.4315e+13$ ways, and our 63 DECTE speakers a truly intractable $1.9076e+29$ ways. Exhaustive search is, in short, not a practical option for general partitional clustering, and ways of avoiding it have been developed. Three conceptually different approaches are described in what follows: projection clustering, proximity-based clustering, and density-based clustering.

4.2.1.1 Projection clustering

The dimensionality reduction methods described in the preceding chapter can be used for clustering by specifying a projection dimensionality of 2 or 3 and then scatter plotting the result. Figure 4.2 shows this for MDECTE using PCAand MDS, with projection dimensionality 2 in both cases; speaker labels have been abbreviated and only a selection of them is shown to avoid clutter.

Both methods show a clear two-cluster structure, and so do those derived via the other dimensionality reduction methods described in the preceding chapter. They all share a problem both with respect to MDECTE and to data in general, however: the intrinsic dimensionality of the data might be greater than 2 or 3, and simply truncating it to enable the data to be plotted runs the risk of losing important information and consequently generating misleading results. In the case of PCA, for example, the degree of information loss is readily seen. Most implementations of it provide a cumulative total of how much variance in the original data is captured or 'explained' by each successive principal component. In the case of Figure 4.2a, the first two components explain only 46.9 percent of the original data variance, that is, about half the original information is thrown away, and the first three are a little but not much better in that they capture 54.6 percent. If the explained variance had been higher, say in the 80–90 percent range, one could have been confident

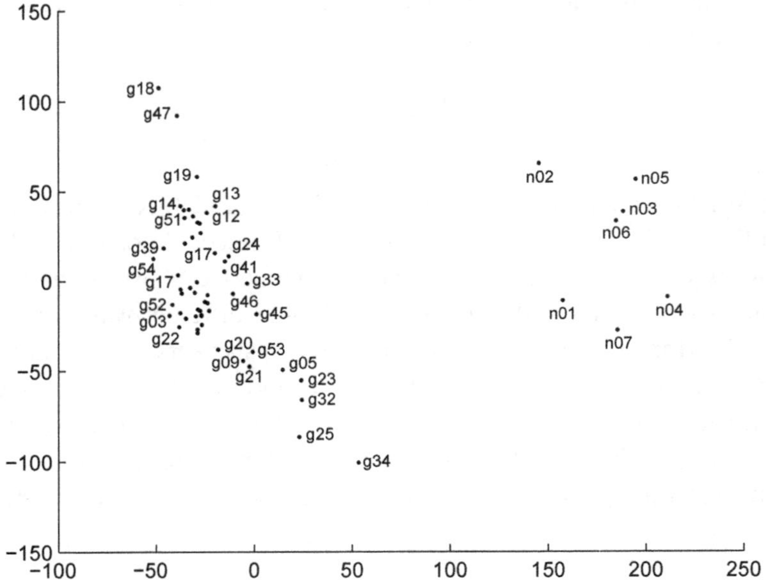

(a) PCA

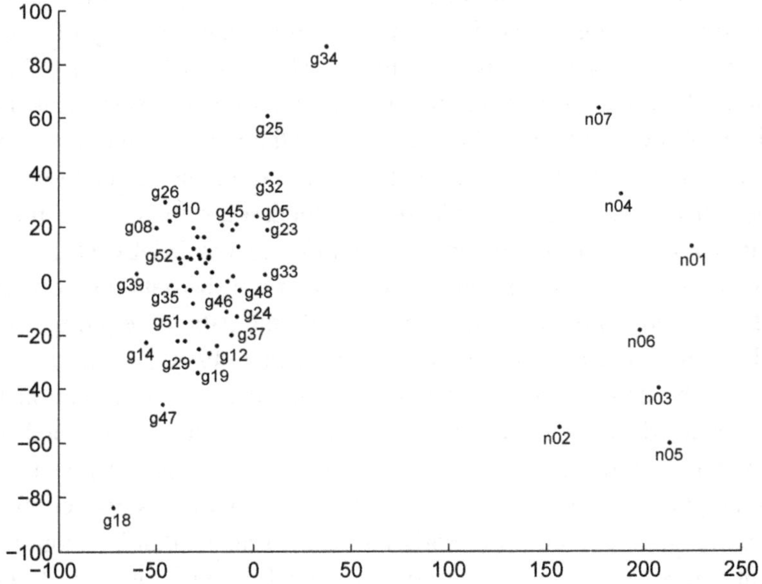

(b) MDS

Figure 4.2: PCA and MDS projection of MDECTE into two-dimensional space in the reliability of the clustering, but under the circumstances it would be unwise to trust the clustering results in Figure 4.2.

An alternative to the dimensionality reduction methods already described is to use a method which projects high-dimensional data into a low-dimensional space without reducing the dimensionality of the original data. The self-organizing map (SOM) is such a method – an artificial neural network that was originally invented to model a particular kind of biological brain organization, but that can also be used without reference to neurobiology as a way of visualizing high-dimensional data manifolds by projecting and displaying them in low-dimensional space, and thereby as a cluster analysis method. It has been extensively and successfully used for this purpose across a wide range of disciplines, and is for that reason described in detail here.

The following account of the SOM is in five parts: the first part describes its architecture, the second exemplifies its use for cluster analysis by applying it to the MDECTE data, the third discusses interpretation of the low-dimensional projection which it generates, the fourth surveys advantages and disadvantages of the SOM for clustering, and the fifth gives pointers to developments of the basic SOM architecture. The standard work on the SOM is Kohonen (2001). Shorter accounts are Haykin (1999: Ch. 9), Van Hulle (2000), Lee and Verleysen (2007: Ch. 5), Izenman (2008: Ch. 12.5), Xu and Wunsch (2009: Ch. 5.3.3); collections of work on the SOM are in Oja and Kaski (1999) and Allinson et al. (2001). For overviews of applications of the SOM to cluster analysis and data analysis more generally see Kohonen (2001: Ch. 7), Kaski, Nikkila, and Kohonen (2000), and Vesanto and Alhoniemi (2000).

A good intuition for how the SOM works can be gained by looking at the biological brain structure it was originally intended to model: sensory input systems (Van Hulle 2000), (Kohonen 2001: Chs. 2 and 4). The receptors in biological sensory systems generate very high dimensional signals which are carried by numerous nerve pathways to the brain. The retina of the human eye, for example, contains on the order of 10^8 photoreceptor neurons each of which can generate a signal in selective response to light frequency, and the number of pathways connecting the retina to the brain is on the order of 10^6 (Hubel and Wiesel 2005). At any time t, a specific visual stimulus v_t to the eye generates a pattern of retinal activation a_t which is transmitted via the nerve pathways to the part of the brain specialized for processing of visual input, the visual cortex, which transmits a transformed version of a_t to the rest of the brain for further processing. It is the response of the visual cortex to retinal stimulation which is of primary interest here. The visual cortex is essentially a two-dimensional region of neurons whose response to stimulation is spatially selective: any given retinal activation at sent to it stimulates not the whole two-dimensional cortical surface but only a relatively small region of it, and activations $a_{t+1}, a_{t+2} \ldots$ similar to a_t stimulate adjacent regions. Similar stimuli in high-dimensional input space are thereby projected to spatially adjacent locations on a two-dimensional output surface. This is the basis for the SOM's use as a clustering method.

Figure 4.3a is a graphical representation of what a highly simplified physical model of a visual input system might look like. Only a very few retinal and cortical 'neurons' are shown. Also, only a few connections between 'retina' and 'cortex' are represented to convey an idea of the pattern of connectivity without cluttering the diagram; each of the cells of the 'retina' is connected to each of the cells in the 'visual cortex' so that if the 'retina' had, say, 10 cells there would be $10 \times (7 \times 4) = 280$ connections, where (7×4) are the dimensions of the 'cortex'.

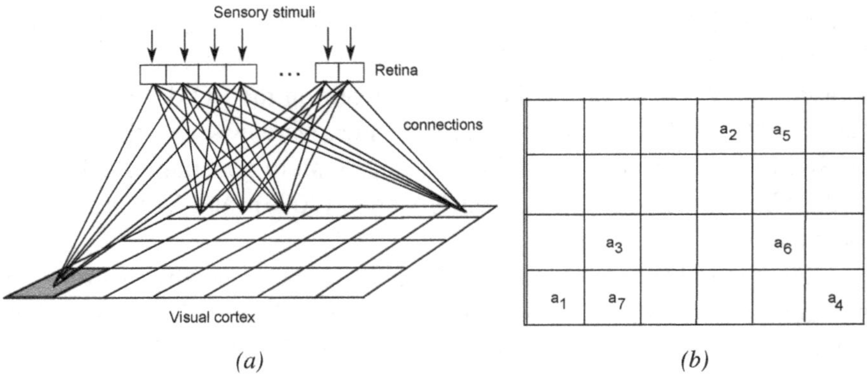

Figure 4.3: A graphical representation of a highly simplified physical model of a biological visual system

Sensory stimuli arrive at the 'retina' and are propagated via the connections to the 'cortex'; for some activation a_t, a single neuron is activated in response. If, say, seven activations $a_1 \ldots a_7$ are input in temporal succession, and if the members of each of three sets a_1, a_3, a_7, a_2, a_5, and a_4, a_6 are similar to one another but different from members of the other two sets, then the 'cortex' is sequentially activated as shown in Figure 4.3b, and these successive activations, when superimposed as they are in 4.3b, show a cluster structure.

The mathematical model corresponding to the above physical one has three components together with operations defined on them:

- An n dimensional input vector R, for some arbitrary n, which represents the retina.
- A $p \times q$ output matrix M which represents the sensory cortex, henceforth referred to as the lattice.
- A $p \times q \times n$ matrix C which represents the connections, where $C_{i,j,k}$ is the connection between the neuron at $M_{i,j}$ (for $i = 1 \ldots p, j = 1 \ldots q$) and the one at R_k (for $k = 1 \ldots n$). Three-dimensional matrices like C have not previously been introduced. They are simply a generalization of the familiar two-dimensional ones, and can be conceptualized as in Figure 4.4, where the value at any of the cells of the two-dimensional matrix is not a scalar but a vector: interpreted in terms of 4.4, the connection from the third component of the input vector R_3 to the output matrix $M_{1,1}$ has the value 0.86.

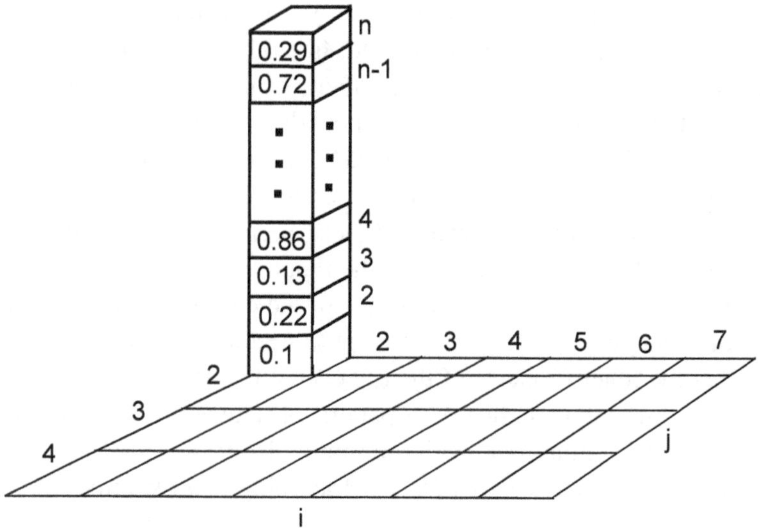

Figure 4.4: Three-dimensional matrix representation of SOM connections

For data clustering a SOM works as follows, assuming an $m \times n$ data matrix D is given. For each row vector D_i (for $i = 1 \ldots m$) repeat the following two steps:

1. Present D_i as input to R.
2. Propagate the input along the connections C to selectively activate the cells of the lattice M; in mathematical terms this corresponds to the inner product of R with each of the connection vectors at $C_{i,j}$. As described earlier, the inner product of two vectors involves multiplication of corresponding elements and summation of the products, yielding a scalar result. For two vectors $v = [a_1, a_2 \ldots a_n]$ and $w = [b_1, b_2 \ldots b_n]$ the inner product p is $p = a_1 b_1 + a_2 b_2 + \ldots + a_n b_n$. The result of the inner product $(R.C_{i,j})$ is stored at $M_{i,j}$: $M_{i,j} = (R.C_{i,j})$.

Once all the data vectors have been processed there is a pattern of activations on the lattice M, and this pattern is the cluster analysis of the data matrix D.

Thus far, an important issue has been left latent but must now be addressed. In biological systems, evolution combined with individual experience of a structured environment determines the pattern of connectivity and cortical response characteristics which implement the mapping from high-dimensional sensory inputs to a two-dimensional representation. An artificially constructed model of such a system must specify these things, but the SOM does not do this explicitly. Instead, like other artificial neural network architectures (Haykin 1999), it learns them from the characteristics of the input data. This learning mechanism is now described.

In terms of the physical model, SOM learning tries to find a pattern of connection strength variation such that similar high-dimensional 'sensory' input signals are mapped to spatially close regions in the 'cortex', that is, so that the similarity relations among the input signals are preserved in their low-dimensional representation. In terms of the corresponding mathematical model, this can be restated as the attempt to find a set of connection vectors C such that the inner products of the $c \in C$ and the set of input vectors $d \in D$ generates in the output lattice M a pattern of activation that represents the neighbourhood relations of D with minimum distortion.

Given a set of input vectors D, SOM learning is a dynamic process that unfolds in discrete time steps $t_1, t_2 \ldots t_p$, where p is the number of time steps required to learn the desired mapping. At each time step t_i, a vector $d_j \in D$ is selected, usually randomly, as input to the SOM, and the connection strength matrix C is modified in a way that is sensitive to the pattern of numerical values in d_j. At the start of the learning process the magnitude of modifications to C is typically quite large, but as the SOM learns via the modification process the magnitude decreases and ultimately approaches 0, at which point the learning process is stopped – the p'th time step, as above. The question, of course, is how exactly the input-sensitive connection strength modification works, and the answer involves looking at the learning algorithm in detail. At the start of learning, the SOM is parameterized with user-specified values:

- Dimensionality of the input vectors R. This is the same as the dimensionality of the data to be analyzed.
- Dimensionality and shape of the output lattice M. In theory the output lattice can have any dimensionality and any shape, though in practice its dimensionality is usually 2 and its shape is usually rectangular or hexagonal. There is evidence that lattice shape can affect the quality of results (Ultsch and Herrmann 2005); 2-dimensional rectangular or hexagonal lattices are assumed in what follows.
- Size of the output lattice M. This is the number of cells in M.
- The following further parameters are explained below: neighbourhood shape, initial neighbourhood size, neighbourhood decrement interval, initial learning rate, and learning rate decrement size and interval.

In addition, the values in the connection matrix C are initialized in such a way that they are non-uniform; uniform connections would preclude the possiblity of learning. This initialization can be random or can use prior information which allows the SOM to learn more quickly (Kohonen 2001: Chs. 3.2 and 3.7). Thereafter, at each time step t_i, the following algorithm is applied until the stopping criterion is met.

1. An input vector $d_k \in D$ is selected.
2. The propagation of the input signal through the connections to the lattice in the physical SOM is represented as the inner product of d_k and the connection vector $C_{i,j}$ for each unit of the lattice. The result of each inner product is stored in the corresponding cell of the lattice matrix $M_{i,j}$, as above. Once all the inner products have been calculated, because the connections strengths in C were initialized to be non-uniform, the result is a non-uniform pattern of activation across the matrix.
3. The lattice matrix M is now searched to identify the cell with the largest numerical activation value. We shall call this cell u_{ij}, where i and j are its coordinates in the lattice.
4. The connection strengths in C are now updated. This is the crucial step, since it is how the SOM learns the connections required to carry out the desired mapping. This update proceeds in two steps:

(a) Update of the connections linking the most highly activated cell u_{ij} in M to the input vector d_k. This is done by changing the connection vector associated with u_{ij} according to Equation (4.2):

$$C_{i,j,k}(t+1) = C_{i,j,k}(t) + (l(t) \times (d_k - C_{i,})) \quad (4.2)$$

where

- t denotes the current time step.
- $C_{i,j,k}(t+1)$ is the new value of the kth element of the connection vector.
- $C_{i,j,k}(t)$ is the current value of the kth element of the connection vector.
- $l(t)$ is the learning rate, a parameter which controls the size of the modification to the connection vector. More is said about this below.
- $(d_k - C_{ij})$ is the difference between the current input vector and the connection vector. This difference is the sum of element-wise subtraction: $\sum_{1...k}(d_k - C_{i,j,k})$, where k is the length of the input and connection vectors.

In essence, therefore, the update to the connection vector C_{ij} associated with the most active cell u_{ij} is the current value of C_{ij} plus some proportion of the difference between C_{ij} and the input vector, as determined by the learning rate parameter. The effect of this is to make the connection vector increasingly similar to the input vector. Figure 4.5 gives an example for some d_k and associated C_{ij}:

Before update the input and connection vectors differ significantly; to make this clearer the difference between vectors is shown as differences in grey-scale components in Figure 4.5. After the first update, the connection vector has moved closer to the input vector. This update in turn means that, the next time the particular input vector d_k is loaded, the inner product of d_k and C_{ij} and thus the activation of M_{ij} will be even greater; ultimately the input and connection vectors will be much more similar than they were at the outset of training. In this way, d_j is associated ever more strongly with a particular unit on the lattice.

(b) Update of the connections linking cells in the neighbourhood of u_{ij} to the input. It is not only the connections associated with the

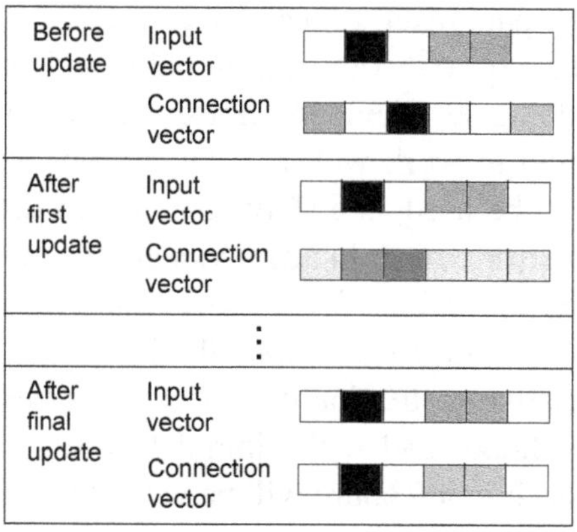

Figure 4.5: Convergence of input and connection vectors by learning

most-activated u_{ij} that are modified. The connections associated with the cells in the neighbourhood of u_{ij} are also modified in the way just described, though the degree of modification decreases with distance from u_{ij}. Figure 4.6a shows a 12×12 SOM lattice on which the most active cell u_{ij} is shown black, and those in its neighbourhood are shown as numerals indicating distance from it.

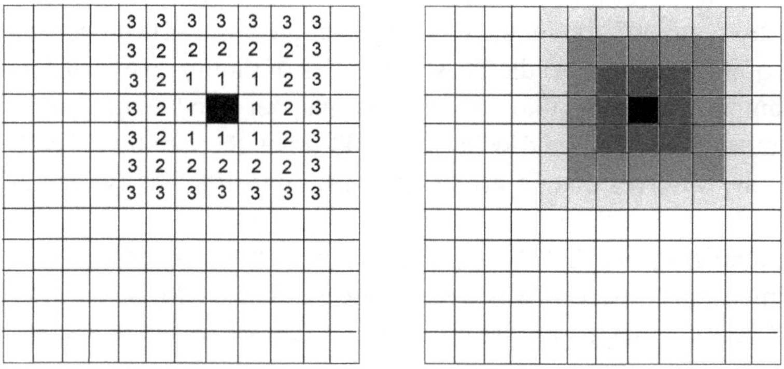

Figure 4.6: Neighbourhood of activation around a most-activated unit u_{ij}

Reference has been made in this step to learning rate and neighbourhood, and, earlier in the discussion, to the need to initialize these as parameters for the learning process. How are appropriate initial parameter values determined? At present, the answer is that there is no theoretically reliable and universally-applicable way of doing so. There are, however, rules of thumb which have been found to work well in a large number of applications and which can be used as a starting point for determining values by trial and error which, in any given application, yield good results (Kohonen 2001: Chs. 3.7 and 3.13). If the initial neighbourhood size is too small, for example, the lattice will not be globally ordered relative to the input data, so it is best to specify a large neighbourhood, and if the initial learning rate is too small the SOM learning procedure will be very slow, so it is best to specify a large learning rate.

5. At appropriate intervals, decrease the size of the neighbourhood and of the learning rate, and return to step 1. Specification of values for the decrement interval and the size of the decrement for both are part of the SOM initialization, as noted, and as with initial neighbourhood size and learning rate are determined heuristically, guided by rules of thumb.

As learning proceeds, the size of the neighbourhood and the learning rate are gradually reduced and approach zero, and changes to the connections slow commensurately. When the magnitude of these connection changes reaches some predefined threshold, the procedure is terminated. How many learning cycles are required? Again, there is no general answer. Much depends on the nature of the data input to the SOM, and rules of thumb give some guidance; for example, Kohonen (ibid.: 112) notes that, for good statistical accuracy, the number of steps should be at least 500 times the number of cells in the lattice, and observes that, in his simulations, up to 100,000 cycles have been used.

Summarizing, the SOM's representation of high dimensional data in a low-dimensional space is a two-step process. The SOM is first trained using the vectors comprising the given data. Once training is complete all the data vectors are input once again in succession, this time without training. The aim now is not to learn but to generate the two-dimensional representation of the data on the lattice. Each successive input vector activates the unit in the lattice with which training has associated it together with neighbouring units, though to an incrementally diminishing degree; when all the vectors have been input, there is a pattern of activations on the lattice, and the lattice is the representation of the input manifold in two-dimensional space.

A SOM with a 11 × 11 output lattice and random initialization of the connections was used to cluster MDECTE, and the results are shows in Figure 4.7. The label in any given cell indicates that that cell was most strongly activated by the data vector associated with the label; DECTE labels have again been abbreviated to avoid clutter.

g18 g47		g02 g19	g13	g12 g55	g24				g23	g25 g32
g14 g51				g46		g04		g53		g34 g05
	g30	g48					g45	g09		g21
g56	g27		g31		g54					g20
								g42		
g29 g37	g07		g50							g01
g33					g15	g17			g10	g41
			g28				g08	g38		
				g06 g16		g40		g36		g11
n03 n05	n02			g03						
n04 n06	n01 n07			g22 g26	g52		g49	g44	g43	g35 g39

Figure 4.7: SOM clustering of MDECTE

The row vectors of MDECTE are unevenly distributed on the lattice, and visual inspection of the distribution should now reveal any cluster structure, but a problem with the distribution is immediately apparent: where are the cluster boundaries? Figure 4.8 shows two possible partitions, but there are also other plausible ones.

Which of these two, if either, is the preferred interpretation of the layout of vector labels on the lattice? Without further information there is no way to decide: the lattice gives no clear indication where the cluster boundaries might be. The eye picks out likely concentrations, but when asked to decide whether a given data item is or is not part of a visually-identified cluster, one is often at a loss. An apparent solution is to fall back on knowledge of the subject domain as a guide. But using a priori knowledge of the data to disambiguate the lattice is not a genuine solution, for two reasons. Firstly, it misses the point of the exercize: the lattice is supposed to reveal the structure of the data, not the other way around. And, secondly, the structure of large,

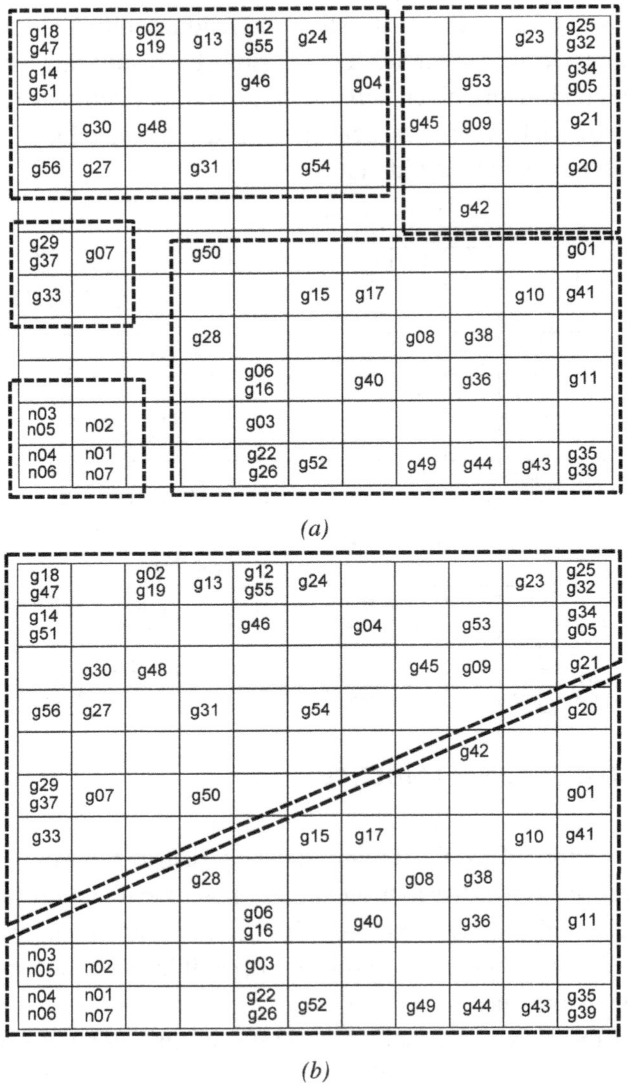

Figure 4.8: Possible cluster boundaries for MDECTE

complex, real-world data sets to which the SOM is applied as an analytical tool is not usually recoverable from mere inspection – if it were, there would be little point to SOM analysis in the first place. To be useful as an analytical tool, the SOM's representation of data structure has to be unambiguously interpretable on its own merits, and the problem is that an activation lattice like that in the above figures does not contain enough information to permit this in the general case. The problem lies in the subjectivity of visual interpretation with which this chapter began: humans want to see pattern, but the pattern any individual sees is determined by a range of personal factors and, among researchers, by degrees of knowledge of the domain which the lattice represents. Some objective interpretative criterion is required, and this is what the following section provides.

Actually, the problem is even worse than it looks. When a SOM is used for cluster analysis, inspection of the pattern of activation on the lattice can not only be subjective but can also be based on a misleading assumption. There is a strong temptation to interpret the pattern spatially, that is, to interpret any groups of adjacent, highly activated units as clusters, and the distance between and among clusters on the lattice as proportional to the relative distances among data items in the high-dimensional input space, as with, for example, MDS. That temptation needs to be resisted. The SOM differs from these other methods in that the latter try to preserve relative distance relations among objects on the data manifold, whereas the SOM tries to preserve the manifold topology, that is, the neighbourhood relations of points on the manifold – cf. Kaski (1997), Verleysen (2003), Lee and Verleysen (2007: Ch. 5). To see the implications of this for cluster interpretation of the output lattice, some additional discussion of the SOM learning algorithm is required.

We have seen that each lattice cell has an associated vector which represents its connections to the input vector. Since the dimensionality of the connection vectors is the same as that of the inputs, and the dimensionality of the inputs is that of whatever n-dimensional input space is currently of interest, the connection vectors are, in fact, coordinates of points in the n-dimensional space. Assume that there is a data manifold in the input space and that the connection vectors have been randomly initialized. In this initial state, there is no systematic relationship between the points specified by the set of connection vectors and the surface of the manifold. By incrementally bringing the connection vectors closer to training vectors taken from the data manifold, a systematic relationship is established in the sense that the connection vectors come to specify points on the manifold; at the end of training, the connection vectors map each of the points on the manifold specified by the training vectors to a particular lattice cell. Moreover, it can and usually does happen that data vectors which are close together on the manifold activate the same unit u_{ij}, as described earlier. In this case, the connection vector for u_{ij} has to be brought closer not only to one but to some number k of input vectors. Since these k vectors are close but not identical, the SOM algorithm adjusts the connection vector of u_{ij} so that it becomes a kind of average vector that specifies a point on the manifold which is intermediate between the k input vectors. In this way, u_{ij} becomes associated not only with a single point on the data manifold, but with an area of the manifold surface containing the k vectors that map to it. This is shown in Figure 4.9, where the shape on the left is a manifold and the square on the right is intended to represent the way in which a SOM trained on this manifold partitions its surface: each dot represents an 'average' vector in the above sense associated with a specific lattice unit u_{ij}, and the lines enclosing a dot represent the boundaries of the area of the manifold containing the k vectors that map to u_{ij}.

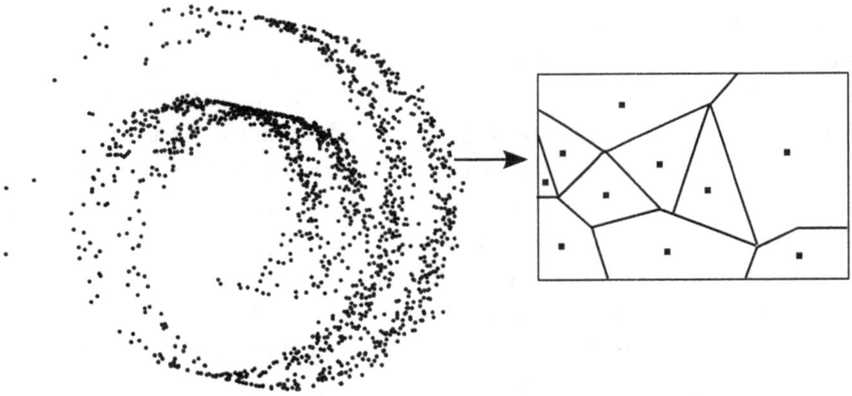

Figure 4.9: Voronoi tesselation of a manifold surface

Mathematically:

1. The process of finding an 'average' vector, or 'centroid', which is intermediate between k vectors was described in Chapter 2 is known as vector quantization: for a set V of k vectors, find a 'reference' vector v_r of the same dimension as those in V such that the absolute difference $d = |v_r - v_i|$ between each $v_i \in V$ and v_r is minimized. The SOM training algorithm quantizes the input vectors, and the connection vectors are the result – cf. Ritter, Martinetz, and Schulten (1992: Ch. 14), Van Hulle (2000: Ch. 2), Kohonen (2001: 59f.).

2. The partition of a manifold surface into areas surrounding reference vectors is a tesselation. TheSOM algorithm implements a particular type, the Voronoi tesselation, in which a reference vector is the centroid, and the area of k associated vectors surrounding the centroid is a neighbourhood; the neighbourhood of a given reference vector v_r in a Voronoi tesselation is defined as the set of vectors closer to v_r than to any other reference vector (ibid.: 59f.).
3. The set of neighbourhoods defined by the Voronoi tesselation is the manifold's topology, as discussed in the preceding chapter.

And because, finally, the SOM algorithm adjusts the connection vector not only of the most-activated unit u_{ij} but also of units in a neighbourhood of gradually diminishing radius, it ensures that adjacent manifold neighbourhoods map to adjacent lattice units.

How does all this relate to cluster interpretation of a SOM lattice? As noted, a Voronoi tesselation is an instance of a topology, that is, a manifold and a discrete collection of subsets of points on the manifold called neighbourhoods. When it is said that a SOM preserves the topology of the input space, what is meant is that it represents the neighbourhood structure of a manifold: when data is input to a trained SOM, the vectors in a given Voronoi neighbourhood are mapped to the same lattice cell, and the vectors in adjoining Voronoi neighbourhoods are mapped to adjacent lattice cells. The result of this topology preservation is that all vectors close to one another in the input space in the sense that they are in the same or adjoining neighbourhoods will be close on the SOM output lattice.

The problem, though, is this: just because active cells are close together on the SOM lattice does not necessarily mean that the vectors which map to them are topologically close in the input space. This apparently-paradoxical situation arises for two reasons – see discussion in, for example, Ritter, Martinetz, and Schulten (1992: Ch. 4).

1. The topology of the output manifold to which the SOM maps the input one must be fixed in advance. In the vast majority of applications the SOM output topology is a two-dimensional plane, that is, a linear manifold, with rectangular or hexagonal neighbourhoods which are uniform across the lattice except for at the edges, where they are necessarily truncated. There is no guarantee that the intrinsic dimensionality of the input manifold is as low as two, and therefore no guarantee that the output topology will be able to represent the input manifold well. In theory, the SOM is not limited to two-dimensional linear topology, and

various developments of it, cited later, propose other ones, but where the standard one is used some degree of distortion in the lattice's representation must be expected – cf. Verleysen (2003), Lee and Verleysen (2007: Ch. 5); the projection is optimal when the dimensionality of the lattice is equal to the intrinsic dimensionality of the data.

2. The dynamics of SOM training do not at any stage use global distance measures. The mapping from input to output space depends entirely on local neighbourhood adjacency. As such, the SOM cannot be expected consistently to preserve proportionalities of distance between individual vectors and vector neighbourhoods.

As a result, the SOM may squeeze its representation of the input topology into the lattice in such a way that units associated with centroids which are far apart on the input manifold may nevertheless be spatially close to one another in the lattice.

How can a SOM lattice be interpreted so as to differentiate cells which are spatially close because they are topologically adjacent in the input space and therefore form a cluster, and cells which are spatially close but topologically more or less distant in the input space? The answer is that it cannot be done reliably by visual inspection alone; interpretation of a SOM lattice by visual inspection is doubly unreliable – a subjective interpretation of an ambiguous data representation.

This is a well known problem with SOMs Kohonen (2001: 165), and a variety of ways of identifying cluster boundaries on the SOM lattice have been proposed; see for example Kaski (1997), Kaski, Nikkila, and Kohonen (2000), Vesanto (1999), Vesanto (2000), Merkl (1997), Merkl and Rauber (1997b), Merkl and Rauber (1997a), Rauber and Merkl (1999), Vesanto and Alhoniemi (2000), Kohonen (2001: Ch. 2.15), Pampalk, Rauber, and Merkl (2002), Pölzlbauer, Rauber, and Dittenbach (2005c), Pölzlbauer, Rauber, and Dittenbach (2005b), Pölzlbauer, Rauber, and Dittenbach (2005a). The most widely used one is the U-matrix – cf. Ultsch (2003a) and Ultsch and Siemon (1990) –, described in what follows.

The U-matrix representation of SOM output uses relative distance between reference vectors to find cluster boundaries. Specifically, given an $m \times n$ output lattice M, the Euclidean distances between the reference vector associated with each lattice cell M_{ij} (for $i = 1..m, j = 1...n$) and the reference vectors of the immediately adjacent cells $M_{i-1,j}$, $M_{i+1,j}$, $M_{i,j-1}$, and $M_{i,j+1}$ are calculated and summed, and the result for each is stored in a new matrix U_{ij} having the same dimensions as M. If the set of cells immediately adjacent to M_{ij} is designated as $M_{adjacent(i,j)}$, and d represents Euclidean distance, then, according to Equation (4.3)

$$U_{ij} = \sum d(M_{adjacent(ij)}, M_{i,j}) \; , \tag{4.3}$$

U is now plotted using a colour coding scheme to represent the relative magnitudes of the values in $U_{i,j}$. Any significant cluster boundaries will be visible. Why? The reference vectors are the coordinates of the centroids of the Voronoi tesselation of the data manifold and thus represent the manifold's topology, as we have seen. Where the sum of distances between the reference vector associated with M_{ij} and those associated with $M_{adjacent(ij)}$ is small, the distance between those centroids on the manifold is small; conversely, a large sum indicates a large distance between centroids on the manifold. Low-magnitude regions in U thus represent topologically close regions on the manifold, and high-magnitude ones topologically distant regions on the manifold. Assuming a grayscale colour coding scheme, therefore, clusters appear on the lattice as regions containing dark gray cells, and boundaries between clusters as regions containing light gray or white ones, or vice versa. Consider, for example, the U-matrix representation of the SOM lattice for the trivial data in Table 4.1.

Table 4.1: A trivial data matrix used to exemplify the U-matrix representation

	v1	v2	v3	v4
item1	1	0	0	0
item2	0	1	0	0
item3	0	0	1	0
item4	0	0	0	1

A SOM with an 11×11 lattice was trained on these data, with the result that the four row vectors in Table 4.1 are mapped to the four corners of the lattice in Figure 4.10

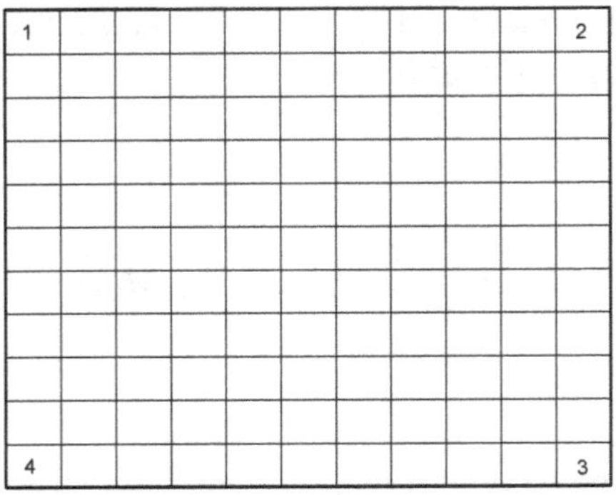

Figure 4.10: SOM lattice of data set in Table 4.1 and U-matrix representation

The U-matrix representation using grayscale colour coding to represent variation in magnitude is shown in Figure 4.11: Figure 4.11a shows the lattice partitioning directly from above using only the grayscale variation to demarcate cluster boundaries, and 4.11b shows a rotation of it which uses the relative numerical magnitudes underlying the map to give a topographic view.

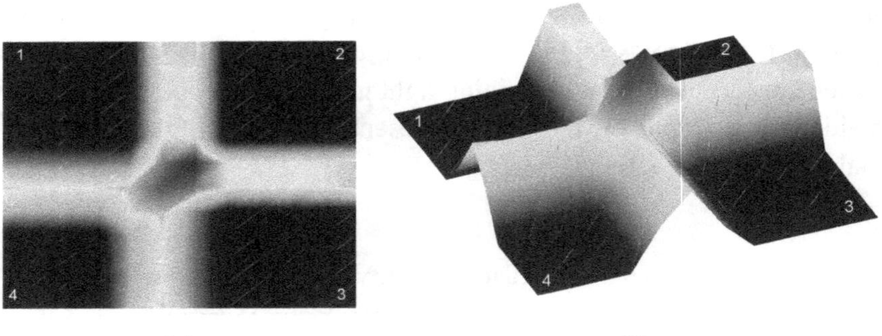

(a) *(b)*

Figure 4.11: U-matrix representation of the SOM lattice in Figure 4.10

Figure 4.12 applies the U-matrix representation to disambiguate the MDECTE lattice of Figure 4.7, and shows the essentially two-cluster structure already encountered with the foregoing projection methods, whereby n01-07 cluster against g01-56.

Software implementations of the SOM typically provide graphical enhancements of the U-matrix display. The two-dimensional grid can, for example, be plotted as a three-dimensional landscape in which the valleys are

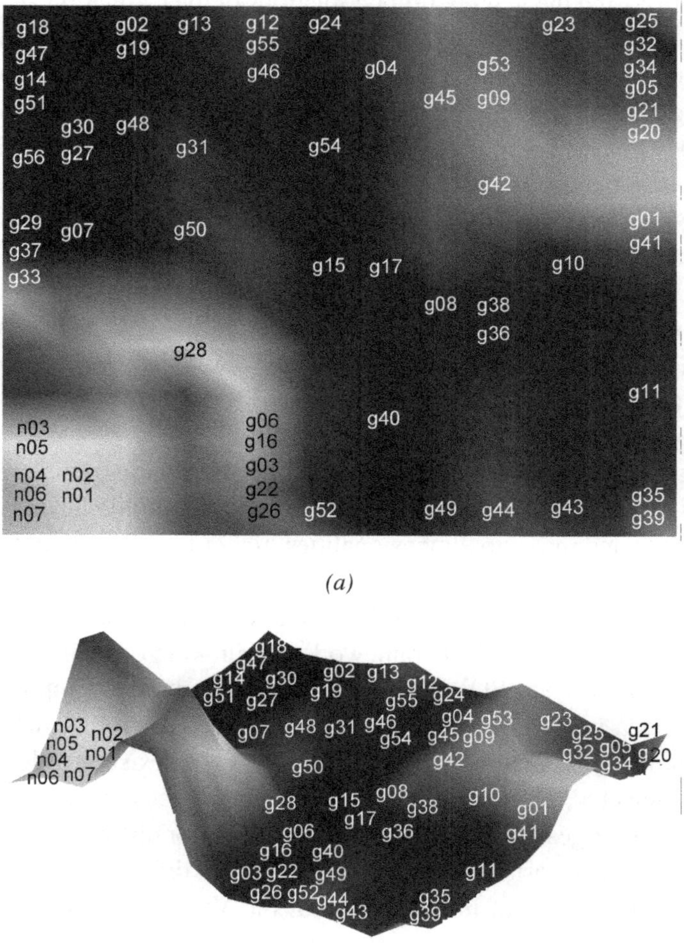

(a)

(b)

Figure 4.12: U-matrix representation of the SOM lattice for MDECTE in Figure 4.7

clusters and the mountains are boundaries separating them, as above. For such enhancements see: Vesanto (1999), Vesanto (2000), Vesanto and Alhoniemi (2000). The U-matrix has also been further developed, for which see: Ultsch (2003b) and Ultsch and Mörchen (2005).

The SOM has three major advantages for cluster analysis. The first and most important is that it takes account of data nonlinearity in its projection: because the Voronoi tesselation follows the possibly-curved surface of the data manifold and the neighbourhood relations of the tesselation are projected onto the output lattice, the SOM captures and represents any nonlinearities present in the structure of the manifold. The second is that, unlike most other clustering methods, the SOM requires no prior assumptions about the number and shapes of clusters: clusters on the SOM lattice represent relative densities of points, that is, of clusters in the input manifold, whatever their number and shape – see: Haykin (1999: 454ff.), Van Hulle (2000: 72ff.), Kohonen (2001: 152ff.)). And, thirdly, the computational time complexity of SOM learning scales linearly with the number of input vectors and quadratically with the number of lattice units – cf. Vesanto (2000), Vesanto and Alhoniemi (2000), Hämäläinen (2002)), which is better than other, more computationally demanding methods like some of those discussed subsequently. Perhaps inevitably, however, the SOM also has drawbacks:

- The default lattice structure, the linear two-dimensional rectangular grid, restricts its ability to represent data manifolds having nonlinear shapes and intrinsic dimensionalities higher than 2 without topological distortion; the foregoing discussion of lattice interpretation shows the consequence of this for SOM-based cluster analysis.
- There is no theoretical framework within which initial values for the fairly numerous SOM parameters can be chosen. Empirical results across a large number of applications have shown that, when parameter values which are 'sensible' in relation to existing rules of thumb are selected, the choice is not crucial in the sense that the SOM usually converges on identical or similar results for different initializations. This is not invariably the case, however, and different combinations, in particular different initializations of the Voronoi centroids and differences in the order of presentation of the training vectors can and do generate different clusterings – cf. Kaski (1997), Cottrell, Fort, and Pages (1998), Cottrell, Bodt, and Verleysen (2001), De Bodt, Cottrell, and Verleysen (2002a). This implies that the result from any single configuration of initial values is not guaranteed to be optimal or even acceptable. Multiple trials using different initializations for the same data are therefore required, followed by selection of the best result. As has just been noted, however, selection of the best result is problematical.
- The training algorithm does not optimize an objective function in the way that MDS, Sammon's Mapping, and Isomap do – cf. Cottrell, Fort, and Pages (1998), Heskes (1999), Kohonen (2001: 148, 356f.), and it has in fact been proven (Erwin, Obermayer, and Schulten 1992) that

such an objective function cannot exist. This makes assessment of the goodness of any given SOM lattice difficult; more is said about this in due course.

- There is no general guarantee that the training algorithm will converge to a stable result.

Numerous developments of the basic SOM described above have been proposed to address specific limitations – see Kaski (1997), Vesanto (2000), Kohonen (2001: Ch. 5) Lee and Verleysen (2007: 142f.), Xu and Wunsch (2009: Ch. 5.3.4)). The main ones are briefly surveyed here.

- Lattice structure: The foregoing discussion has noted that the default SOM lattice structure, the linear two-dimensional rectangular grid, restricts its ability to represent data manifolds of higher intrinsic dimensionality and nonlinear shape directly without topological distortion. One type of development addresses this restriction by proposing more flexible lattice structures which can better represent the input topology – cf. Fritzke (1999), Kaski (1997), Kohonen (2001: Ch. 5), Vesanto (2000), Lee and Verleysen (2007: 142f.) –, and more specifically to alter the shape of the lattice as learning proceeds. Some of the many such proposals are: Growing Cell Structures (Fritzke 1993), Growing Grid (Fritzke 1995), Growing SOM (Bauer and Villmann 1997), Hierarchical SOM (Blackmore and Miikkulainen 1995), Growing Hierarchical SOM (Rauber, Merkl, and Dittenbach 2002), and Neural Gas (Martinetz, Berkovich, and Schulten 1993).
- Theoretical tractability: The Generalized Topographic Mapping (GTM) has been proposed by Bishop, Svensen, and Williams (1998) to address the theoretical limitations of the SOM, and is intended as a mathematically principled replacement for it. GTM is a nonlinear latent variable model, and so is based on the idea that the n observed variables in data describe a natural process which is equally or almost equally well described by a smaller number k of variables, where $k < n$; another latent variable model, factor analysis (FA), was briefly described in the preceding chapter. GTM goes a step further than FA, however, in hypothesizing a probability density model for the lower-dimensional process and relating the higher-dimensional observed data to that model. Details of how GTM works are given in Bishop, Svensen, and Williams (ibid.); for present purposes the important thing to note is that it is trained by optimizing an objective function, that the algorithm is guaranteed to converge to the objective function, and that statistical

methods can be used to identify suitable initial parameter values for the model.

Clustering results from the GTM and the SOM are typically very similar, which is unsurprising given their close similarity. What the GTM offers, however, is on the one hand an understanding of results in terms of a well developed probability theory, and on the other an objective measure for assessing the goodness of those results.

Finally, a fairly recent development of projection clustering must be mentioned: subspace clustering. The foregoing discussion of dimensionality reduction has described linear and nonlinear ways of reducing data of observed dimensionality n to an approximation of its intrinsic dimensionality k, where k is less than n. This assumes that all the data objects are best described using the same number k of latent variables, which is not necessarily the case. Subspace clustering groups variables in accordance with the optimal number of latent variables required to describe them, or, put another way, of the i dimensional subspace (for $i = 1 \ldots n$) of the original n-dimensional data space in which they are embedded. This approach to clustering has in recent years found extensive application in areas like computer vision, motion segmentation, and image processing, and there is now a substantial literature devoted to it. Recent surveys are available in Parsons, Hague, and Liu (2004), Agrawal et al. (2005), Gan, Ma, and Wu (2007: Ch. 15), Kriegel, Kröger, and Zimek (2009), and Vidal (2011).

4.2.1.2 Proximity-based clustering

Nonhierarchical proximity-based approaches treat clustering as a mathematical optimization problem, where only a small subset of all possible partitions is examined in the hope of finding the optimal one. An initial k-cluster partition is defined and an iterative procedure is used in which, at each step, individual data points are moved from cluster to cluster to form a new partition and the result is evaluated in relation to the objective function f: if the value of f shows an improvement over the preceding one the new partition is retained, and if not it is discarded and another one is tried. Such iterative procedures are widely used and are known as gradient descent or gradient ascent procedures depending on whether optimality of the objective function is defined by a minimum or maximum value. Ideally, the procedure will gradually converge on a partition for which no change leads to an improvement in the value of f, at which point the partition is taken to be optimal. This assumption does not always hold, however, because gradient procedures can and often do converge on local maxima or minima, that is, where further iteration produces no improvement in the value of f but the true maximum or minimum has not been reached. Figure 4.13 shows this for gradient descent.

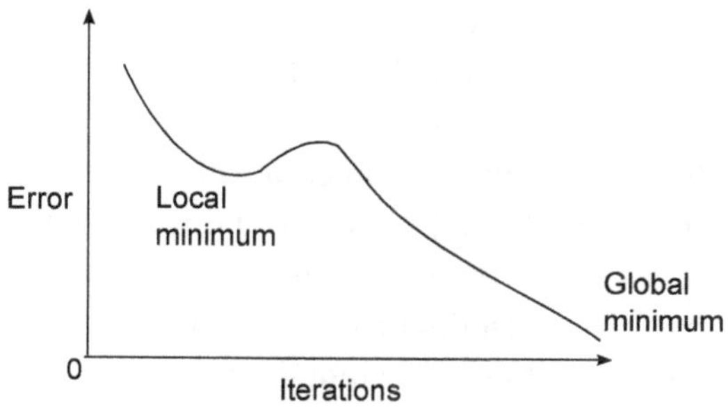

Figure 4.13: Gradient descent with local minimum

If the gradient descent procedure enters a local minimum it cannot escape and the objective function value will be a local optimum; the obverse is the case for gradient ascent. There are ways of escaping local minima, and they can be applied when necessary. The difficulty lies in knowing when it is necessary; anyone using this type of optimization must always be aware of the possibility that the supposedly optimal clustering is not in fact so.

Since it was first proposed in the mid-1960s (Forgy 1965), k-means clustering has become the most frequently used proximity-based non-hierarchical clustering method. The first part of the following account describes the standard k-means algorithm, the second identifies its main advantages and some problems associated with it, the third applies it to the MDECTE data, and the fourth outlines developments of the standard algorithm. Because it has been and continues to be so widely used, k-means is discussed in greater or lesser detail in virtually all cluster analysis, multivariate analysis, and data mining textbooks: for example Jain and Dubes (1988: Ch. 3.3), Jain, Murty, and Flynn (1999), Berkhin (2006), Xu and Wunsch (2005: Ch. 4.3), Tan, Steinbach, and Kumar (2006: Ch. 8), Gan, Ma, and Wu (2007: Ch. 9), Xu and Wunsch (2009: Ch. 4.3), Mirkin (2011: Ch. 6.2), and Mirkin (2013: Ch. 3).

k-means is based on the idea that, for a given set of data objects O, each cluster is represented by a prototype object, and a cluster is defined as the subset of objects in O which are more similar to, or in distance terms closer to, the prototype than they are to the prototype of any other cluster. An objective function is used find a set of clusters each of which optimally meets this criterion. For a data set O comprising m n-dimensional data points, O is partitioned into k prototype-centred clusters by the following iterative procedure:

1. Initialize the procedure by selecting k n-dimensional prototype locations in the data space; these can in principle be anywhere in the space, so that they might correspond to data points but need not. The prototypes are the initial estimate of where the clusters are centred in the space, and their locations are refined in subsequent steps. Placement of initial prototypes and selection of a value for k, that is, of the number of required clusters, is problematical, and is further discussed below.
2. Assign each of the m data points to whichever of the k prototypes it is closest to in the space using a suitable proximity measure. This yields k clusters.
3. Calculate the centroid of each of the k clusters resulting from (2). Each centroid becomes a new cluster prototype.
4. Repeat (2) and (3) until the objective function is optimized, that is, until the centroids stop changing their locations in the space.

This procedure is visualized in Figure 4.14 for 30 data points in two-dimensional data space, though it extends straightforwardly to any dimensionality. There are three visually obvious clusters, labelled A – C, and the object is for the above procedure to find them by identifying the cluster centroids in a way that optimizes the objective function.

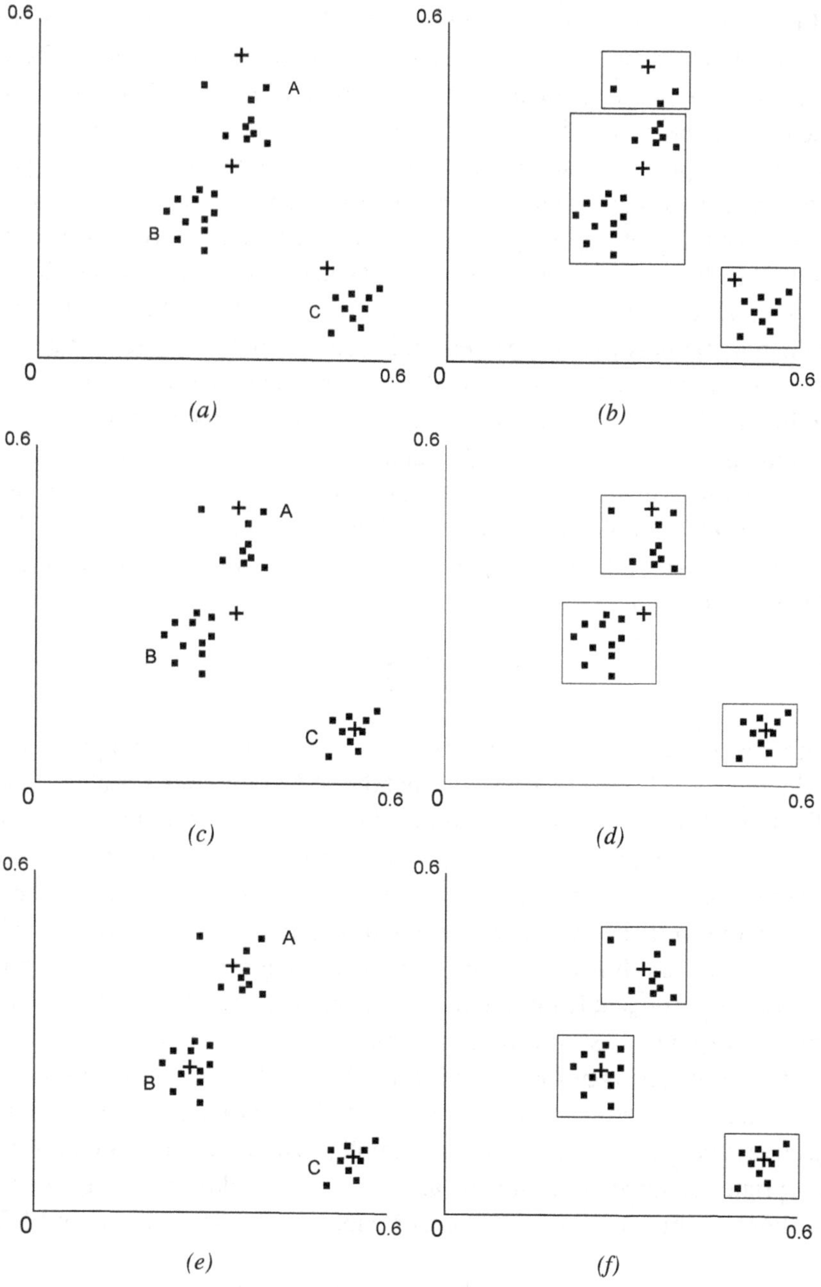

Figure 4.14: Identification of clusters using the *k*-means procedure

Figure 4.14a shows a scatter plot of the 29 data points and k = 3 prototypes randomly selected in the space and represented as crosses. The data points are assigned to their nearest prototypes, which results in the clusters indicated by the rectangles in Figure 4.14b. Cluster centroids are now calculated; their positions, shown in 4.14c, reflect the means of the data points in the clusters of 4.14b, and they become the new prototypes. Because the positions of the prototypes have changed, so does the clustering, as shown in 4.14d. The cluster centroids are once again calculated; their new positions are shown in 4.14e and the associated clustering in 4.14f. Further iterations will not change the positions of the centroids, and the procedure terminates. The data points associated with the final prototypes correspond to the visually-identifiable clusters in Figure 4.14a, and the k-means procedure can therefore be claimed to have partitioned the data set into three disjoint subsets. The further claim is that, for k = 3, the partition is optimal in the sense that it has minimized an objective function. Where Euclidean distance is used, this function is usually the sum of squared errors (SSE), described earlier and here defines by Equation (4.4):

$$SSE = \sum_{i=1...k} \sum_{x \in C_i} (x - p_i)^2 \tag{4.4}$$

where x is a data point, C_i is the i'th of k clusters, and p_i is the prototype of the i'th cluster. This expression says that the SSE is the sum, for all k clusters, of the Euclidean distances between the cluster prototypes and the data points associated with each prototype. For k-means to have optimized this function, the prototypes have to be placed in the data space so that the Euclidean distances between them and their associated data points is globally minimized across all clusters. It is easy to see that this is what k-means does: the procedure converges on stable cluster centroids, and a centroid is by definition the minimum distance from the all the points on which it is based.

Use of k-means is not restricted to Euclidean distance, though this is the most frequently used measure. A variety of different measures an associated objective functions can be used. For example, cosine similarity might be more appropriate for some kinds of data, and in that case a different objective function shown in Equation (4.5), Total Cohesion, can be used instead of SSE:

$$SSE = \sum_{i=1...k} \sum_{x \in C_i} cosine(x, p_i) \tag{4.5}$$

Cosine similarity is an attractive alternative to Euclidean distance when the data has not been normalized, as described earlier, because by basing proximity measurement solely on the angles between pairs of vectors the magnitudes of vector values (or, equivalently, vector lengths) are eliminated as a factor in clustering. The implication of using cosine similarity, however, is that vector length doesn't matter. There are undoubtedly applications where it does not, in which case cosine proximity is the obvious alternative to data normalization, but there are also applications where it does. With respect to MDECTE, for example, use of cosine proximity implies that all phonetic segments, from very frequent to very infrequent, are equally important in distinguishing speakers from one another, but as the foregoing discussion of data has argued, a variable should in principle represent something which occurs often enough for it to make a significant contribution to understanding of the research domain. The frequent segments in the DECTE interviews are prominent features which any attempt to understand the phonetics of Tyneside speech must consider, whereas the very infrequent ones tell one little about Tyneside speech and may well be just noise resulting from speaker mispronunciation or transcription error. Cosine proximity measurement eliminates the distinction, and is therefore unsuitable in this application. This observation applies equally to the use of cosine proximity measurement with other clustering methods as an alternative to measures which take vector magnitude into account.

Relative to the selection criteria for inclusion in this discussion, *k*-means is a prime candidate: it is intuitively accessible in that the algorithm is easy to understand and its results are easy to interpret, it is theoretically well founded in linear algebra, its effectiveness has repeatedly been empirically demonstrated, and computational implementations of it are widely available. In addition,

- Its computational time complexity grows with data space size as $O(nkdt)$, where n is the number of data vectors, k is the number of clusters, d is the data dimensionality, and t is the number of iterations. This means that *k*-means essentially grows linearly with data size, unlike other clustering methods to be considered in what follows, and is therefore suitable for clustering very large data sets in reasonable time – cf. Jain, Murty, and Flynn (1999), Manning, Raghavan, and Schütze (2008: Ch. 16.4), Xu and Wunsch (2009: Ch. 4.3).
- It is guaranteed to converge on a solution, though on this see further below.

The procedure of k-means also has several well known problems, however.

– *Initialization.* k-means requires two user-supplied parameter values: the number of clusters k and the locations of the k initial centroids $c_1 \ldots c_k$ in the data space. These values crucially affect the clustering result. On the one hand, if the value chosen for k is incompatible with the number of clusters in the data, then the result is guaranteed to mislead because k-means will deliver k clusters whatever the actual number of clusters intrinsic to the data, including none. For example, Figure 4.15 shows the cluster structure from Figure 4.14, but with $k = 2$ and $k = 4$: in both cases k-means fails to identify the visually-obvious cluster structure.

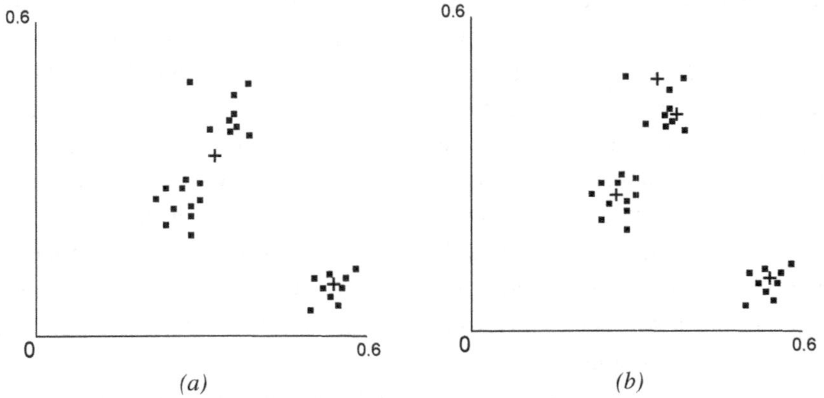

Figure 4.15: k-means cluster solutions for $k = 2$ and $k = 4$

On the other hand, different sets of initial prototypes can lead to different final ones, and thereby to different partitions of the data into clusters. In Figure 4.16 initial prototypes are shown as asterisks and final ones as crosses; given the correct $k = 3$, one prototype initialization led to a clustering compatible with visual intuition, and the other did not.

Various ways of selecting an initialization compatible with the intrinsic cluster structure of the data exist. The obvious one is to base the initialization on reliable a priori knowledge of the domain from which the data comes, where available. Failing this, a projection clustering method can be used to visualize the data and thereby to gain some insight into its cluster structure, or one of the range of initialization heuristics proposed in the literature can be applied (Xu and Wunsch

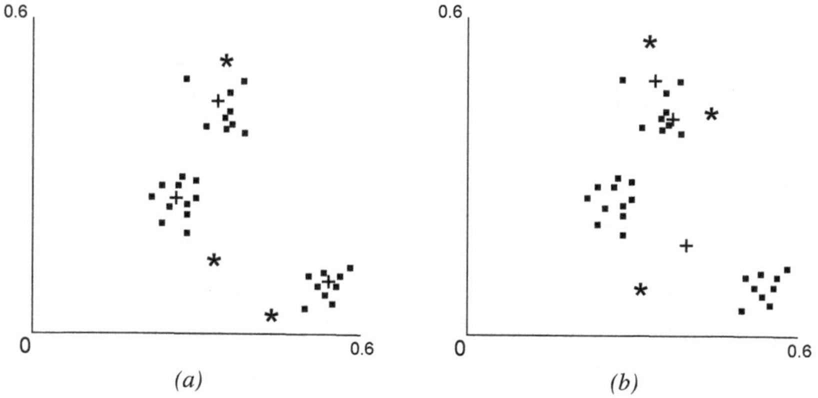

Figure 4.16: Effect of initial prototype location on *k*-means cluster solutions

2009: Ch. 4.3; Mirkin 2011: Ch. 6.2.7). Finally, a common approach is to conduct multiple analyses on the same data using different initializations and to select the best one, given some definition of "best"; for more on this approach see the discussion of cluster validation at the end of the present chapter.

- *Convergence. k*-means is guaranteed to converge on a stable solution, though that solution might be local; convergence to a global optimum is not guaranteed (Xu and Wunsch 2009: Ch. 4.3). The usual way to forestall this possibility is to conduct multiple analyses of the same data using different initializations and then to select the resulting consensus result.
- *Outliers.* Because *k*-means is based on centroids, it is strongly affected by the presence of outliers which distort the location of the centroids in the data space. Outliers should therefore be identified and eliminated prior to analysis.
- *Cluster shape. k*-means is limited in the range of clusters it can identify. This is demonstrated with reference to the clusters in Figure 4.17, where (a)–(c) are simplified abstractions of the cluster shapes in Figure 4.1b–4.1d respectively, and (d) is an additional shape often used in the literature to exemplify a particularly challenging cluster structure, where one cluster is completely enclosed by another.

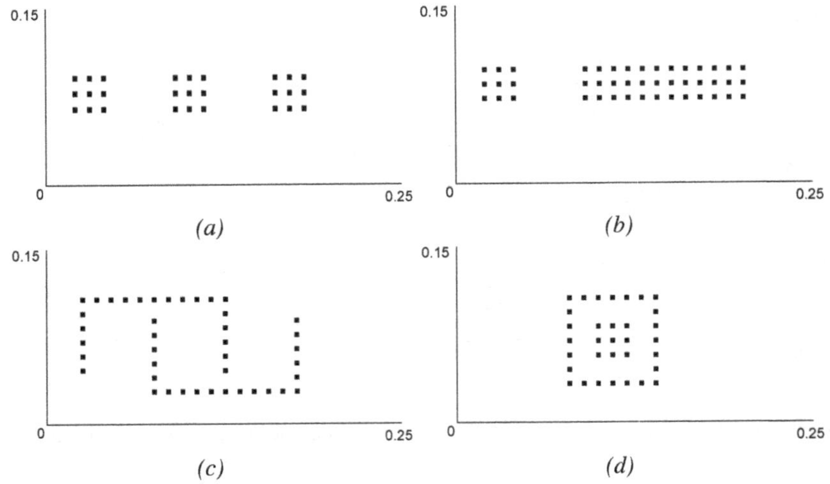

Figure 4.17: A range of cluster shapes for *k*-means analysis

The *k*-means analysis of the data underlying Figures 4.17a-c gave the results shown in Figure 4.18.

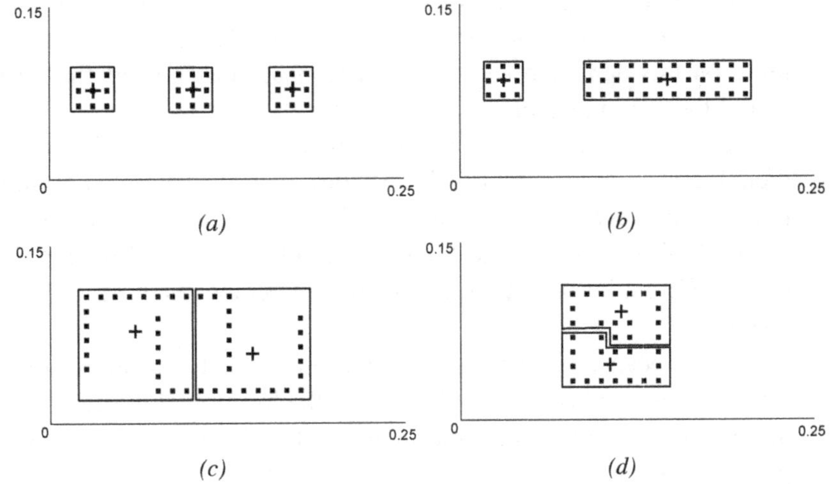

Figure 4.18: *k*-means solutions for clusters in Figure 4.17

In Figure 4.18a–4.18d the consensus prototypes arrived at by multiple initializations of *k*-means are shown by crosses, and the data points associated with each prototype are enclosed by boxes. The partitions of 4.18a and 4.18b accord with visual intuitions about cluster structure. However, those of 4.18c and 4.18d do not; this failure is symptomatic of a fundamental limitation on the clustering capability of *k*-means.

k-means partitions the data points in an *n*-dimensional metric space into *k* regions such that the points in each region are closer to their centroid than they are to the centroid of any other region; mathematically this is a Voronoi partition, already introduced in the discussion of the SOM. It does this purely on the basis of a linear metric such as Euclidean distance and, crucially, without reference to the density structure or, put another way, without reference to the shape of the data. Where the partition happens to coincide with the density structure, as in 4.18a and 4.18b, *k*-means successfully identifies the cluster structure, and where not, as in Figures 4.18c and 4.18d, it fails. The condition for success, that is, for coincidence of *k*-means partition and density structure, is that the dense regions of the data space be linearly separable, which means that the dense regions can be separated from one another by straight lines in two dimensions, or by planes or hyperplanes in higher dimensions; for a recent discussion of linear separability see Elizondo (2006). Linear separability and its coincidence with *k*-means partitioning is shown in Figures 4.19a and 4.19b, where the lines indicate the separability. There is, however, no way of linearly separating the intertwined clusters of 4.19c and 4.19d no matter how straight lines are drawn, some points from the upper and lower clusters will be in the same partition. Placement of the *k*-means prototypes indicates that the partition will cut across the data density structure and give the results in 4.18c and 4.18d.

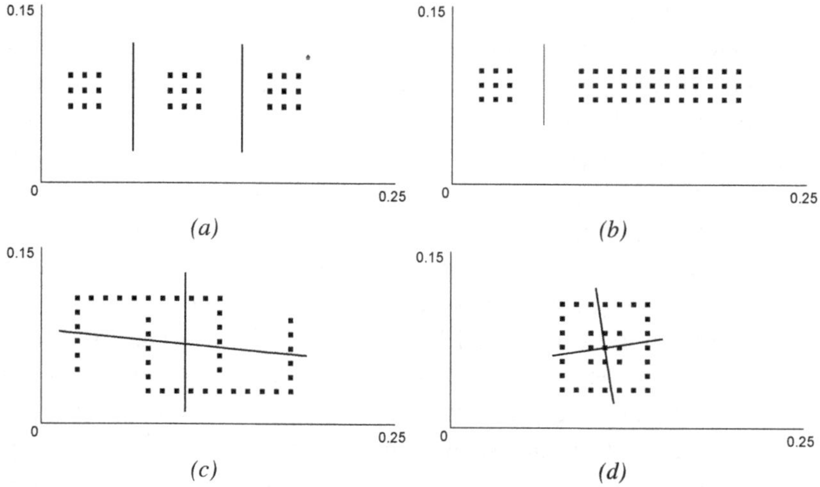

Figure 4.19: *k*-means and linear separability

The moral, therefore, is that *k*-means can only be relied upon reliably to identify clusters in data whose dense regions are linearly separable.

The PCA and MDStwo-dimensional visualizations of MDECTE derived earlier in Figure 4.2 indicate, on the one hand, that *k*-means can be used because the data density structure is linearly separable, and on the other suggest a suitable value for *k*. Table 4.2 shows the *k*-means results for $k = 2$; ten different prototype initializations were tried, and all gave the same result, which is compatible with both the PCA and MDS ones.

Table 4.2: *k*-means clusterings of MDECTE for $k = 2$

Cluster 1	n57	n58	n59	n60	n61	n62	n63			
Cluster 2	g01	g02	g03	g04	g05	g06	g07	g08	g09	g10
	g11	g12	g13	g14	g15	g16	g17	g18	g19	g20
	g21	g22	g23	g24	g25	g26	g27	g28	g29	g30
	g31	g32	g33	g34	g35	g36	g37	g38	g39	g40
	g41	g42	g43	g44	g45	g46	g47	g48	g49	g50
	g51	g52	g53	g54	g55	g56				

Because of its simplicity and computational efficiency, *k*-means continues to be developed in the hope of eliminating or at least mitigating its disadvantages. These developments are reviewed in Gan, Ma, and Wu (2007), Jain (2010), Jain, Murty, and Flynn (1999), Kogan (2007), Mirkin (2011, 2013), and Xu and Wunsch (2005, 2009); those most obviously relevant to the present discussion are outlined below.

- *Initialization.* No general and reliable method for selecting initial parameter values for the number of clusters and placement of prototypes in known, and given the crucial role that these play in determining the *k*-means result, it is unsurprising that initialization remains a research focus. Several approaches to the problem have already been mentioned. For initial prototype placement one of these approaches was to use of various parameter value selection heuristics. The simplest of these heuristics is random selection; others use a variety of criteria derived from the data. (Pena, Lozano, and Larranaga 1999) compared random selection to those proposed by (Forgy 1965), (MacQueen 1967) and (Kaufman and Rousseeuw 1990) and concluded that random selection and Kaufman's method were superior to the other two, with preference given to Kaufman's; more recent ones are (Bradley and Fayyad 1998; Likas, Vlassis, and Verbeek 2003). With respect to selection of the number of clusters k, (Ball and Hall 1967)'s ISODATA algorithm sidesteps the problem by providing a mechanism for merging and splitting clusters so that the value of k can increase or decrease in the course of the iterative k-means procedure: a cluster is split when its variance is above a specified threshold, and two clusters are merged when the distance between their centroids is below another specified threshold. This can provide an optimal number of partitions irrespective of the initial specification of k, and so is an attractive solution in principle. To work well in practice, however, Isodata requires optimization of no fewer than six threshold parameters, and the extra complication might not be thought worthwhile relative to simply finding an optimal value of k empirically, as described above.

- *Convergence.* As noted, the standard k-means procedure does not guarantee convergence to a global optimum. Stochastic optimization techniques like simulated annealing, genetic algorithms, and neural networks (Pham and Karaboga 2011) can do this, but at a very heavy computational cost. For applications of this type see (Krishna and Murty 1999; Patane and Russo 2001).
- *Outliers.* Outliers are a problem for k-means where centroids are used for cluster prototypes because, when an outlier is included in a cluster, the averaging pulls the cluster centroid away from where the other points would place it and towards the outlier. Using medoids instead of centroids eliminates this effect. Relative to a given cluster, a centroid is an abstraction which only coincidentally corresponds to any one of the actual data points in the cluster. A medoid, on the other hand, is an actual data point, and more specifically the point which is closest to the centroid and therefore best represents the centre of the cluster. When medoids instead of centroids are used as cluster prototypes, outliers cannot affect the prototypes because no averaging is involved. Examples of algorithms that use medoids are PAM (Partitioning Around Medoids) (Kaufman and Rousseeuw 1990), CLARA (ibid.), and CLARANS (Ng and Han 2002).
- Detection of non-linearly-separable clusters Because the standard k-means method uses a linear proximity measure between data points and the definition of centroids is a linear operation, it is a linear clustering method which, as we have seen, means that it can only be relied upon to identify clusters corresponding to linearly-separable regions of data density. If, however, it were possible to separate dense regions nonlin-

early, as in Figure 4.20, this fundamental limitation could be overcome.

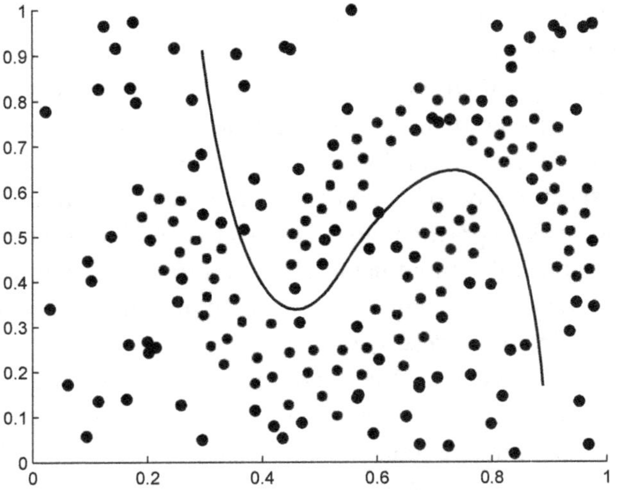

Figure 4.20: Nonlinear separation of intertwined clusters

The kernel *k*-means algorithm makes this possible. Like kernel methods generally, this development of *k*-means is based on Cover's Theorem (Cover 1965) in computational learning theory which says, in essence, that when clusters not linearly separable the data can be transformed by projection into a higher dimensional space using a suitable nonlinear function so that the clusters become linearly separable in the higher dimensional space; linear methods like standard *k*-means can then be used to cluster the transformed data (Dhillon, Guan, and Kulis 2004, 2005; Shawe-Taylor and Cristianini 2004).

4.2.1.3 Density-based clustering

Standard *k*-means fails to identify non-linearly-separable clusters because it partitions the data into clusters without reference to its density structure. The obvious solution is to take account of density, and this is what density-based clustering methods do; for general discussions see: Jain and Dubes (1988: Ch. 3.3.5), Han and Kamber (2001), Berkhin (2006), Tan, Steinbach, and Kumar (2006: Ch. 9.4), Gan, Ma, and Wu (2007: Ch.13), Everitt et al. (2011: Ch. 8.2). This section first describes Dbscan, currently the best established and most popular density-based clustering method, and then goes on to look more briefly at other approaches to density clustering. Dbscan was proposed by Ester et al. (1996), and is discussed in Tan, Steinbach, and Kumar (2006: Ch. 8.4), Gan, Ma, and Wu (2007: Ch. 13), and Everitt et al. (2011: Ch. 8.3).

Dbscan is based on a topological view of data manifolds, which was introduced in the discussion of Isomap in the preceding chapter. On this view, a data manifold is defined not in terms of the positions of its constituent points in an n-dimensional space relative to the n basis vectors, but rather as a set of adjacent, locally-Euclidean neighbourhoods. The essential idea is that clusters are collections of sufficiently dense adjacent neighbourhoods, and that neighbourhoods which are insufficiently dense are noise, given some definition of 'sufficiently dense'. This is shown in Figure 4.21.

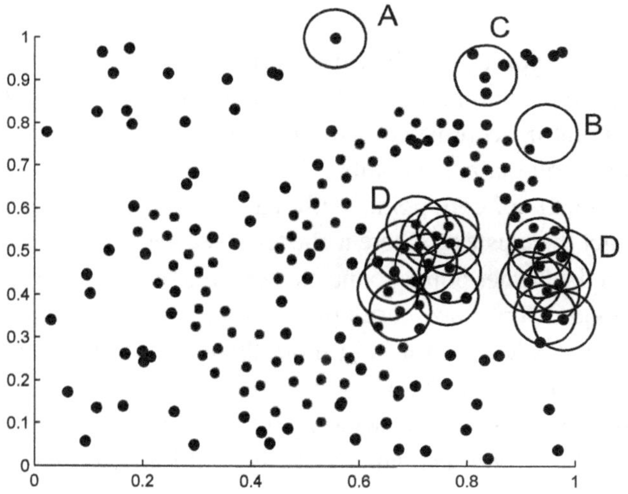

Figure 4.21: Dense and noise neighbourhoods in Dbscan

Each point p in Figure 4.21 has a neighbourhood, and a neighbourhood is defined as a radius r centred on p; density is the number of points in a neighbourhood, and sufficient density is some threshold number of neighbourhood points. In Figure 4.21 a sample of radii is shown as circles centred on points, and the threshold density number is taken to be 5. The point labelled A is visually not a member of the two main clusters, and Dbscan would regard it as noise because its neighbourhood contains only one member, itself, and is thus insufficiently dense. B is denser, but with only two members insufficiently so, and is also regarded as noise by Dbscan, which is again visually confirmed. C

contains four members and is nearly but not quite sufficiently dense in terms both of intuition and of Dbscan. The collection of neighbourhoods labelled *D* are all sufficiently dense and adjacent, and so belong to their respective clusters; continuation of the series of circles would eventually cover the two visually-identifiable intertwined clusters.

To implement this idea Dbscan requires predefinition of two parameters: the radius *r*, called *Eps*, which defines the size of the neighbourhood, and the threshold number of points for sufficient density, called *MinPts*. Relative to these parameters, three types of point are distinguished:

- Core points, whose neighbourhood contains *MinPts* or more points.
- Border points, whose neighbourhood contains fewer than *MinPts* but which are themselves in the neighbourhood of one or more core points.
- Noise points, which are all points that are not either core or border points.

Assuming *m* data points, the Dbscan algorithm is as follows:

1. Visit each data point m_i, $i = 1 \dots m$, labelling each as a core, border, or noise point in accordance with the above definitions
2. Eliminate all the noise points.
3. Link all pairs of core points within a radius *Eps* of one another.
4. Abstract the clusters, where a cluster is the set of all linked core points.
5. Assign the border points to the clusters. If a border point is in the neighbourhood of only one core point, assign it to the cluster to which the core point belongs. If it is in more than one neighbourhood, assign it arbitrarily to one of them.

Like *k*-means, Dbscan was selected for inclusion because it is a easy to understand and interpret, is mathematically well founded, has an established user base, and is readily available in software implementations. It has important advantages over *k*-means, however. One is that Dbscan does not require and in fact does not permit prespecification of the number of clusters, but rather infers it from the data; selection of *k* is one of the main drawbacks of *k*-means, as we have seen. Another is that Dbscan can find non-linearly-separable clusters, which extends its range of applicability beyond that of *k*-means. The *k*-means procedure was, for example, able to identify the linearly separable clusters in Figures 4.18a and 4.18b (see p. 188), but not the non-linearly-separable ones of 4.18c and 4.18d; as Figure 4.22 shows, Dbscan is able to identify them all.

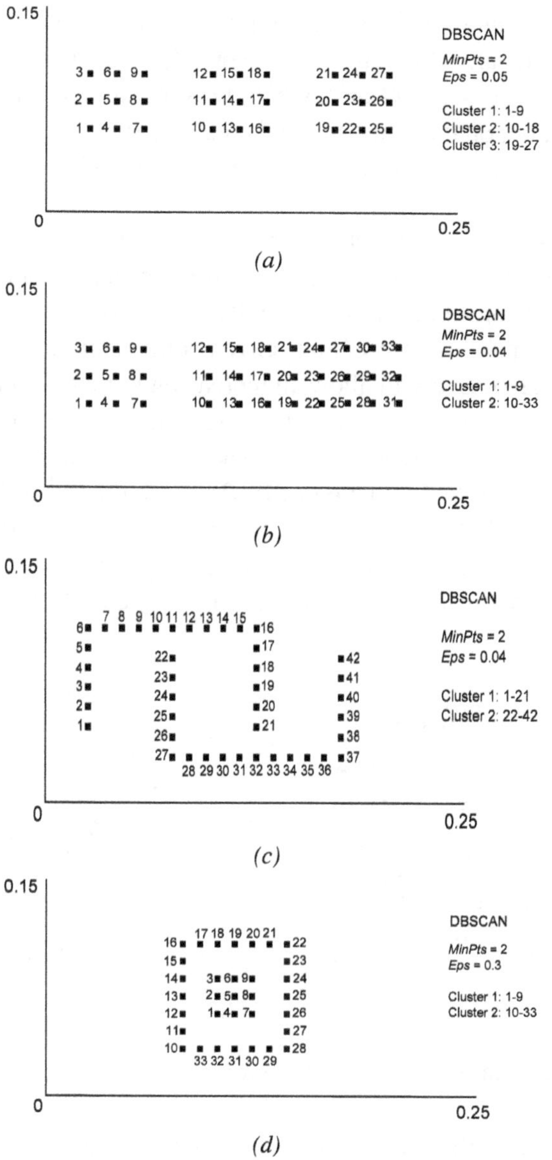

Figure 4.22: Dbscan clustering of manifolds for which *k*-means fails

Dbscan's computational time complexity is somewhat more demanding than that of k-means, though still reasonably moderate in ranging from $O(mlogm)$ to a worst-case $O(m^2)$, where m is the number of data points (Ester et al. 1996), but on the other hand it is highly resistant to noise and outliers when the clusters are uniformly dense, though not otherwise, on which see further below.

Perhaps inevitably, however, Dbscan has its problems.

- *Selection of parameter values.* As with k-means, selection of suitable parameter values strongly affects the ability of Dbscan to identify the intrinsic data cluster structure. Given some value for $MinPts$, increasing the size of Eps will allow an increasing number of points to become core points and thereby include noise points in the clusters, as in Figure 4.23a, and decreasing Eps makes it more and more difficult for any neighbourhood to achieve $MinPts$, with the result that many points which actually belong to a cluster become border or noise points, as in 4.23b. Conversely, given some value for Eps, increasing $MinPts$ makes it more difficult for points to become core, and decreasing it makes it easier.

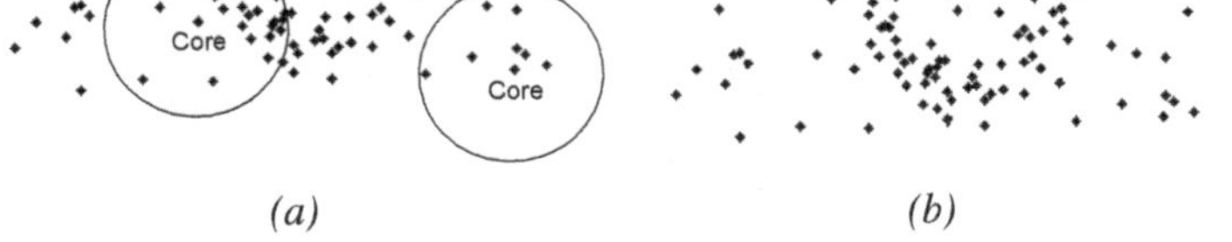

Figure 4.23: The relationship of $MinPts$ and Eps

The question, therefore, is how to select optimal Eps and $MinPts$ values for any given data. Dbscan does not optimize a global objective function, so identification of suitable Eps and $MinPts$ values by trial and error is not a reliable option. Heuristic methods for parameter value selection exist (Daszykowski, Walczak, and Massart 2001; Ester et al. 1996) and appear to work well for at least some kinds of data, but the fundamental problem remains: in the absence of a clear definition of 'best', how does one know which value combination gives the best result? Direct visualization of the shape of the data, as in Figure 4.23, provides an objective check, but where the data dimensionality precludes this there is no obvious way of knowing whether the specific parameter value selection has given the best or even a good result.

– *Variation in data density.* Dbscan has difficulty with data in which the density of the clusters varies substantially. Using heuristically-determined parameters values $Eps = 1.8565$ and $MinPts = 4$ it had no problem finding the two clusters in Figure4.24a.

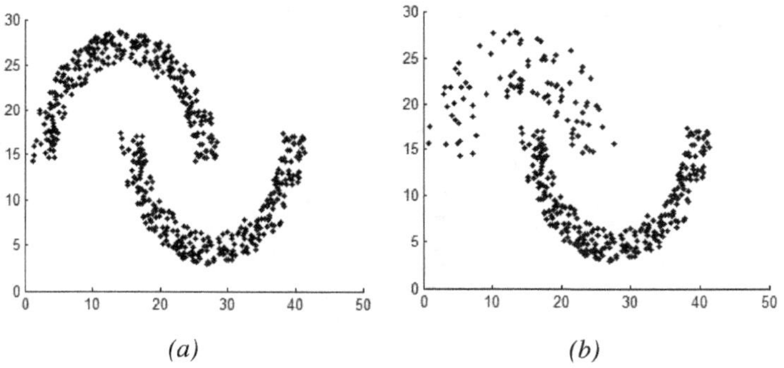

Figure 4.24: Clusters with similar and dissimilar densities

In Figure 4.24b the upper cluster was sparsified by adding random values to the underlying data, with the result that Dbscan was unable to identify the two-cluster structure in a way that agrees with visual intuition; in 4.25a the same parameter values as for 4.24a find the bottom cluster 5 as before, but cause the sparse cluster to be partitioned into four smaller clusters 1–4. A small increase of Eps in 4.25b to 2.5 produces a 3-cluster result by partitioning the sparse cluster into two, and a tiny further increase in 4.25c to 2.52 produces a two-cluster solution by combining the dense cluster with most of the sparse one. A further increase of Eps to 2.7 merges all the data points into a single cluster. The problem is that if Eps is small enough to keep the lower cluster separate from the upper one then it is too small to allow the upper one to be identified as a single cluster, and if Eps is large enough for the upper one to be identified as a single cluster then it is too large to keep the upper cluster separate from the lower one.
These problems are related. Various combinations of Eps with different $MinPts$ additional to those shown in Figure 4.25 were tried, but none allowed Dbscan to find the visually-obvious cluster structure.

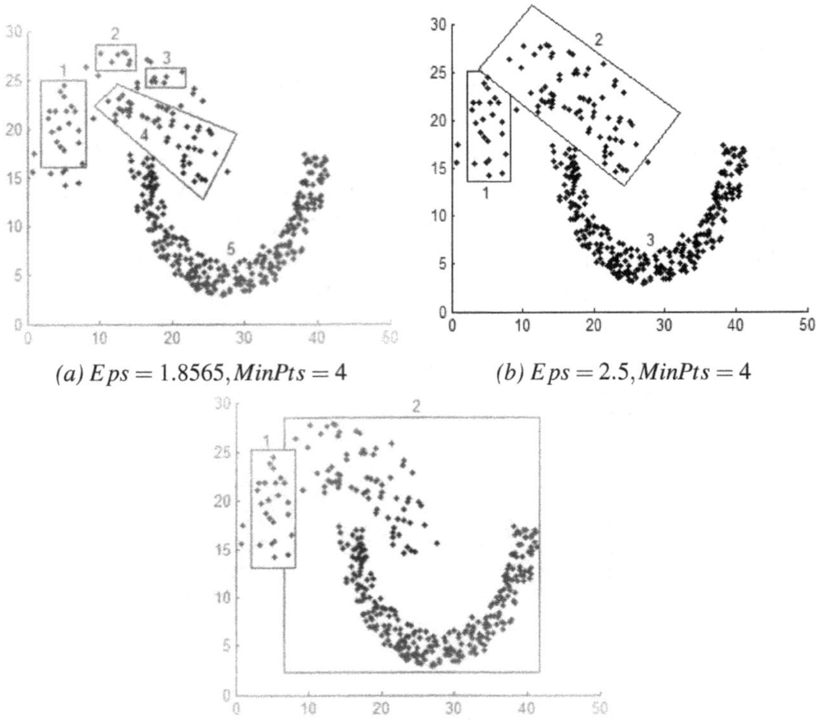

(a) $Eps = 1.8565, MinPts = 4$

(b) $Eps = 2.5, MinPts = 4$

(c) $Eps = 2.52, MinPts = 4$

Figure 4.25: Dbscan clusters with various Eps values

With sufficient searching a parameter value combination which does identify the visually-obvious structure might be found, but this should not obscure the fundamental point: it is only when the correct cluster structure is known a priori, as here, that one can judge whether or not Dbscan has found it and, if not, can adjust the parameter values until it does. Where such a priori knowledge is lacking, any Dbscan-generated clustering result might be suboptimal or just plain wrong because of a poor choice of parameters values, or because the data clusters vary in density, or both.

Application of Dbscan to MDECTE exemplifies the foregoing comments. Two experiments were conducted. In the first, $MinPts$ values in the range 2-50 were used, in each case using an Eps value determined by one of the above-mentioned heuristic methods; these values were in the range 77.73–82.22, which is small in relation to the scale of the data. For *MinPts* 2-35 Dbscan returned one cluster with a relatively few noise points, and from *MinPts* = 36 onwards there were only noise points without any clusters. In the second experiment *MinPts* was held constant at 4 and *Eps* values in the range 5–85 were used, in increments of 5: from 5 to 65 there were only noise points and no clusters, and from 70 onwards there was one cluster with a relatively few noise points. A reasonable conclusion would be that MDECTE has only one cluster, and therefore no interesting cluster structure. We know from previous analyses using different clustering methods that this is not the case, however. The two-cluster structure shown in Figures 4.2 (p. 160) and 4.12 (p. 177) is significant: there is one relatively dense cluster and one relatively sparse one, and, as expected in the light of the foregoing discussion, Dbscan was unable to identify it; in all trials the points in the smaller sparse cluster were treated as noise.

Because Dbscan can identify a superset of data shapes identifiable by *k*-means, it is tempting simply to dispense with *k*-means and to use Dbscan as the default analytical method. The foregoing discussion of problems with Dbscan and its application to MDECTE show, however, that this would be ill-advised. Where it is known or strongly suspected that the data density structure is linearly separable, the more reliable *k*-means method should be used, and if the data is non-linearly separable then results from Dbscan should, in view of its initialization and sparsity problems, be corroborated using some other clustering method or methods.

That variation in data density is a problem for Dbscan was quickly recognized, and proposals for addressing it have appeared, including Gdbscan (Sander et al. 1998), Optics (Ankerst et al. 1999), Snn (Ertöz, Steinbach, and Kumar 2003), Vdbscan (Liu, Zhou, and Wu 2007), Dvbscan (Ram et al. 2010), and most recently modifications to Dbscan by Dawoud and Ashour (2012) and Serdah and Ashour (2012). These are effective to varying degrees and have their own problems, but all improve on Dbscan with respect to the cluster density problem. The remainder of this section briefly present other approaches to density-based clustering.

A fairly obvious approach to identification of density in data is to cover the data space with a grid and count the number of data objects in each cell, as shown in Figure 4.26: regions of the grid where contiguous cells contain relatively many points are clusters, regions where they contain relatively few are noise, and empty regions contain no data.

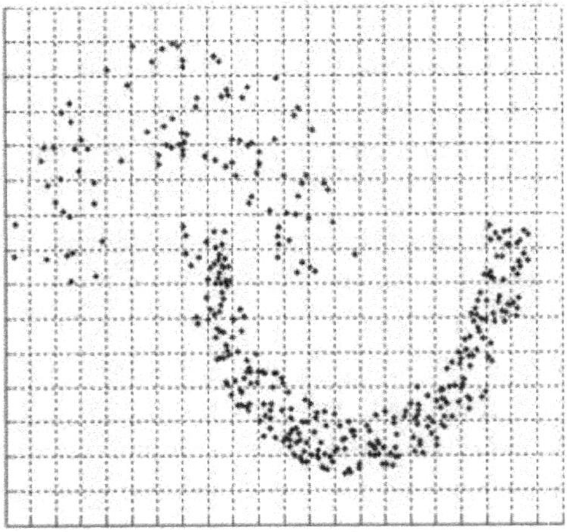

Figure 4.26: Grid covering the data in Figure 4.24b

Jain and Dubes (1988) in fact suggested such an approach, but realized that high dimensionality would be a problem for reasons given earlier with respect to dimensionality reduction: the number of cells grows very rapidly with increasing data dimensionality and the distribution of data points becomes so sparse that fewer and fewer cells contain even a single point, making it increasingly difficult to identify variation in density. An additional problem is specifying cell resolution, where the smaller the cell the lower the number of points in each, on average, giving a noisy and indistinct representation of the data density, and the larger the cell the larger the average number of points in each and the coarser the density representation. This is analogous to the Dbscan neighbourhood parameter Eps: like Eps, it is not obvious what the optimum cell size should be, and, also as with Dbscan, it may not be possible to find a single cell size that deals adequately with clusters of different densities.

For general discussions of grid-based clustering see: Jain and Dubes (ibid.: Ch. 3.3.5), Berkhin (2006), Tan, Steinbach, and Kumar (2006: Ch. 9.3.1), Gan, Ma, and Wu (2007: Ch. 12). Some specific methods which address the issues of dimensionality and parameter selection are Gridclus (Schikuta 1996), Bang (Schikuta and Erhart 1997), Sting (Wang, Yang, and Muntz 1997), Dbclasd (Xu et al. 1998), Optigrid (Hinneburg and Keim 1999), Clique (Agrawal et al. 1998, 2005), and (Yue et al. 2008).

Kernel-based clustering is based on concepts from statistical density function estimation, the aim of which is to find a mathematical function that generates some given data distribution (Dhillon, Guan, and Kulis 2004, 2005; Shawe-Taylor and Cristianini 2004). To do this, the contribution of each data point to the overall density function being estimated is expressed by a kernel probability density function, and the overall function is the sum of the individual kernel functions. A frequently used kernel function is the multivariate normal one, an example plot of which is given in Figure 4.27, where 4.27a shows its shape in three dimensions, and 4.27b in two dimensions as seen from above.

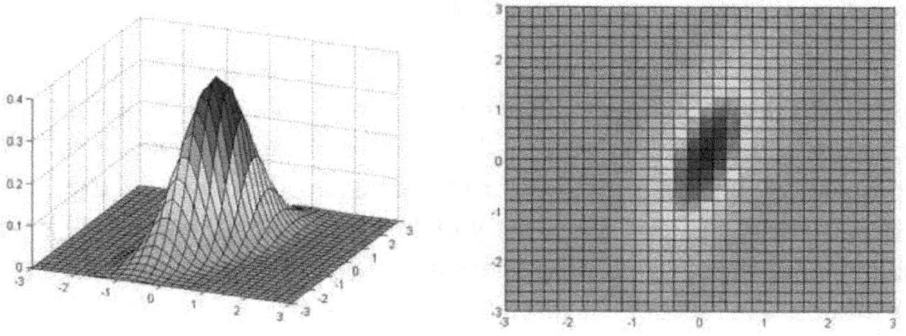

(a) *(b)*

Figure 4.27: Plot of a multivariate normal probability density function

Figure 4.28 shows a normal kernel function as concentric circles analogous to Figure 4.27b around a representative selection of points in a data distribution: dense functional areas like *A* constitute clusters, and less dense ones like those labelled *B* are regarded as noise.

The basic idea here is very like that of the neighbourhood density of Dbscan or the cell density of grid-based clustering, but in this case the important parameter is the shape of the kernel function. The standard kernel-based density clustering method is Denclue (Hinneburg and Gabriel 2007; Hinneburg and Keim 1998, 2003), for which see also the discussion in Tan, Steinbach, and Kumar (2006: Ch. 9.3.3).

4.2.2 Hierarchical clustering

Given an $m \times n$ data matrix D which represents m objects in n-dimensional space, hierarchical analysis does not partition the m objects into k discrete

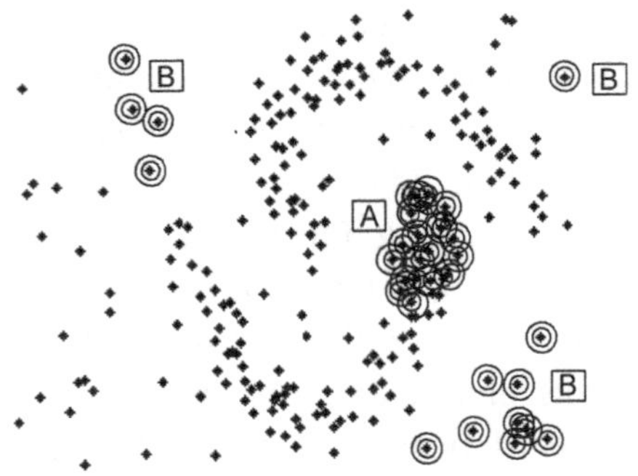

Figure 4.28: Clustering with kernel probability density functions

subsets like the clustering methods described so far. Instead, it constructs a constituency tree which represents the distance relations among the m objects in the space and leaves it to the user to infer a partition from the tree.

Hierarchical clustering is very widely used, and so is covered in most accounts of cluster analysis, multivariate analysis, and related disciplines like data mining. A selection of discussions is Jain and Dubes (1988: Ch. 3.2), Jain, Murty, and Flynn (1999), Tan, Steinbach, and Kumar (2006: Ch. 8), Gan, Ma, and Wu (2007: Ch. 7), Izenman (2008: Ch. 12), Xu and Wunsch (2009: Ch. 3), Everitt et al. (2011: Ch. 4), Mirkin (2011: Ch. 7), Mirkin (2013: Ch. 4).

4.2.2.1 The method

Construction of a hierarchical cluster tree is a two-step process: the first step abstracts a proximity table from the data matrix to be analyzed, and the second constructs the tree by successive transformations of the table. An intuition for how tree construction proceeds is best gained by working through an example; the example presented in what follows is based on MDECTE, and more specifically on a subset of MDECTE small enough to render illustrative tables and figures tractable for graphical representation. The concept of proximity among data objects has already been described in the foregoing discussion of data geometry. Euclidean distance is used to exemplify its application to construction of a proximity table for the first 6 of the full 63 rows of MDECTE. The Euclidean distances between all possible pairings of these 6 rows were calculated and stored in a 6×6 matrix D, shown in Table 4.3.

Table 4.3: Euclidean distance matrix for the first six row of MDECTE

	g01	g02	g03	g04	g05	g06
g01	0	116.9	59.0	82.6	103.8	69.7
g02	116.9	0	113.2	98.8	124.4	116.8
g03	59.0	113.2	0	79.4	112.9	69.3
g04	82.6	98.8	79.4	0	108.8	78.9
g05	103.8	124.4	112.9	108.8	0	113.0
g06	69.7	116.8	69.3	78.9	113.0	0

The Euclidean distance from g01 to g02 is 116.9, from g01 to g03 it is 59.0, and so on. To further simplify the discussion to follow, it is observed that the table is symmetrical on either side of the zero-values on the main diagonal because the distance between any two row vectors in the data matrix is symmetrical –the distance from g02 to g03 is the same as the distance from g03 to g02. Since the upper-right triangle simply duplicates the lower-left one it can be deleted without loss of information. The result is shown in Table 4.4.

Table 4.4: Euclidean distance matrix for the first six rows of MDECTE with superfluous entries removed

	g01	g02	g03	g04	g05	g06
g01	0					
g02	116.9	0				
g03	59.0	113.2	0			
g04	82.6	98.8	79.4	0		
g05	103.8	124.4	112.9	108.8	0	
g06	69.7	116.8	69.3	78.9	113.0	0

Given m objects to be clustered, construction of a cluster tree begins with m clusters each of which contains a different object. Thereafter, the tree is constructed in a sequence of steps in which, at each step, two clusters are joined into a superordinate cluster and the distance matrix D is transformed so as to incorporate the newly created cluster into it. The sequence ends when only one cluster, the tree itself, remains and D is empty. The joining of clusters requires some criterion for deciding which of the clusters available at any given step in the tree construction process should be selected for combination. The literature contains a variety of different criteria, and these will be presented in due course; the one chosen for detailed description joins the two clusters with the smallest distance between them in the distance matrix. The following sequence of cluster joins and matrix transformations exemplifies this.

Initially, each row vector of the data matrix is taken to be a cluster on its own; clusters here and henceforth are shown in brackets. Table 4.5a shows the original distance matrix of Table 4.4. It is searched to find the smallest distance between clusters. This is the distance 59.0 between row 1, that is, g01 and row 3, that is, g03, shown bold-face. These are combined into a new composite cluster (1,3).

Table 4.5: Joining the two nearest single-speaker clusters into a composite cluster

	g01	g02	g03	g04	g05	g06
g01	0					
g02	116.9	0				
g03	59.0	113.2	0			
g04	82.6	98.8	79.4	0		
g05	103.8	124.4	112.9	108.8	0	
g06	69.7	116.8	69.3	78.9	113.0	0

(a)

	(1,3)	(2)	(4)	(5)	(6)
(1,3)	0				
(2)	**113.2**	0			
(4)	**79.4**	98.8	0		
(5)	**103.8**	124.4	108.8	0	
(6)	**69.3**	116.8	78.9	112.9	0

(b)

Cluster1	Cluster2	Joining distance
1	3	59.0

(c)

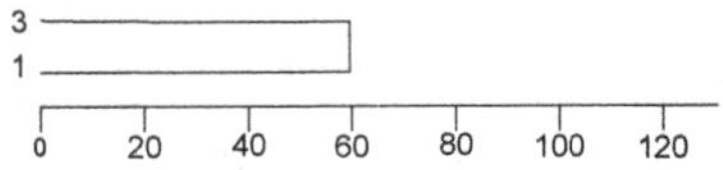

(d)

Table 4.5a is now transformed into the one in Table 4.5b:

1. Rows and columns (1) and (3) are removed from the Table 4.5a and replaced in 4.5b with a single blank row and column to accommodate the new (1,3) cluster; 0 is inserted as the distance from (1,3) to itself.
2. Into the blank cells of the (1,3) row and column of Table 4.5b are inserted the minimum distances from (1,3) to the remaining clusters (2), (4), (5), and (6). What does this mean? Referring to Table 4.5a, the distance between (1) and (2) is 116.9 and between (3) and (2) it is 113.2; the minimum is 113.2, and that value is inserted into the cell representing the minimum distance between (1,3) and (2) in Table 4.5b. The distance between (1) and (4) in Table 4.5a is 82.6 and between (3) and (4) is 79.4; the latter value is inserted into Table 4.5b as the minimum distance between (1,3) and (4). By the same procedure, the minimum distances between (1,3) and 5 and between (1,3) and 6 are inserted. The resulting table is smaller by one row and one column; inserted values are shown in bold-face, and the remaining ones are unchanged.

Table 4.5c is a list showing the sequence of cluster joins together with the distance at which they were combined; this list is the basis for the graphical representation of the cluster tree shown in Table 4.5d. The scale line below the tree in 4.5d allows the joining distance to be read from the graphical representation; one can, for example, see that (1) and (3) are joined just short of 60, that is, at 59.0. The reduced matrix of Table 4.5b is used as the basis for the next step, shown in Table 4.6a.

Table 4.6: Joining the composite cluster in Table 4.5 to the nearest single-speaker cluster

	(1,3)	(2)	(4)	(5)	(6)
(1,3)	0				
(2)	113.2	0			
(4)	79.4	98.8	0		
(5)	103.8	124.4	108.8	0	
(6)	**69.3**	116.8	78.9	112.9	0

(a)

	((1,3),6)	(2)	(4)	(5)
((1,3),6))	0			
(2)	**113.2**	0		
(4)	**78.9**	98.8	0	
(5)	**103.8**	124.4	108.8	0

(b)

Cluster1	Cluster2	Joining distance
1	3	59.0
(1,3)	6	69.3

(c)

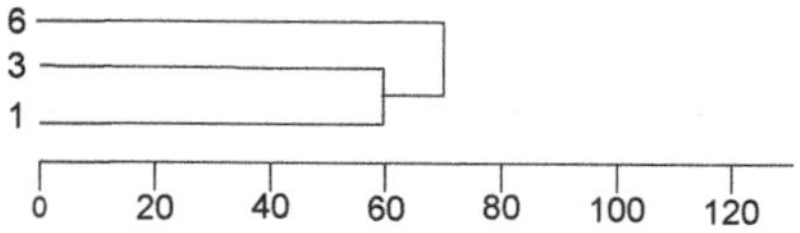

(d)

The matrix in Table 4.6a is searched to find the smallest distance between clusters. This is 69.3 between (1,3) and (6), and these are combined into a composite cluster ((1,3),6). The matrix is transformed into the one in 4.6b as in the previous step:

1. Rows and columns (1,3) and (6) are removed and replaced with a single blank row and column to accommodate the new ((1,3),6) cluster, with 0 inserted as the distance from ((1,3),6) to itself.
2. Into the blank cells of the ((1,3),6) row and column are inserted the minimum distance from ((1,3),6) to the remaining clusters (2), (4), and (5). Referring to Table 4.6a, the distance between (1,3) and (2) is 113.2 and between (6) and (2) it is 116.8; the minimum is 113.2, and that value is inserted into the cell representing the minimum distance between ((1,3),6) and (2). The distance between (1,3) and (4) is 79.4 and between (6) and (4) is 78.9; the latter value is inserted into the matrix as the minimum distance between ((1,3),6) and (4). By the same procedure, the minimum distance between ((1,3),6) and 5 is inserted. The

resulting table is again smaller by one row and one column; inserted values are highlighted, and the remaining ones are again unchanged.

Table 4.6d shows the tree after the second step together with the distance between (1,3) and (6). The reduced matrix of Table 4.6b is used as the basis for the next step, shown in Table 4.7a.

Table 4.7: Joining the composite cluster in Table 4.6 to the nearest single-speaker cluster

	((1,3),6)	(2)	(4)	(5)
((1,3),6)	0			
(2)	113.2	0		
(4)	**78.9**	98.8	0	
(5)	103.8	124.4	108.8	0

(a)

	(((1,3),6),4)	(2)	(5)
(((1,3),6),4)	0		
(2)	**98.8**	0	
(5)	**103.8**	124.4	0

(b)

Cluster1	Cluster2	Joining distance
1	3	59.0
(1,3)	6	69.3
((1,3),6)	4	78.9

(c)

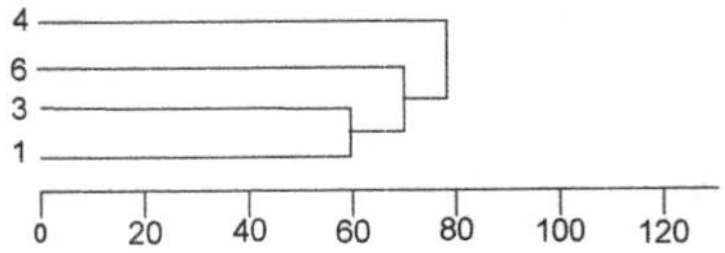

(d)

The matrix in Table 4.7a is searched to find the smallest distance between clusters. This is 78.9 between ((1,3),6) and (4), and these are combined into a composite cluster (((1,3),6),4). The matrix is transformed into the one in 4.7b as in the previous step:

1. Rows and columns ((1,3),6) and (4) are removed and replaced with a single blank row and column to accommodate the new (((1,3),6),4) cluster, with 0 is inserted as the distance from (((1,3),6),4) to itself.
2. Into the blank cells of the (((1,3),6),4) row and column are inserted the minimum distance from (((1,3),6),4) to the remaining clusters (2) and (5). Referring to Table 4.7a, the distance between ((1,3),6) and (2) is 113.2 and between (4) and (2) it is 98.8; the minimum is 98.8, and that value is inserted into the cell representing the minimum distance between (((1,3),6),4) and (2). The distance between ((1,3),6) and (5) is 103.8 and between (4) and (5) is 108.8; the former value is inserted into the matrix as the minimum distance between (((1,3),6),4) and (5). The resulting table is again smaller by one row and one column; inserted values are highlighted, and the remaining one is again unchanged.

Table 4.7d shows the tree after the third step together with the distance between ((1,3),6) and (4). The reduced matrix of Table 4.7b is used as the basis for the next step, shown in Table 4.8a.

Table 4.8: Joining the composite cluster in Table 4.7 to the nearest single-speaker cluster

	(((1,3),6),4)	(2)	(5)
(((1,3),6),4)	0		
(2)	**98.8**	0	
(5)	103.8	124.4	0

(a)

	((((1,3),6),4),2)	(5)
((((1,3),6),4),2)	0	
(5)	**103.8**	0

(b)

Cluster1	Cluster2	Joining distance
1	3	59.0
(1,3)	6	69.3
((1,3),6)	4	78.9
(((1,3),6),4)	2	98.8

(c)

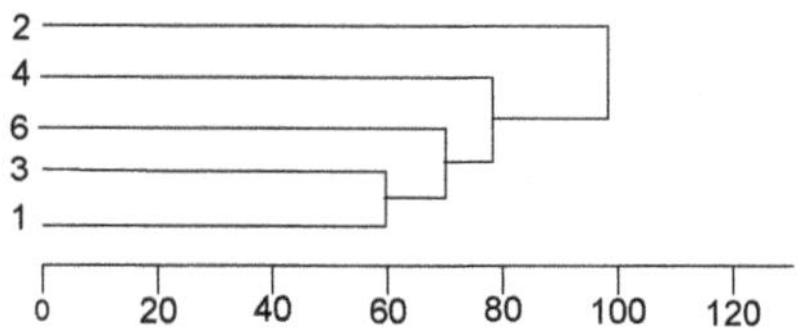

(d)

The matrix in Table 4.8a is searched to find the smallest distance between clusters. This is 98.8 between (((1,3),6),4) and (2), and these are combined into a composite cluster ((((1,3),6),4),2). It is transformed into the one in 4.8b as in the previous step:

1. Rows and columns ((((1,3),6),4) and (2) are removed and replaced with a single blank row and column to accommodate the new ((((1,3),6),4),2) cluster, with 0 is inserted as the distance from ((((1,3),6),4),2) to itself.
2. Into the blank cells of the ((((1,3),6),4),2) row and column is inserted the minimum distance from ((((1,3),6),4),2) to the remaining cluster (5). Referring to Table 4.8a, the distance between (((1,3),6),4) and (5) is 103.8 and between (2) and (5) it is 124.4; the minimum is 103.8, and that value is inserted into the cell representing the minimum distance between ((((1,3),6),4),2) and (5). The resulting table is again smaller by one row and one column; the inserted value is highlighted.

Table 4.8d shows the tree after the fourth step together with the distance between (((1,3),6),4) and (2). The reduced matrix of Table 4.8b is used as the basis for the next step, which is trivial because only one cluster remains. This remaining cluster is combined with the existing composite one, yielding a single final cluster (((((1,3),6),4),2),5) and completing the tree, as in

Table 4.9.

Table 4.9: Joining the composite cluster in Table 4.8 to the nearest single-speaker cluster

	((((1,3),6),4),2)	(5)
((((1,3),6),4),2)	0	
(5)	**103.8**	0

(a)

	(((((1,3),6),4),2),5)
(((((1,3),6),4),2),5)	0

(b)

Cluster1	Cluster2	Joining distance
1	3	59.0
(1,3)	6	69.3
((1,3),6)	4	78.9
(((1,3),6),4)	2	98.8
((((1,3),6),4),2)	5	103.8

(c)

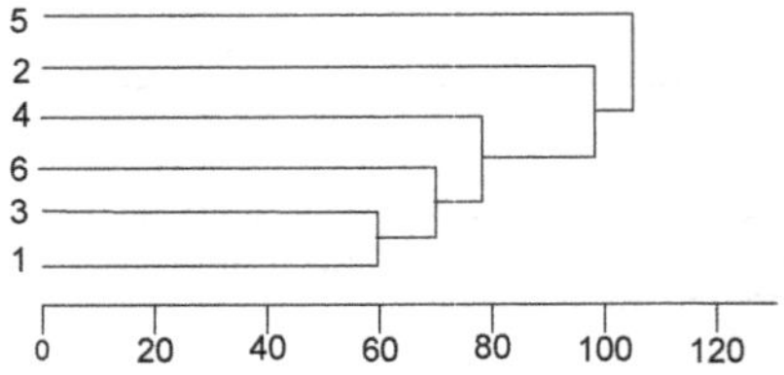

(d)

4.2.2.2 *Variants*

For a matrix with m rows there will at any step in the above tree-building sequence be a set of p clusters, for p in the range $2 \ldots m$, available for joining, and two of these must be selected. At the first step in the clustering sequence, where all the clusters contain a single object, this is unproblematical: simply choose the two clusters with the smallest distance between them. At subsequent steps in the sequence, however, some criterion for judging relative proximity between composite and singleton cluster pairs or between composite pairs is required, and it is not obvious what the criterion should be. The one exemplified in the foregoing sequence is such a criterion, known as Single Linkage, but there are various others (Jain and Dubes 1988: ch.3.2), (Tan, Steinbach, and Kumar 2006: Ch. 8.3), (Gan, Ma, and Wu 2007: Ch. 6.8), (Manning, Raghavan, and Schütze 2008: Ch. 17), (Xu and Wunsch 2009: Ch. 3), (Everitt et al. 2011: Ch. 4). Some of the more commonly used criteria are described in what follows.

For simplicity of exposition, it is assumed that a stage in the tree building sequence has been reached where there are $p = 3$ clusters remaining to be joined. This is shown in Figure 4.29: 4.29a shows a scatterplot of the data being clustered, and 4.29b the current state of tree construction.

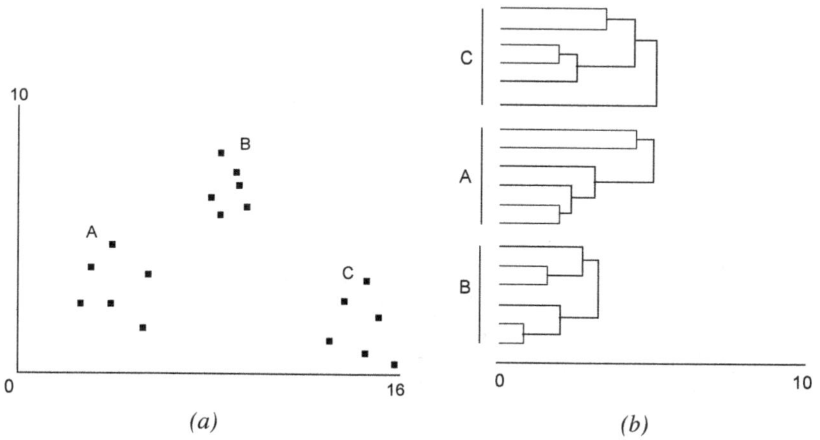

Figure 4.29: An intermediate stage in a hierarchical cluster tree construction

Which pair of subtrees should be joined next? Based on visual examination of the scatterplot, the intuitively obvious answer is the pair of clusters closest to one another, that is, *A* and *B*. Where the data are higher-dimensional and cannot be directly plotted, however, some explicit specification of closeness is required. This is what the various cluster-joining criteria referred to above provide.

- The Single Linkage criterion defines the degree of closeness between any pair of clusters (X,Y) as the smallest distance between any of the data points in X and any of the data points in Y: if there are x vectors in X and y vectors in Y, then, for $i = 1 \ldots x, j = 1 \ldots y$, the Single Linkage distance between X and Y is defined as in Equation (4.6).

$$SingleLinkageDistance(X,Y) = min(dist(X_i,Y_j)) \qquad (4.6)$$

where $dist(X_i,Y_j)$ is the distance between the i'th vector in X and the j'th vector in Y stated in terms of whatever metric is being used, such as Euclidean distance. The Single Linkage distances between all unique pairs of the p vectors remaining to be clustered are calculated, and the pair with the smallest distance is joined. This is exemplified for the three clusters of Figure 4.29 in Figure 4.30.

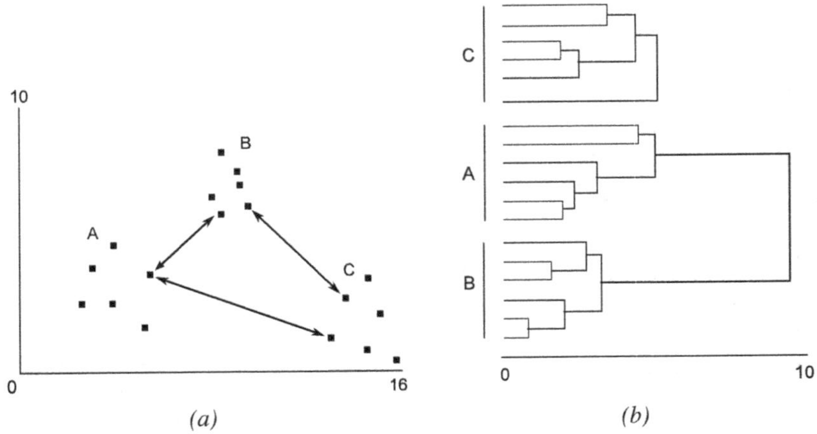

Figure 4.30: Single Linkage

The arrowed lines in Figure 4.30a represent distances between the points closest to one another in cluster pairs (A,B), (A,C), and (B,C); the one between A and B is shortest, so these two clusters are joined, as in Figure 4.30b. Single Linkage is also commonly known as Nearest Neighbour clustering, for self-evident reason, and is the linkage criterion exemplified in detail in the foregoing discussion.

- The Complete Linkage criterion defines the degree of closeness between any pair of clusters (X,Y) as the largest distance between any of the data points in X and any of the data points in Y: if there are x vectors in X and y vectors in Y, then, for $i = 1 \ldots x$, $j = 1 \ldots y$, the Complete Linkage distance between X and Y is defined as in Equation (4.7)

$$CompleteLinkageDistance(X,Y) = max(dist(X_i, Y_j)) \qquad (4.7)$$

where $dist(X_i, Y_j)$ is the distance between the ith vector in X and the jth vector in Y stated in terms of whatever metric is being used, such as Euclidean distance. The Complete Linkage distances between all unique pairs of the p vectors remaining to be clustered are calculated, and the pair for which the Complete Linkage distance is smallest is joined. This is exemplified for the three clusters of Figure 4.29 in Figure 4.31.

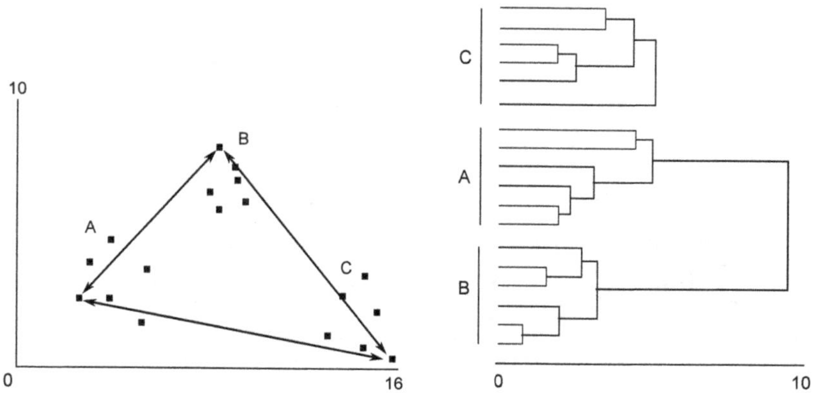

(a) (b)

Figure 4.31: Complete Linkage

The arrowed lines in Figure 4.31a represent distances between the points furthest from one another in cluster pairs (A,B), (A,C), and (B,C); the one between A and B is shortest, so these two clusters are joined, as in Figure 4.31b. The intuition behind this joining criterion may not be immediately obvious, but is does make sense: finding and joining the cluster pair with the smallest maximum distance between their members creates a cluster with the smallest diameter at that stage in the clustering procedure, and therefore the most compact cluster. Complete Linkage is also commonly known as Furthest Neighbour clustering, again for self-evident reason.

– The Centroid Linkage criterion defines the degree of closeness between any pair of clusters (X,Y) as the distance between their centroids, as in Equation (4.8):

$$CentroidDistance(X,Y) = dist(centroid(X), centroid(Y)) \quad (4.8)$$

where *dist* is defined as above. The centroid distances between all unique pairs of the p vectors remaining to be clustered are calculated,and the pair for which the distance is smallest is joined.
The centroid distances for all unique pairings of the p clusters are calculated using the proximity matrix, and the pair for which $centroiddistance(A,B)$ is smallest is joined. This is exemplified for the three clusters of Figure 4.29 in Figure 4.32.

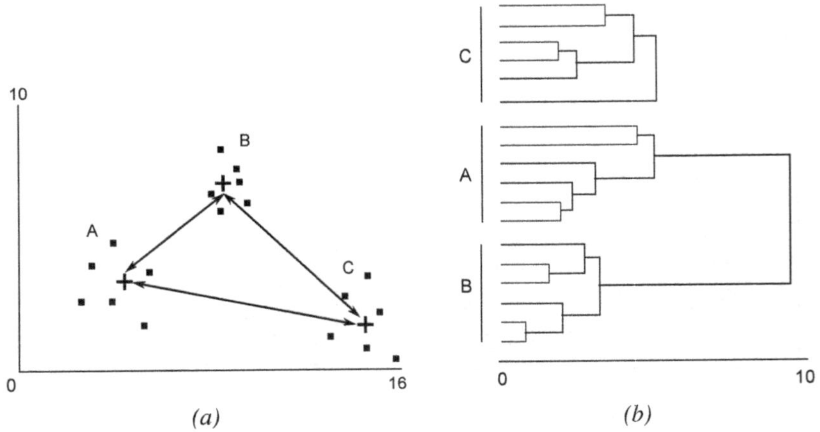

Figure 4.32: Centroid Linkage

The arrowed lines in 4.32a represent distances between centroids in cluster pairs (A,B), (A,C), and (B,C), which are shown as crosses; the one between A and B is shortest, so these two clusters are joined, as in Figure 4.32b.

- The Average Linkage criterion defines the degree of closeness between any pair of clusters (X,Y) as the mean of the distances between all ordered pairs of objects in the two different clusters: if X contains x objects and Y contains y objects, this is the mean of the sum of distances (X_i,Y_j) (where $X_i \in X, Y_j \in Y, i = 1\ldots x, j = 1\ldots y$, as in Equation (4.9):

$$AverageDistance(X,Y) = \frac{\sum_{i=1\ldots x, j=1\ldots y} dist(X_i,Y_j)}{xy} \qquad (4.9)$$

where *dist* is defined as previously; note that distances of objects to themselves are not counted in this calculation, and neither are symmetric ones on the grounds that the distance from, say X_i to Y_j is the same as the distance from Y_j to X_i.

- Increase in Sum-of-Squares Linkage (Ward's Method) defines the degree of closeness between any pair of clusters (X,Y) in terms of minimization of variability using an objective function. To describe it, two measures need to be defined. The error sum of squares (ESS) is the sum of squared deviations of the vectors in A from their centroid. If A contains m vectors, then ESS is defined by Equation (4.10):

$$ESS(A) = \sum_{i=1...m} (A_i - centroid(A))^2 \tag{4.10}$$

The total error sum of squares (TESS) of a set of p clusters is the sum of the error sum of squares of the p clusters. At each step in the tree-building sequence, the ESS for each of the p clusters available for joining at that step is calculated. For each unique combination of cluster pairs the increase in TESS is observed, and the pair which results in the smallest increase in TESS is joined.

Finally, hierarchical variants are standardly divided into agglomerative vs. divisive methods. Agglomerative tree construction was exemplified above: it begins by partitioning the set of data objects so that each member of the set is a cluster on its own, and then builds the tree incrementally by joining pairs of clusters at each step until no more pairs remain and the tree is complete. Because it incrementally builds trees of increasing complexity from simpler components, agglomerative clustering is also called bottom-up clustering. Divisive tree construction begins with a single cluster consisting of all the data objects, and builds the tree incrementally by partitioning that cluster into subtrees at each step until each cluster contains a single data object and no more division is possible, at which point the tree is complete; because it incrementally subdivides a set into subsets, divisive clustering is also known as top-down clustering. Divisive clustering, described in detail by Xu and Wunsch (2009: 37ff.), is less often used than agglomerative clustering – see: Everitt and Dunn (2001: 67ff.), Izenman (2008: 411), Everitt et al. (2011: 84ff.) – because, on the one hand, the latter is computationally more tractable with respect to large data matrices, and on the other because divisive methods are not always available in hierarchical clustering software implementations. For these reasons, the remainder of this section deals with agglomerative methods only.

Though the above tree-building sequence makes it obvious, it nevertheless seems worth making explicit the distinction between the tree generated by a hierarchical analysis and its graphical representation: the tree is the table of joining distances (c) in Table 4.9 (p. 208), more commonly known as the agglomeration schedule, and its representation in Table 4.9d is a dendrogram. The subcluster constituency shown by the dendrogram represents the order in which subclusters were joined in the tree-building sequence and the lengths of the lines joining subclusters represent the inter-cluster distances at which they were joined, as recorded in the agglomeration schedule.

Because a cluster tree represents constituency only, the sequential ordering of constituents has no interpretative significance. Given the tree in Table 4.9d, for example, any pair of constituents can be rotated about its axis, thereby reversing the sequencing, without affecting its constituency structure, as in Figure 4.33.

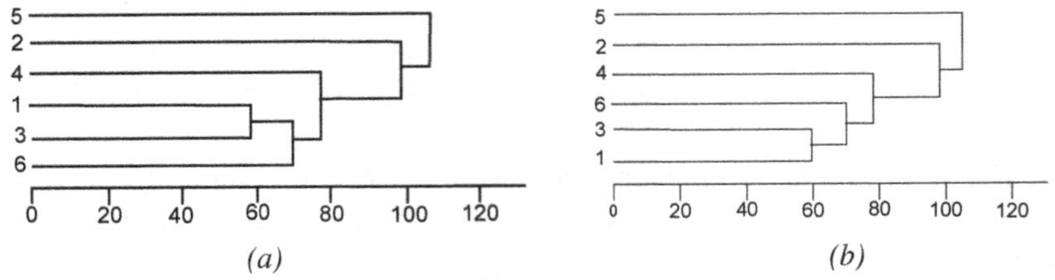

Figure 4.33: Different sequential orderings of the same constituency struture

In Figure 4.33b clusters (1) and (3) and clusters (1,3) and 6 are reversed without affecting the constituency structure; in fact, for any given constituency structure, there are 2^{n-1} trees with different sequencings of constituents (Everitt et al. 2011: Ch. 4). Algorithms for optimizing the sequencing of constituents relative to various definitions of optimality have been devised; see for example Everitt et al. (ibid.: Ch. 4).

Note also that dendrograms are more often shown sideways, than in the 'icicle' or downward-facing format more familiar to linguists from phrase structure trees. This is a purely practical matter: an icicle format rapidly broadens out as the number of data objects grows, making it impossible to display on a page.

4.2.2.3 Issues

The main and considerable advantage of hierarchical clustering is that it provides an exhaustive and intuitively accessible description of the proximity relations among data objects, and thereby provides more information that a simple partitioning of the data generated by the non-hierarchical methods covered thus far. It has also been extensively and successfully used in numerous applications, and is widely available in software implementations. There are, however, several associated problems.

- *How many clusters?* Given that a hierarchical cluster tree provides an exhaustive description of the proximity relations among data objects, how many clusters do the data 'really' contain? As already noted, it is up to the user to decide. Looking at a dendrogram like the one in Figure 4.34 the answer seems obvious: there are two clusters *A* and *B*; each of these itself has some internal cluster structure, but that structure is insignificant compared to the main A/B partition. This intuition is based on the relative lengths of the lines joining subclusters or, equivalently, on the relative values in the agglomeration schedule: unusually large intervals between successive merges is taken as an indication that the subclusters in question constitute 'main' clusters.

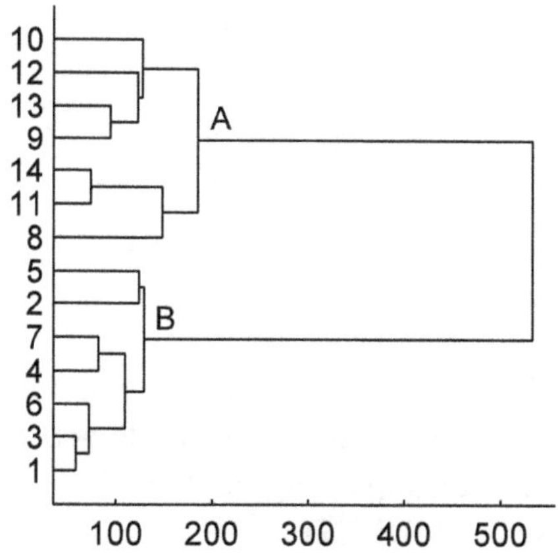

Figure 4.34: Hierarchical tree showing a clear two-cluster structure

What about a structure like the one in Figure 4.35, however? There are no obvious main clusters, and depending on the agglomeration level one selects, or, as it is usually expressed, where one cuts the tree, two, three, four or more clusters can be identified.

In Figure 4.35a the cut is placed so that subclusters below a threshold of 100 are not distinguished, yielding two clusters *A* and *B*. In 4.35b the cut is at 90, yielding three clusters $A-C$, and in 4.35c there are four clusters $A-D$ for a threshold of 73, and in 4.35d there are five clusters $A-E$. Which is the best cut, that is, the once that best captures the cluster structure of the data? There have been attempts to formalize selection of a best cut (ibid.: Ch. 4), but the results have been mixed, and the current position is that the best cut is the one that makes most sense to experts in the subject from which the data comes.

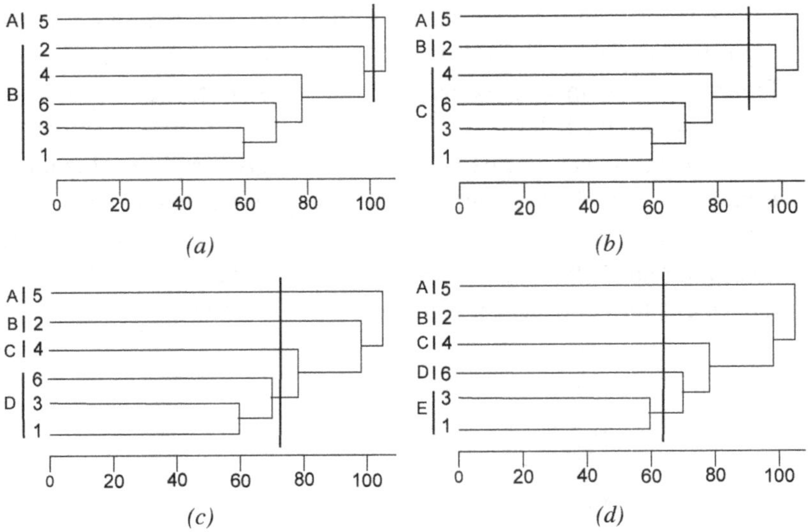

Figure 4.35: Different 'cuts' of the same dendrogram

– *Which tree?* It is universally recognized in the literature that different cluster joining criteria can and typically do generate different trees for the same data. For example, Figure 4.36 shows dendrograms for two different analyses of MDECTE.
Both trees show a two-cluster structure consisting of the Newcastle speakers 57–63 at the top of each, and the remaining ones, but the structuring of the latter differs greatly. This is hardly surprising. The various joining criteria articulate different views of how data points should be combined into clusters, and these views find their expression in different cluster trees relative to the same data. It does, however, raise two questions: given several hierarchical analyses of the same data generated by different joining criteria, which analysis should be preferred, and why?
The traditional answer is, again, that an expert in the domain from which the data was taken should select the analysis which seems most reasonable in terms of what s/he knows about the research area. The obvious objection to this is that it is subjective. It runs the risk of reinforcing preconceptions and discounting the unexpected and potentially productive insights which are the prime motivation for use of cluster analysis in hypothesis generation (Handl, Knowles, and Kell 2005);

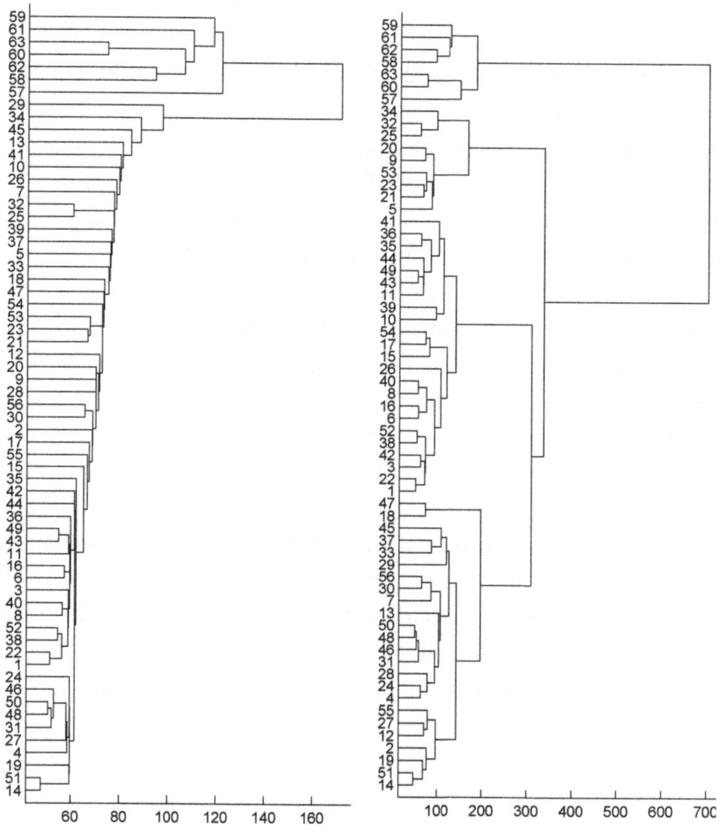

(*a*) Single Linkage (*b*) Ward's Method

Figure 4.36: Dendrograms for MDECTE generated by different linkage criteria

given a range of different analyses, one might well subconsciously look for what one wants to see. The aesthetics of tree structuring might also become a selection factor. Looking at the trees in Figure 4.36, for example, one might find the clear cluster structure of the Ward's Method tree more appealing than the uninformative chained structure of the Single Linkage one. Ultimately, of course, all human interpretation is subjective, but to be scientifically convincing that subjectivity needs to be constrained as much as possible.

One way of constraining tree selection is to observe that there is a fundamental difference between the Single Linkage criterion and the others listed above: the latter are distance-based clustering methods, and the former is a density-based one. Complete Linkage, Average Linkage, Centroid Linkage, and Ward's Method all build clusters on the basis of linear distance between data points and cluster centres relative to the coordinates of the metric space in which the data is embedded. The result is a tree each level of which is a Voronoi partition of the data space: starting at the root, the first level divides the space into two partitions each of which contains the data points closer to their centre than they are to the centre of the other partition, the second level divides the space into four such partitions, and so on. This partitioning is, moreover, done without reference to the density structure of the data. Single Linkage, on the other hand, builds clusters solely on the basis of local neighbourhood proximity and without reference to cluster centres; it is to the other kinds of hierarchical clustering, therefore, as Dbscanis to *k*-means. As such, the expectation is that the non-Single Linkage group will, like *k*-means, correctly identify the cluster structure of data when its dense regions are linearly separable but not otherwise, whereas Single Linkage will be able to identify non-linearly separable clusters. Figure 4.37 shows a scatterplot of linearly separable clusters taken from the earlier discussions of *k*-means and Dbscan.

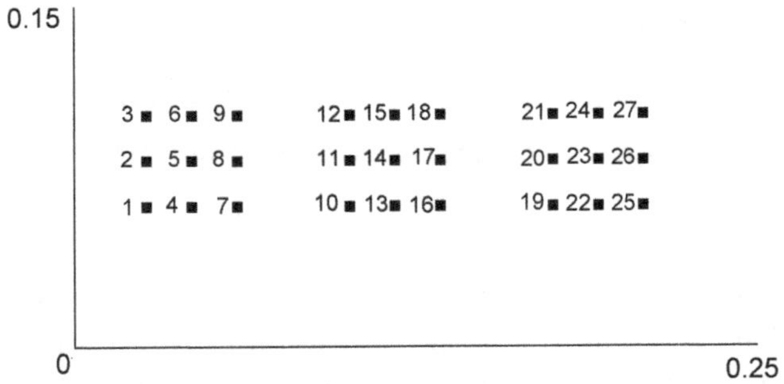

Figure 4.37: Scatterplot of two-dimensional data showing three linearly separable clusters

These clusters are correctly identified by both Single Linkage and Average Linkage clusters in Figure 4.38, where the latter is used as a representative for the non-Single-Linkage varieties.

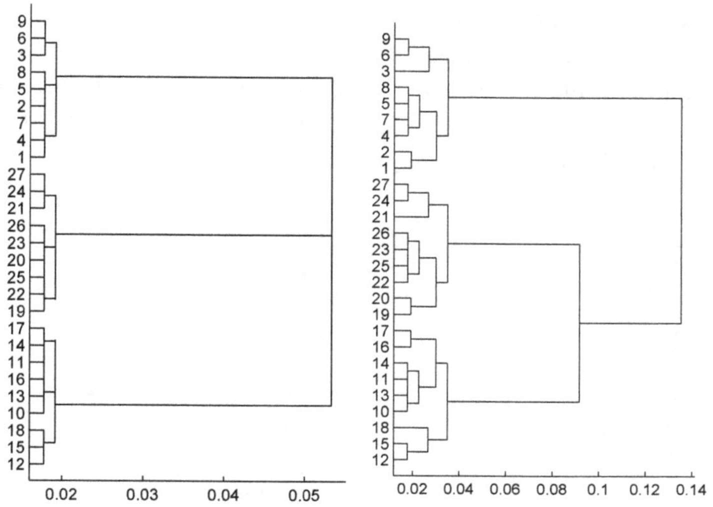

(a) Single Linkage (b) Average Linkage

Figure 4.38: Single linkage and average linkage dendrograms for the data underlying Figure 4.37

For the two-dimensional data underlying the scatterplot in Figure 4.39, however, the Average Linkage cluster tree in Figure 4.40 partitions the data points exactly as *k*-means did for this data earlier on in Figure 4.18c, cutting across the data density structure, whereas the Single Linkage tree identifies the data density structure just as Dbscan did in Figure 4.22c (p. 195).

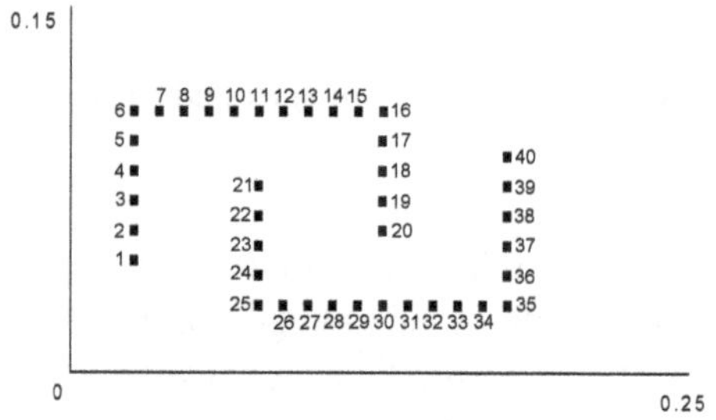

Figure 4.39: Scatterplot of two-dimensional data showing two non-linearly separable clusters

This difference between Single Linkage clustering and the others underlies the commonly-made observation in the literature that the non-Single Linkage criteria have a strong predisposition to find roughly spherical clusters in data even where clusters of that shape are known not to be present or indeed where the data are known not to have any meaningful cluster structure at all; see for example Dalton, Ballarin, and Brun (2009), Everitt et al. (2011: Ch. 4), whereas Single Linkage can identify clusters of arbitrary shape.

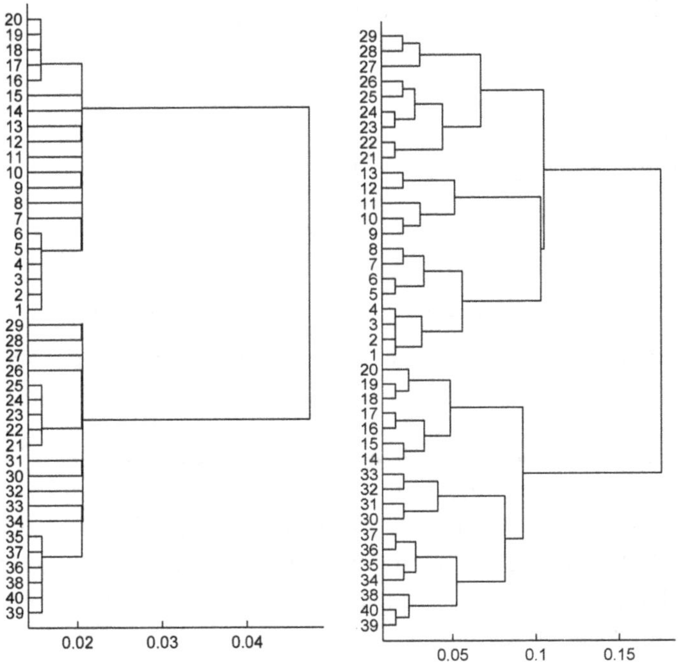

(a) Single Linkage *(b)* Average Linkage

Figure 4.40: Single linkage and average linkage dendrograms for the data underlying Figure 4.39

The ability of Single Linkage to identify a superset of the cluster structures identifiable by the other linkage methods implies that it is more likely than the others correctly to identify the structure latent in any given data, and that it is therefore the most authoritative of the linkage methods: in principle, no matter how aesthetically attractive a non-Single Linkage tree might be, or how much it accords with expert expectation, if it differs substantively from the Single Linkage tree, then the suspicion must be that the difference is an artefact of the linkage criterion rather than a reflection of intrinsic cluster structure. In practice there is a caveat, however. Single Linkage does not make a distinction between cluster structure and noise in data, and this can generate spurious structures; more is said about the effect of noise below. When the given data is known to have a non-linearly separable cluster structure and not to contain much or any noise, Single Linkage is authoritative. Where this is not known, selection of the best cluster tree, that is, the one which best captures the intrinsic cluster structure of the data, must be guided by the cluster validation methods discussed later in this chapter.

- *Outliers and noise.* All the joining criteria are affected by outliers and noise to different degrees and for different reasons; see for example Manning, Raghavan, and Schütze (2008: 350ff.). Outliers are not a problem for Single Linkage because it simply represents them as one-member clusters distant from the other clusters in the tree; this characteristic in fact makes Single Linkage clustering a good way to check for outliers. It is, however, much affected by noise, which results in chaining like that in Figure 4.36a. Noise is less of a problem for the remaining criteria but outliers are more of one: because Complete Linkage, Centroid Linkage, Average Linkage, and Ward's Method are all based in one way or another on the notion of a cluster having a centre of gravity, an outlier artificially pulls that centre away from its natural location among the other data points, thereby affecting calculation of the centre and consequently distorting the structure of the tree. The effect of outliers in particular is seen as instability in the cluster structure. A stable clustering is one in which removal of a small number of data objects, or addition of a small number drawn from the same source as the original data, has a correspondingly small effect on the structure of a cluster tree; removal or addition or one or more outliers has a disproportionately large effect, and can often completely change the structure. When using hierarchical analysis, therefore, it is important to identify outliers and to eliminate them or at least to be aware of their presence when interpreting results.
- *Computational complexity.* In the general case, the time complexity of hierarchical agglomerative methods is $O(n^3)$, where n is the number of data objects, and even in optimized cases it is $O(n^2 log(n))$ except for Single Linkage, where further optimization to $O(n^2)$ is possible (Jain, Murty, and Flynn 1999), (Manning, Raghavan, and Schütze 2008: p.353ff). Even with the optimizations, however, this level of

computational complexity limits the ability of hierarchical methods to process large data sets to the extent that, in data mining where data sets grow ever larger, they are now considered obsolete by some researchers (Tan, Steinbach, and Kumar 2006). Size is a relative matter, however, and this might not be a problem for corpus linguistic applications.

4.2.2.4 Developments

Hierarchical methods are not parameterized, and so initialization is not a problem, but cluster shape, data size, noise, and outliers are.

- *Cluster shape.* As noted, all the hierarchical methods except Single Linkage are predisposed to find linearly separable clusters in data even where the data do not actually contain that structure or indeed any cluster structure at all. The obvious solution is to use Single Linkage instead, but as we have seen this has a predisposition to generate uninformative chained structures when the data contains noise. Ideally, one would want a hierarchical method which is not limited to linearly separable structures on the one hand and not compromised by chaining on the other. CURE (Guha, Rastogi, and Shim 1998) and CHAMELEON (Karypis, Han, and Kumar 1999) are two such methods: both are density rather than proximity-based, both can find a greater range of cluster structures than the standard methods described above, and both are able to deal with substantial differences in cluster size and density, which can also be problems for the standard methods. For discussions of CURE and CHAMELEON see Berkhin (2006), Ertöz, Steinbach, and Kumar (2003), Tan, Steinbach, and Kumar (2006).
- *Data size.* The computational complexity of the standard agglomerative methods limits their ability to scale up to ever-larger data. CURE and CHAMELEON are computationally less demanding; BIRCH (Zhang, Ramakrishnan, and Livny 1996) was developed specifically to deal with large data sets, and can achieve a time complexity of $O(n)$, that is, the time required grows linearly with the number of data objects.
- *Noise and outliers.* Noise and outliers can adversely affect all the standard methods in the ways described above. Ideally, a clustering method would be resistant to the effect of outliers and noise, thereby offering stable clustering; all three methods mentioned above provide this to varying degrees.

- *Linear separability.* The non-single linkage hierarchical methods are limited to data with a linearly-separable density structure because the Minkowski distance measures that the literature associates with them are linear. These methods neither know nor care how the values in the distance matrices on the basis of which they construct cluster trees were derived, and there is no obstacle in principle to using values generated by a nonlinear metric like the geodesic one described in the discussion of data geometry. This allows non-linearly separable regions to be separated nonlinearly and removes the linear separability limitation on non-single linkage hierarchical methods.

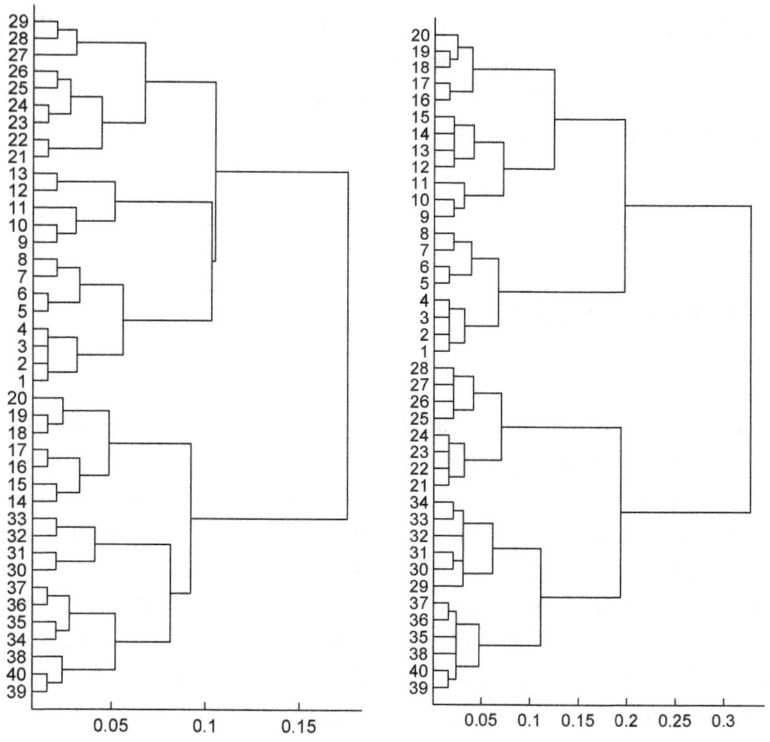

(a) Average Linkage, Euclidean distance

(b) Average Linkage, geodesic distance

Figure 4.41: Average Linkage hierarchical clustering applied to non-linearly separable data using Euclidean and geodesic distance measures

Compare, for example, the Average Linkage trees for Euclidean and geodesic distance measures in Figure 4.41.
The Euclidean distance-based tree in Figure 4.41a shows the effect of the linear separability constraint already observed in the foregoing discussion, but the geodesic distance-based one in 4.41a overcomes this by identifying clusters compatible with the density structure of the data, like Single Linkage does in Figure 4.41a.
As usual, however, there is a caveat. The foregoing discussion of nonlinearity detection pointed out a potential disadvantage of the graph approximation to geodesic distance measurement: that it does not make the distinction between model and noise which the regression-based approach makes, and treats the data matrix as a faithful representation of the domain from which the data was abstracted. Because the graph distance-based approach includes noise, whether random or systematic, in its calculations, this may or may not be a problem in relation to the application in question.

4.3 Cluster validation

The fundamental purpose of a cluster analysis method is to identify structure that might be present in data, that is, any non-regular or non-random distribution of points in the n-dimensional space of the data. It is, however, a commonplace of the cluster analysis literature that no currently available method is guaranteed to provide this with respect to data in general, and the foregoing discussion of a selection of methods confirms this: projection methods based on dimensionality reduction can lose too much information to be reliable, and the linear ones together with k-means and linear hierarchical methods fail to take account of any nonlinearity in the data; the reliability of the SOM, k-means, and Dbscan depends on correct parameterization; different hierarchical joining criteria can assign data points to different clusters and typically impose different constituency structures on the clusters. In addition, some methods impose characteristic cluster distributions on data even when the data are known not to contain such distributions or indeed to have any cluster structure at all.

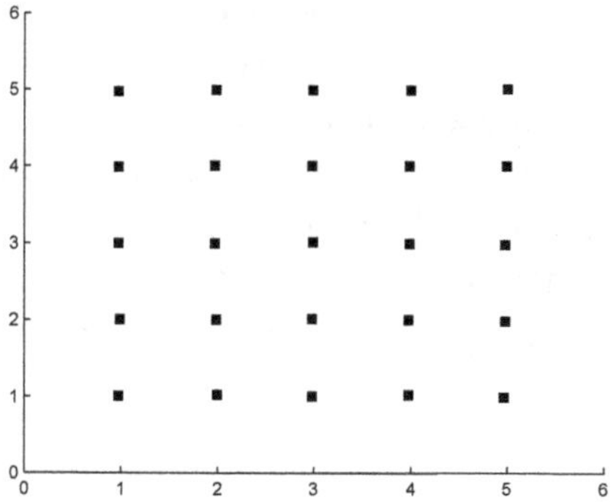

Figure 4.42: Regular distribution

This is shown in Figure 4.43 for the regular distribution in Figure 4.42, where the distance between each point and its immediate neighbours is constant.

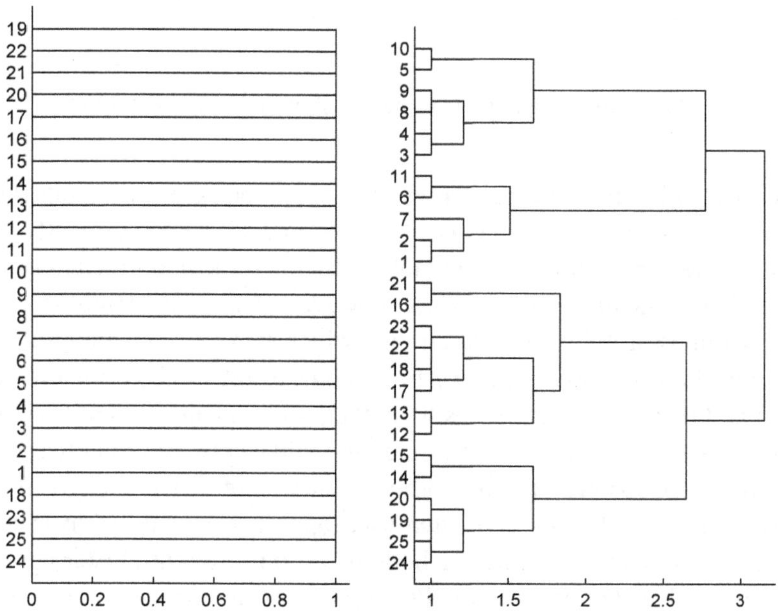

(a) Single linkage *(b)* Average Linkage

Figure 4.43: Hierarchical analyses of the regular distribution in Figure 4.42

Except for the limiting case where each point is regarded as a cluster on its own, the distribution in Figure 4.42 has no meaningful cluster structure, and the shape of the Single Linkage analysis reflects this. The Average Linkage tree, however, shows a rich and complex structure, and Complete Linkage and Ward's Method generate almost identical trees; if the data were higher dimensional and thus not amenable to graphical confirmation, these latter results would be very misleading. k-means gives similarly misleading results, as Figure 4.44 shows.

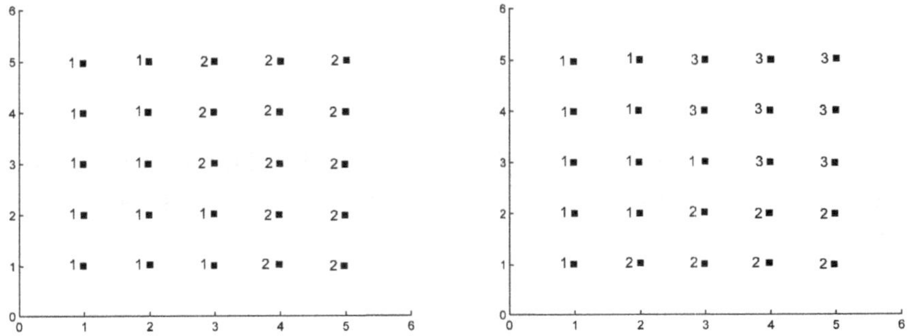

(a) $k = 2$ (b) $k = 3$

Figure 4.44: Two k-means analyses of the distribution in Figure 4.42

It is not difficult to understand why these methods give the results they do: all are based on the notion of cluster centres, and all are consequently predisposed to find convex linearly-separable clusters, as the literature has often observed. Understanding why does not, however, change the demonstrated fact that, by imposing this structure on data which lacks any meaningful cluster structure, these methods can give misleading results. Where the data distribution is directly visualizable, as here, the problem can be identified, but for higher-dimensional data, which cannot, it could easily pass unnoticed.

Given these sources of unreliability, validation of cluster analytical results is required. One obvious and often-used approach to validation is to generate a series of results using methods based on different clustering criteria in the hope that they will mutually support one another and converge on a consistent solution: if a range of methods based on dimensionality reduction, topology preservation, proximity, and density give identical or at least compatible results, the intuition is that the reliability of the solution is supported by consensus. It would, however, be useful to supplement such a consensus with one or more alternative validating criteria. And, of course, there might not be a consensus, in which case a selection must be made, which implies selection criteria. This section presents a range of quantitative ones.

The discussion is in two parts. The first part considers ways of determining the degree to which any given data has a cluster structure prior to application of clustering methods, known in the literature as 'clustering tendency'. The motivation here is the observation that, if data is known to contain little or no cluster structure, then there is no point to attempting to analyze it, and, if an analysis is carried out despite this, then the result must be an artefact of the method. The second part then presents a range of validation criteria for results from application of different analytical methods to data known to contain cluster structure.

4.3.1 Clustering tendency

Figure 4.45a shows a scatterplot 100 2-dimensional vectors, and 4.45b a Ward's Method hierarchical analysis of them, using squared Euclidean distance.

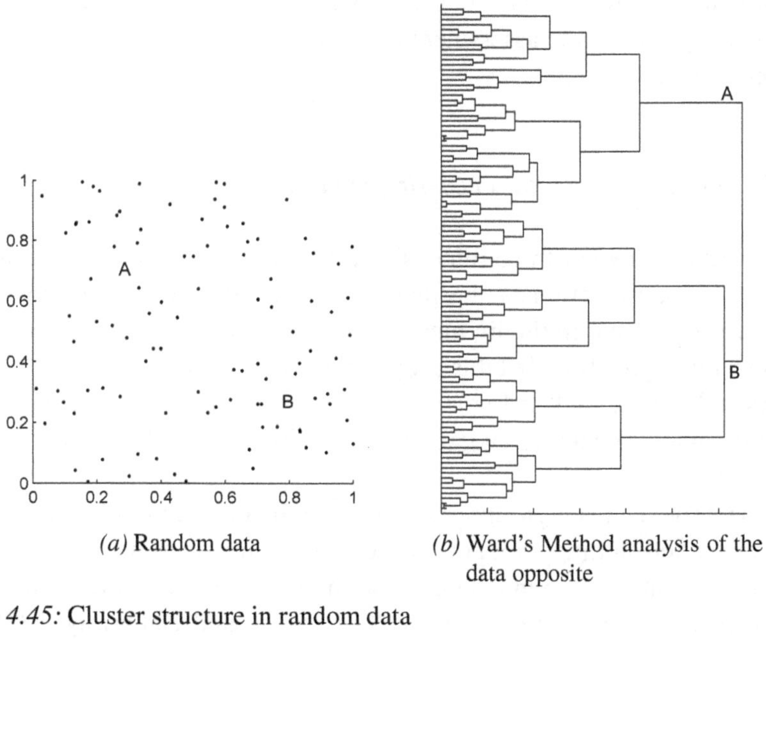

(a) Random data

(b) Ward's Method analysis of the data opposite

Figure 4.45: Cluster structure in random data

The eye picks out regularities in the scatterplot – there appear, for example, to be two dense regions in the upper left and lower right triangles marked *A* and *B*, separated by an diagonal low-density area – and the tree confirms the intuition: there are two well-separated clusters, and each of these has its own internal structure which, if one looks hard enough, can be seen as variations of point density in the scatterplot. However, the data were produced by a random number generator, and so one knows a priori that they are random. The obvious conclusion is that that the fallibility of subjective visual interpretation of graphical information has once again been demonstrated, and that the tree is an artefact of the clustering method. On the other hand, one might observe that finite sets of random numbers are typically not uniformly distributed but rather have local variations in density (Chaithin 2001), as is readily observable in the sequence of heads and tails as a coin is flipped, and argue that, the randomness of the data generator in the present case notwithstanding, it is a matter of empirical fact that there is observable structure in the data which the cluster tree has captured. Which interpretation is correct? Research into ways of answering the question is known as 'assessment of clustering tendency' (Jain and Dubes 1988: 201ff.), the aim of which is essentially to decide whether data are sufficiently nonrandom to merit investigation of the hypothesis that it contains meaningful cluster structure, though without attempting to identify whatever cluster structure might be present. There are two main approaches to assessment of clustering tendency, one graphical and the other statistical.

4.3.1.1 Graphical tests for clustering tendency

Where data are two or three-dimensional they can be scatter-plotted directly, and visual interpretation of the plot will reveal the presence or absence of structure. It is usual in the literature to express distrust in the subjectivity of graphical interpretation, but this subjectivity is a matter of degree. It would be unreasonable to deny that Figure 4.46a demonstrates the presence of cluster structure in the data underlying it, whereas 4.46b, repeated from 4.45, is more equivocal.

Where the data is high-dimensional it can be reduced to dimensionality 2 or 3 for scatter-plotting, but, depending on the intrinsic dimensionality of the data, this might lose too much information to provide a reliable result. Visual Assessment of cluster Tendency (VAT) (Bezdek and Hathaway 2002) pro-

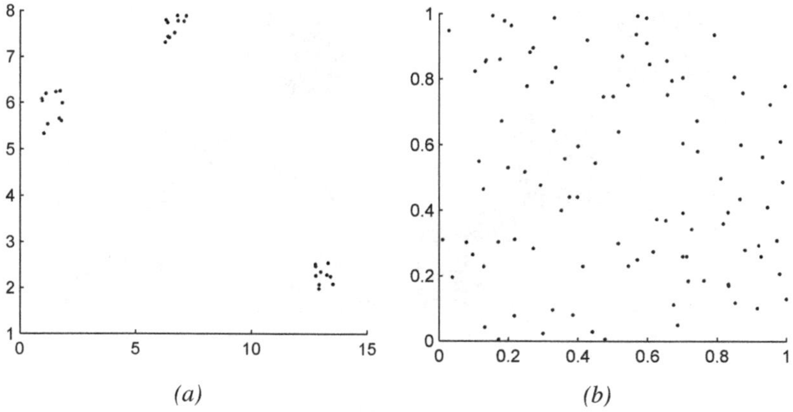

Figure 4.46: Visual identification of cluster structure

vides an alternative method for graphical representation of structure which can be applied to numerical data of any dimensionality. Given an *m* x *n* data matrix *M*, where *m* is the number of data objects and *n* the dimensionality, VAT is based on an *n* x *n* distance matrix *D* abstracted from *M*; any distance measure can be used, but Euclidean is assumed. The entries in *D* are rearranged so that the closer any two objects are to one another in the data space, the more spatially adjacent they are to one another in *D*. *D* is then represented as an *n* x *n* two-dimensional grid *G* such that each grid cell G_{ij} represents the value in D_{ij} as a grey-scale colouring, where a value of 0 in *D* is represented in *G* as black, the maximum distance value as white, and points in between as regularly spaced intervals of grey shading. Data points which are close to one another appear as dark blocks on the main diagonal surrounded by lighter-coloured cells, and represent clusters. Figure 4.47 shows VAT graphs corresponding to the scatterplots 4.46.

The very distinct clusters of Figure 4.46a show as black blocks in 4.47a, and the far less distinct clustering of 4.46b is reflected in the less easily interpretable blocking of 4.47b. In every VAT image there is a line of single-cell dark blocks in the main diagonal reflecting the fact at every object is at distance-0 from itself, and for cluster identification this is without value; the larger blocks on the diagonal in 4.47b indicate some degree of clustering, but these are not nearly as distinct as those of 4.47a, reflecting the less definite clustering visible in 4.46b.

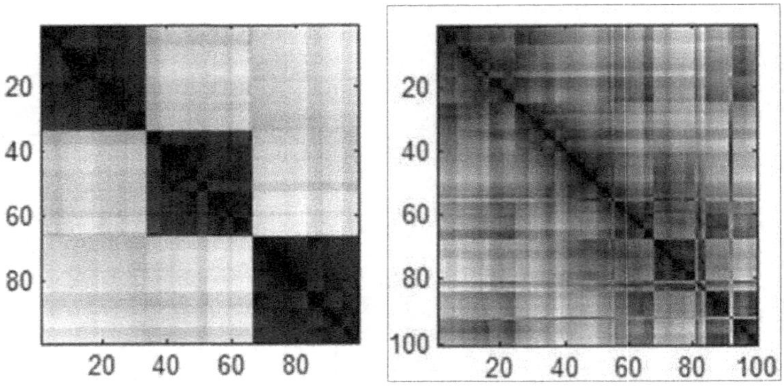

(a) (b)

Figure 4.47: VAT graphs of the data underlying Figure 4.46

Applied to the two-dimensional data on which the Figure 4.46 plots are based, the VAT graphs of Figure 4.47 take us no further forward. The utility of VAT comes with higher dimensionality. Figure 4.48 shows the VAT graph for MDECTE, where a small block at the upper left and a much larger block on the lower right indicate two main clusters, and the larger block contains some evidence of subclustering.

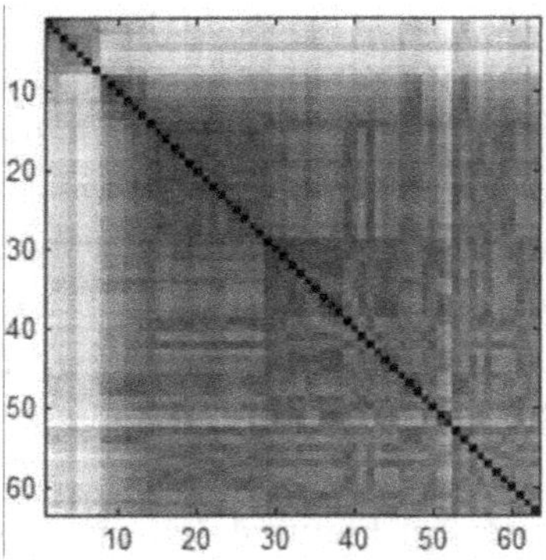

Figure 4.48: VAT graph of MDECTE

Specifics of the algorithm used to rearrange the distance matrix are given in (Bezdek and Hathaway 2002), and enhancements to VAT in (Bezdek, Hathaway, and Huband 2006), (Hu and Hathaway 2008), (Wang et al. 2009), (Havens and Bezdek 2012), (Hu 2012).

4.3.1.2 Statistical tests for clustering tendency

Statistical identification of clustering tendency in data is based on testing the null hypothesis that the data are random and therefore have no cluster structure. This testing is done by comparing the data to what they would look like if they were random relative to some criterion. If the criterion indicates that the data deviate sufficiently from randomness, then the null hypothesis is falsified and the alternative hypothesis, that the data have cluster structure, is adopted with some specified degree of confidence. Otherwise, the null hypothesis stands and the conclusion is that the data very probably contain no cluster structure.

There are several ways of defining randomness in data (Jain and Dubes 1988: 144, 186ff. Gordon 1999). Relative to the geometric view of data on which the discussion has thus far been based, randomness is understood as the positioning of vectors in an n-dimensional space by a Poisson process, introduced earlier in the discussion. The dataset is compared to what it would have looked like if its had been generated by a Poisson process, and the comparison is based on the random position hypothesis, which says that data containing m points in n-dimensional space are random if all sets of $k < m$ points in the space are equally likely to occur (Jain and Dubes 1988: 144f., 202ff.). To test for spatial randomness a statistic is derived from the data using some structure-sensitive measure, and if that statistic is sufficiently different from its value for random data, then the null hypothesis is rejected and the conclusion is that the data are nonrandom and therefore clustered in some way. Various such structure-sensitive tests exist – see Jain and Dubes (ibid.: 211ff.), Gordon (1999: 188ff.), Everitt et al. (2011: Ch. 9.3). The one presented here, the Hopkins statistic, is selected both on account of its intuitive accessibility and demonstrated effectiveness (Jain and Dubes 1988: 218; Lawson and Jurs 1990; Banerjee and Dave 2004; Tan, Steinbach, and Kumar 2006: 547f.).

Relative to some m x n matrix M, where m is the number of data objects and n the dimensionality, the Hopkins statistic determines whether the distribution of the m objects in the n-dimensional space deviates significantly from a random distribution of objects in that space. It does this on the basis of two measurements. The first measurement randomly selects $k < m$ row vectors of M, and for each of these k vectors the distance to one or more of its nearest neighbours in M is calculated using some distance metric; the k nearest-neighbour distances are then summed and designated p. The second measurement generates k vectors randomly distributed in the data space, and the nearest-neighbour distances between each of these k vectors to the row vectors of M are calculated and summed, and designated q. The Hopkins statistic H is then defined by Equation (4.11):

$$H = \frac{q}{p+q} \tag{4.11}$$

H is calculated multiple times and summed, each time using a new random selection of k data vectors and a new set of k random vectors, and the mean of the sum is taken to be an indication of how far the distribution is from randomness. The reasoning here is as follows. If the data distribution is random then p and q will, on average, be the same, and the above formula gives 0.5; this is the baseline value for random data. If the data distribution is not random then p will on average be smaller than q because the nearest-neighbour distances between clustered points in a data space are on average smaller than between randomly-distributed ones. In such a case the value of the $(p+q)$ term is smaller than when $p = q$, which gives increasingly larger H-values as p approaches 0 up to a maximum $H = 1$.

The mean value of H was calculated for both the distributions of Figure 4.46 and for MDECTE, in each case summing 50 H-values: for 4.46a $Hmean = 0.95$, for 4.46b $Hmean = 0.56$, and for MDECTE $Hmean = 0.87$. The H-value for 4.46a confirms the graphical impression of strong clustering, the H-value for 4.46b disambiguates the plot and the VAT graph, confirming that the data is indeed very nearly random, and the H-value for MDECTE indicates substantial non-randomness.

For general discussions of clustering tendency see Jain and Dubes (1988: Ch. 4.6), Gordon (1999: Ch.7), Tan, Steinbach, and Kumar (2006: Ch. 8.5.6), Everitt et al. (2011: Ch. 9.3).

4.3.2 Validation

We have seen that clusters can have a variety of shapes, and that clustering methods do not cope equally well with identifying different varieties. Because each method has different criteria for identifying clusters, each has a predisposition to find certain kinds of structure, and this predisposition can prevent it from identifying other kinds of structure or cause it to impose the structure it is designed to find on data (Gordon 1998, 1999; Handl, Knowles, and Kell 2005). For example, k-means builds clusters on the basis of distance from specified centers and so has a strong predisposition to find spherical clusters; where the data's intrinsic cluster structure is spherical, or at least roughly so, this is exactly what is required, but where not *k*-means gives a distorted representation of the intrinsic structure. Single Linkage hierarchical clustering, on the other hand, is good at finding irregularly-shaped clusters but obscures sphericity with chaining where these are not well separated or where the data is noisy. And so on. In general, therefore, a good match between intrinsic data structure and clustering method yields good insight into the data, and a poor match is misleading (Jain 2010). The intrinsic structure of data is, however, usually not known in advance. The aim of cluster analysis, after all, is to identify it, and if it were known there would be no point to an analysis. As such, there is no obvious a priori way of matching the given data to the appropriate clustering method, and consequently no a priori way of knowing how well a clustering result represents the intrinsic cluster structure.

One possible response to this is that it doesn't matter when, as here, cluster analysis is used as a tool for hypothesis generation, because the aim is to stimulate ideas which lead to hypotheses about the research domain from which the clustered data was abstracted, and a poor hypothesis based on a poor analysis can be falsified and eliminated in subsequent hypothesis testing. There is no obvious objection to this in principle, but in practice it is inefficient. A more efficient alternative is to base hypothesizing on clustering results in whose accuracy one can have a reasonable degree of confidence. Cluster analysts have increasingly taken the latter view and have developed a general methodology for providing that confidence, whereby a given data set is multiply analyzed using different clustering methods and different parameterizations of each method, and the result which best represents its intrinsic cluster structure is selected. The problem, of course, is knowing which is best; this section outlines the current state of cluster validation strategies and methods designed for that purpose.

A traditional and commonly-used criterion for cluster validation is domain knowledge, whereby a subject expert selects the analysis which seems most reasonable in terms of what s/he knows about the research area. This has some validity, but it is also highly subjective and runs the risk of reinforcing preconceptions and discounting the unexpected and potentially productive insights which are the prime motivation for use of cluster analysis in hypothesis generation, as already noted. Domain knowledge needs to be supported by objective criteria for what counts as 'best'; the remainder of this section outlines such criteria.

The foregoing account of clustering tendency has already dealt with part of the problem of cluster validation: that the given data are random, and that any clustering result is therefore an artefact of the analytical method used. The discussion to follow assumes that the result to be validated is based on data known to be sufficiently non-random to be meaningfully clustered, and that the aim of validation is to determine how close it has come to identifying the non-random structure, that is, to finding the correct number of clusters on the one hand, and to assigning the correct data objects to each of the clusters on the other. The literature contains numerous methods for doing this. Decades ago (Milligan and Cooper 1985) assessed the performance of no fewer than 30 of them, and many have been proposed since, so there is no hope of being able to cover them comprehensively, and a selection strategy is required. The literature distinguishes a class of "external" methods which assume the existence of a priori reliable information about what the structure found by cluster analysis should look like, and which thereby provides an independent criterion relative to which a clustering can be assessed. When external information is available the sensible course is to use it. However, where cluster analysis is used for hypothesis generation, as here, the implication is that it is so used because no clear prior information about the structure of the data is available. As such, the external approach to cluster validation is only tangentially relevant here, and nothing more is said about it; for further information see (Gan, Ma, and Wu 2007; Halkidi, Batistakis, and Vazirgiannis 2001; Halkidi, Vazirgiannis, and Batistakis 2000; Jain and Dubes 1988: Ch. 4). What remains is a still-sizable number of validation methods. These are categorized by the approach they take to validation and exemplified with a small selection of methods. The result is an introduction to a complicated topic, and the further readings cited in the course of discussion are an indispensable adjunct in any serious research application.

Note that the methods described in what follows apply only to so-called 'crisp' clustering, where each data object is assigned to one and only one cluster, and not to fuzzy clustering, where a data object may belong to one or more cluster; fuzzy clustering (Gan, Ma, and Wu 2007: Ch. 8) is not covered in this book.

4.3.2.1 Cluster validation with respect to cluster compactness and separation

The validation methods in this category assess a given clustering in terms of how well it adheres to the proximity-based cluster definition with which this chapter began: a cluster is an aggregation of points in the test space such that the distance between any two points in the cluster is less than the distance between any point in the cluster and any point not in it. They are therefore applicable only when the underlying data density structure is linearly separable. The two criteria most often used are compactness and separation – cf., for example, Halkidi, Batistakis, and Vazirgiannis (2001), Handl, Knowles, and Kell (2005), and Liu et al. (2010) .

- *Compactness.* Compactness is the degree to which the members of a proposed cluster are similar to one another. The *Root Mean Square Standard Deviation* (RMSSTD) validity index – cf. Sharma (1996), Halkidi, Batistakis, and Vazirgiannis (2001), Gan, Ma, and Wu (2007: Ch. 17.2.6) – measures cluster compactness as the degree of variability among the component vectors of a cluster, and more specifically as the mean deviation of the members of the cluster from the cluster centroid expressed as a standard deviation: the smaller the deviation, the more compact the cluster. Relative to a cluster C containing k n-dimensional vectors, the sum of squared deviations (SS) from the cluster centroid is first calculated using Equation (4.12):

$$SS = \sum_{i=1...k} \sum_{j=1...n} (C_{i,j} - centroid_j)^2 \tag{4.12}$$

  The RMSSTD is then the square root of the mean SS, as in Equation (4.13):

$$RMSSTD = \sqrt{\frac{SS}{kn}} \tag{4.13}$$

  Figure 4.49 exemplifies the application of this measure to three clusters of varying visual compactness.

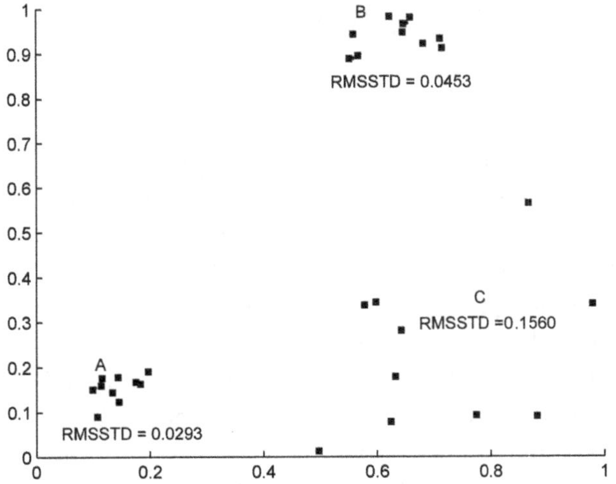

Figure 4.49: The Root Mean Square Standard Deviation (RMSSTD) for clusters of varying compactness

Cluster *B* is slightly less compact than *A* and its RMSSTD is correspondingly slightly larger, whereas the far less compact cluster *C* has a much larger RMSSTD than either *A* or *B*. Because the index is expressed as a standard deviation and is thus a value in the range $0 \ldots 1$, moreover, its significance is readily interpretable.

– *Separation.* Separation is the degree to which compact clusters are distant from one another, and *RS* – cf. Sharma (1996), Halkidi, Batistakis, and Vazirgiannis (2001), Gan, Ma, and Wu (2007: Ch. 17.2.7) – is a measure of separation. It is based on the sum of squared differences (*SS*) as described in the discussion of RMSSTD. Assume a data matrix *M* in which the *m* rows have been partitioned into *k* clusters $c_1, c_2 \ldots c_k$. Three types of *SS* are distinguished. Firstly, the *SS* for each of the clusters c_j is calculated and the *k SS* values are summed to yield *SSw*, where '*w*' stands for 'within cluster'. Secondly, the *SS* for the whole matrix *M* is calculated to give *SSt*, where '*t*' stands for 'total data'. Finally, the between-cluster sum of squared differences *SSb* is calculated by observing that $SSt = SSw + SSb$, so that $SSt - SSw = SSb$. On this formulation, the smaller the value of *SSw* the larger the *SSb*. But from the discussion of RMSSTD it follows that the smaller the *SSw* the more compact the clusters. The larger the *SSb*, therefore, the more compact the clusters and the larger the separation between them. *SSb* is thus a measure of cluster separation. *RS* expresses this measure as a ratio of SSb/SSt to give a readily interpretable range $0 \ldots 1$, so that larger *RS* values in this range indicate greater cluster separation and smaller ones lesser separation. This is exemplified in Figure 4.50. The clusters in 4.50a are more compact and better separated than those in 4.50b; the *SSb* and the SSb/SSt ratios are commensurately larger for 4.50a than for 4.50b.

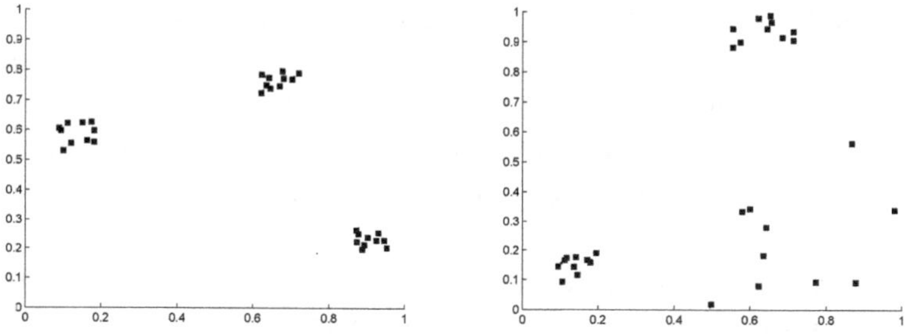

(a) (b)

(c)

$SS_w(A)$: 0.0329
$SS_w(B)$: 0.0262
$SS_w(C)$: 0.0232
$SS_w(A,B,C) = 0.0329 + 0.0262 + 0.0232 = 0.0823$
SS_t: 0.2780
$SS_b = 0.2780 - 0.0823 = 0.1957$
$SS_b / SS_t = 0.1957 / 0.2780 = 0.7040$

(d)

$SS_w(A)$: 0.0293
$SS_w(B)$: 0.0453
$SS_w(C)$: 0.1560
$SS_w(A,B,C) = 0.0293 + 0.0453 + 0.1560 = 0.2306$
SS_t: 0.3209
$SS_b = 0.3209 - 0.2306 = 0.0903$
$SS_b / SS_t = 0.0903 / 0.3209 = 0.2814$

Figure 4.50: Measures of separation between clusters

Most of the validation methods in this category combine compactness and separation measures into a single index. A long-established one still much used today is Dunn's index (Dunn 1974), the essential idea of which is that the ratio of the smallest distance between any two clusters to the size of the largest cluster in any clustering is a useful measure of its compactness and separation. Given a data matrix *D* of *m* *n*-dimensional row vectors and a partition of the D_i into a set of *k* clusters $C = c_1, c_2 \ldots c_k$, Dunn's index is defined as in Equation (4.14):

$$Dunn(C) = \frac{min_{i,j=1\ldots k, i \neq j}(dist(c_i, c_j))}{max_{i=1\ldots k}(size(c_i))} \tag{4.14}$$

where *dist* measures the cluster separation and size the compactness. The value of $Dunn(C)$ depends on the selection of measures used to calculate these. For *dist* the various linkage methods discussed in relation to hierarchical analysis, such as Single and Complete Linkage, can be used, and the size of a cluster can be calculated using, for example, average distance among cluster members or average distance to the cluster centroid (Brun et al. 2007). The essence of the measure remains the same across these specifics, however. The smaller the distance between clusters and the larger the size of the largest cluster, the less compact and well-separated the clusters are and the smaller the value of $Dunn(C)$. Conversely, the larger the distance between clusters and the smaller the size of the largest cluster the more compact and well separated they are, and the larger the value of $Dunn(C)$. Using average Euclidean distance between and within clusters to measure both cluster separation and compactness, the Dunn index value for the compact and well-separated clusters of Figure 4.50a is 4.804 and for the less compact and well separated ones of 4.50b is 1.103.

This index can be used to select the best from a set of candidate clusterings by applying it to each of the candidates and identifying the one with the highest index value. For example, MDECTE was k-means clustered for values of k in the range $2 \ldots 12$ and the Dunn index values calculated and plotted, as in Figure 4.51. This indicates that the optimum number of clusters is 2.

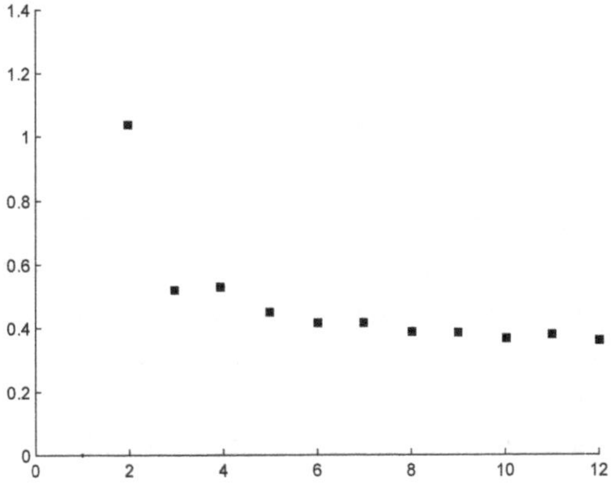

Figure 4.51: Dunn index values for k-means clustering of MDECTE, k in the range $2 \ldots 12$

Various modifications to Dunn's original formulation have been proposed. Bezdek and Pal (1998), for example, observed that it is highly sensitive to noise and outliers and can therefore produce misleading results in data where these are present, and proposed modifications to mitigate this. Where these are present or suspected of being present, therefore, the reformulation by Bezdek and Pal (ibid.) should be used. For further discussion of the Dunn index see Stein, Eissen, and Wissbrock (2003), Gan, Ma, and Wu (2007: Ch. 17.2.3).

Another traditional index that combines cluster compactness and separation is that proposed by Davies and Bouldin (1979), defined for the above D and C as in Equation (4.15):

$$DB(C) = \frac{\sum_{i,j=1...k, i \neq j} max(\frac{\delta_i + \delta_j}{distance(c_i, c_j)})}{k} \tag{4.15}$$

where δ_i and δ_j are the average distances of all vectors in clusters c_i and c_j from their respective cluster centroids, and *distance* is the distance between the centroids of c_i and c_j. The δ terms measure the compactness of two clusters in terms of the degree of dispersion of their members, and the Davies-Bouldin (DB) index is therefore the average cluster dispersion to cluster distance ratio. The greater the dispersion and the smaller the distance the less compact and well-separated the clusters, and the larger the value of $DB(C)$. And, conversely, the smaller the dispersion and the larger the distance the more compact and better-separated the clusters, and the smaller the index value. Applied to the data underlying the plots in Figure 4.50, the Davies-Bouldin index value for the compact and well-separated clusters of 4.50a is 0.048 and for the less compact and well separated ones of 4.50b is 0.145.

Like the Dunn index, *DB* can be used to select the best from a set of candidate clusterings by applying it to each of the candidates and identifying the one with the smallest index value. This was done for *k*-means clustering of the data underlying the scatterplots in Figure 4.50 as for Dunn, and the results are plotted in Figure 4.52; as before, the indication is that the optimum number of clusters is 2.

Methods similar to Dunn and Davies-Bouldin are reviewed in Halkidi, Batistakis, and Vazirgiannis (2001), Handl, Knowles, and Kell (2005), and Gan, Ma, and Wu (2007: Ch.17); more recent developments of the cluster compactness / separation approach are *SD* Halkidi, Vazirgiannis, and Batistakis (2000) and *CDbw* (Halkidi and Vazirgiannis 2008). Banerjee and Dave (2004) take a different approach by adapting the Hopkins statistic used for clustering tendency to assessment of cluster compactness and separation.

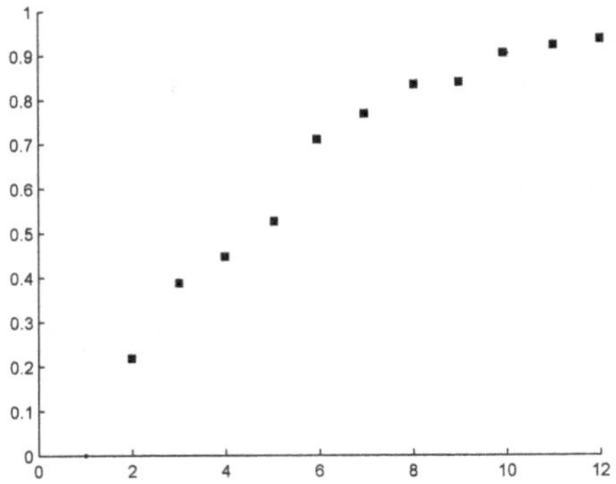

Figure 4.52: DB index values for k-means clustering of MDECTE, k in the range $2 \ldots 12$

4.3.2.2 Cluster validation of constituency structure

Like proximity-based nonhierarchical clustering methods, hierarchical ones partition data objects into discrete subsets. Where the partitional methods represent proximity relations among data objects in high-dimensional space as relative proximity among them in low-dimensional space, however, hierarchical methods represent these relations as a constituency structure which can be graphically represented as a dendrogram. As we have seen, different joining criteria used by the various hierarchical methods for building such constituency structures are based on different views of what constitutes a cluster, and, with respect to any given data matrix, the consequent expectation is that they will generate different structures. This expectation is routinely confirmed empirically. The question of which structure best captures the proximity relations among the data objects therefore arises, and the answer requires a validation index which, unlike the ones presented so far, is able to evaluate such structure.

The cophenetic correlation coefficient ((Sokal and Rohlf 1962), (Sneath and Sokal 1963), (Rohlf 1974), (Baker and Hubert 1975)) measures the degree of consistency between the distance matrix underlying the cluster tree and another matrix, the ‘ultrametric’ or ‘cophenetic’ matrix derived from the table of joins, and is standardly used in validation of hierarchical clustering, and so is described in what follows. Given a data matrix M and a hierarchical clustering of M, the cophenetic correlation coefficient for the clustering is calculated using two matrices: the distance matrix D derived from M on which the clustering is based, and the cophenetic distance matrix. The first of these is familiar. Construction of the second is easier to exemplify than to describe. The example is taken from the foregoing discussion of hierarchical clustering, where a small subset of the first 6 rows of the full 63 rows of MDECTE was used as data.

Table 4.10 shows the Euclidean distance matrix abstracted from the data.

Table 4.10: Euclidean distance matrix D

	g01	g02	g03	g04	g05	g06
g01	0					
g02	116.9	0				
g03	59.0	113.2	0			
g04	82.6	98.8	79.4	0		
g05	103.8	124.4	112.9	108.8	0	
g06	69.7	116.8	69.3	78.9	113.0	0

Table 4.11 shows the table of joins constructed by the clustering algorithm.

Table 4.11: Table of joins

Cluster1	Cluster2	Joining distance
1	3	59.0
(1,3)	6	69.3
((1,3),6)	4	78.9
(((1,3),6),4)	2	98.8
((((1,3),6),4),2)	5	103.8

Figure 4.53 shows the cluster tree corresponding to the table of joins.

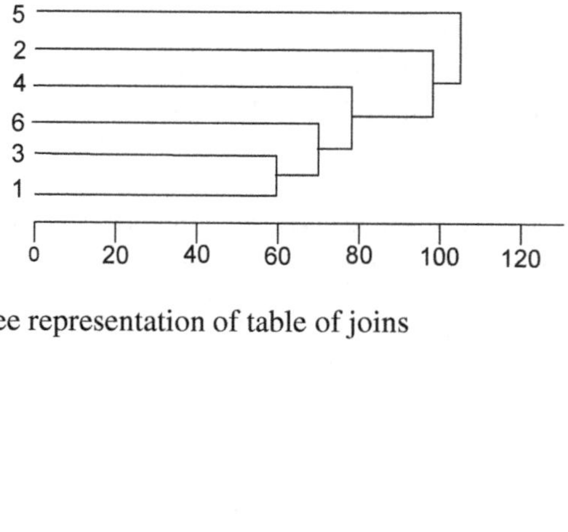

Figure 4.53: Tree representation of table of joins

The cophenetic distance between two rows D_i and D_j is the distance at which they become members of the same cluster for the first time. These distances are found in the table of joins, and, using them, a matrix C containing the cophenetic distances between D_i and D_j is constructed. Referring to Table 4.12, the construction procedure is as follows.

Table 4.12: Cophenetic proximity matrix C for tree in Table 4.11 and Figure 4.53

	(1)	(2)	(3)	(4)	(5)	(6)
(1)	0					
(2)	98.8	0				
(3)	59.0	98.8	0			
(4)	78.9	98.8	78.9	0		
(5)	103.8	103.8	103.8	103.8	0	
(6)	69.3	98.8	69.3	78.9	103.8	0

Starting at the top of Table 4.11 and working downwards, (1) and (3) become part of the same cluster at cophenetic proximity 59.0; this value is entered at the corresponding coordinates in Table 4.12. (1,3) and (6) become part of the same cluster at cophenetic proximity 69.3; this value is entered at coordinates ((1),(6)) and ((3),(6)) in Table 4.12. ((1,3),6) and (4) become part of the same cluster at cophenetic proximity 78.9; this value is entered at coordinates ((1),(4)), ((3),(4)), and ((6),(4)) in Table 4.12, and so on. When complete, the cophenetic proximity matrix shows, for each pair of data objects, the proximity value at which they become part of the same cluster.

Once the cophenetic distance matrix is constructed it can be compared to the Euclidean distance matrix to see how similar they are and thereby the degree to which the clustering from which the cophenetic matrix was derived preserves the distance relations among the data objects as represented by the Euclidean distance matrix. This can be done in various ways, the simplest of which is to row-wise linearize the two matrices and then to calculate their Pearson correlation coefficient, which gives the cophenetic correlation coefficient; the linearization is shown in Table 4.13 broken in the middle to fit on the page.

Table 4.13: Row-wise linearized distance (D) and cophenetic (C) matrices

D	116.9	59.0	113.2	82.6	98.8	79.4	103.8	124.4
C	98.8	59.0	98.8	78.9	98.8	78.9	103.8	103.8
	...(D)	112.9	108.8	69.7	116.8	69.3	78.9	112.9
	...(C)	103.8	103.8	69.3	98.8	69.3	78.9	103.8

The Pearson correlation coefficient of the vectors in Table 4.13 is 0.96, a high correlation which shows that the cluster structure in Figure 4.53 well represents the distance relations between data objects.

The cophenetic correlation coefficient can be used to select the best from a set of hierarchical analyses of the same data. Figure 4.54 shows three different analyses of the first 12 rows of MDECTE.

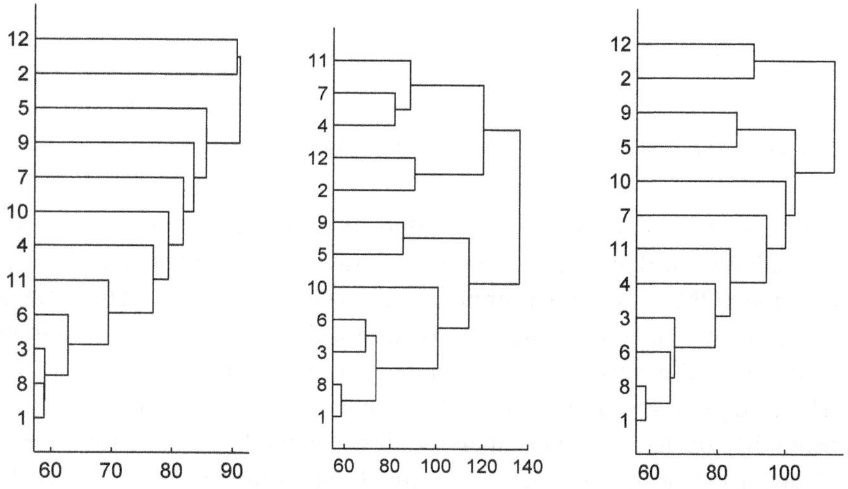

(a) Single Linkage *(b)* Complete Linkage *(c)* Average Linkage

Figure 4.54: Three hierarchical analyses of a matrix containing 12 rows randomly selected from MDECTE

As expected, the tree structures in Figure 4.54 are all different, so the problem of selection arises. Their cophenetic coefficients in descending order of magnitude are: Average Linkage 0.81, Single Linkage 0.78, and Complete Linkage 0.61. The Complete Linkage tree is the most intuitively appealing in terms of its separation of the data into well-defined nested structures, but in terms of the degree to which it distorts the Euclidean distance structure of the data it is the worst. Average Linkage and Single Linkage, on the other hand, both show chaining but also substantially less distortion. It is known that Complete Linkage tends to impose well-separated spherical clusters on data whatever its structure; this observation, together with the numbers, means that the cophenetic correlation coefficient must overrule aesthetics, and the indication is that the data really does contain the chaining structure which Average Linkage and Single Linkage display.

The cophenetic correlation coefficient is a measure of the strength of the linear relationship between distance and cophenetic matrices, but though it is widely used its validity for assessing the relative goodness of hierarchical clustering results has been disputed. Gordon (1996), for example, considers that it "does not seem a very relevant measure", suggesting instead the Goodman-Kruskal lambda measure (Goodman and Kruskal 1954), and warns against "the incautious use of criteria that almost always produce misleading results".

4.3.2.3 Cluster validation of topology preservation

Preservation of manifold topology differs from preservation of linear distances among the data points which constitute it, as explained earlier, and as such the SOM or any other topology-preserving clustering method cannot be assessed by validation indices designed for distance-preserving methods. This section outlines two alternative indices for the SOM.

The foregoing discussion of the SOM noted that, where the intrinsic dimensionality of data is greater than the dimensionality of the output lattice, some distortion of the topology of the input manifold is to be expected. The discussion also noted that different selections of initial parameter values such as the locations of the Voronoi centroids, lattice size and shape, and different sequencings of training data items can generate different cluster results, and this calls the status of the result as a reliable representation of the intrinsic data structure into question. One approach to SOM assessment is to attempt to identify an objective function which the SOM optimizes and, using that function, to select the optimal result in any specific application, as with MDS, Sammon's Mapping, and Isomap. It has been shown that a general objective function applicable across all SOM parameterizations does not exist (Erwin, Obermayer, and Schulten 1992); an objective function *does* exist for every SOM with a specific lattice dimensionality, lattice size, and neighbourhood size, relative to which an error value can be calculated, but error values for differently-parameterized SOMs cannot be reliably compared to assess their relative goodness.

Two alternative criteria that have been developed are quantization error and topology preservation error; for these and others see De Bodt, Cottrell, and Verleysen (2002a,b).

- *Quantization error.* As noted in the discussion of the SOM, the process of finding a centroid vector which is intermediate between the k vectors of a Voronoi partition in the course of SOM training is known as vector quantization: for a set V of k vectors, find a 'reference' vector v_r of the same dimension as those in V such that the absolute difference $d = |v_r - v_i|$ between each $v_i \in V$ and v_r is minimized. The SOM training algorithm quantizes the input vectors, and the reference vectors are the result (Ritter, Martinetz, and Schulten 1992: ch.14), (Van Hulle 2000: ch.2), (Kohonen 2001: pp.59-60), (De Bodt, Cottrell, and Verleysen 2002a), (De Bodt, Cottrell, and Verleysen 2002b). Quantization error is a measure of how close the reference vectors are to the data vectors, or, in other words, how well the SOM has learned the data, and is calculated as the average of the sum of Euclidean distances between each data vector and the reference vector which the SOM training has most closely associated with it.
- *Topology preservation error.* Unlike the PCA projection, which tries to preserve as much of the data variance as possible, and unlike the projections generated by MDS and its derivatives, which try to preserve relative distances among points in the data space, the SOM projection aims to preserve the topology of the data, and the topology preservation error is a measure of how well it has done this in any particular application. We have seen that the topology of an n-dimensional data manifold D containing m points is the set of neighbourhoods of each point D_i in the manifold, for $i = 1 \ldots m$, where a neighbourhood of D_i is the set of other points within some distance of it, and distance is defined relative to a metric like the Euclidean either as a radius centred on D_i or some number k of nearest neighbours of D_i. A projection of D into a reduced-dimensionality space D' is topology-preserving if the neighbourhoods in D and D' are the same, that is, D_i and the point in D' to which it projects have the same neighbours. Various ways of measuring deviations from topology preservation have been proposed for the SOM: Bauer and Pawelzik (1992), Villmann, Der, and Martinetz (1994), Kaski and Lagus (1996), Kiviluoto (1996), Villmann et al. (1997), Kaski (1997), Venna and Kaski (2001), De Bodt, Cottrell, and Verleysen (2002b)); review in Pölzlbauer (2004).

  The one proposed in Kiviluoto (1996), called 'topographic error', is selected for description here on account of its intuitive simplicity. It assesses topology preservation as the degree to which the mapping from the input manifold to the SOM lattice is continuous, where continuity means that points which are adjacent in the manifold are also adjacent in its projection. In a SOM trained to project D onto its lattice, every point D_i is associated with a lattice cell standardly referred to as the 'best matching unit'. Where the mapping is continuous, the second-best matching unit will be adjacent to the best-matching one; where it is not adjacent, there is a local failure in continuity and with it of topology preservation. Topographic error is defined as the proportion of all data points D_i for which the first and second best-matching units are not adjacent: the lower the topographic error, the better the SOM preserves the topology of D. The topographic error function is shown in Equation (4.16):

$$TopographicError = \frac{\sum_{i=1...m} u(D_i)}{m} \tag{4.16}$$

where u is a function that returns 0 if the best matching unit and second best matching unit are adjacent in the lattice and 1 otherwise.

Selecting the best from a set of SOM analyses generated using different parameter values is a matter of choosing the one that minimizes both the quantization and topology preservation errors. This is exemplified in Figure 4.55, which plots the behaviour of the quantization and topology preservation errors as the dimension of a square SOM lattice is increased in increments of 1 from 2×2 to 60×60, in each case training the SOM on MDECTE with random selection of training vectors. The indication of Figure 4.55 is that a lattice size of about 25×25 is optimal for projection of the MDECTE data.

4.3.2.4 Stability assessment

Stability assessment successively applies a given clustering scheme, that is, some combination of clustering method and parameterization, to a given data matrix and to multiple versions of that matrix which have been perturbed in some way, and observes the behaviour of the cluster result with respect to the original matrix and to each of the perturbed ones. If the cluster result does not change it is stable, and if it does then the result is unstable to proportion to the degree of change. The relative stability of the result is on statistical grounds taken to indicate the extent to which the cluster result captures the intrinsic cluster structure of the data (Ben-Hur, Elisseeff, and Guyon 2002;

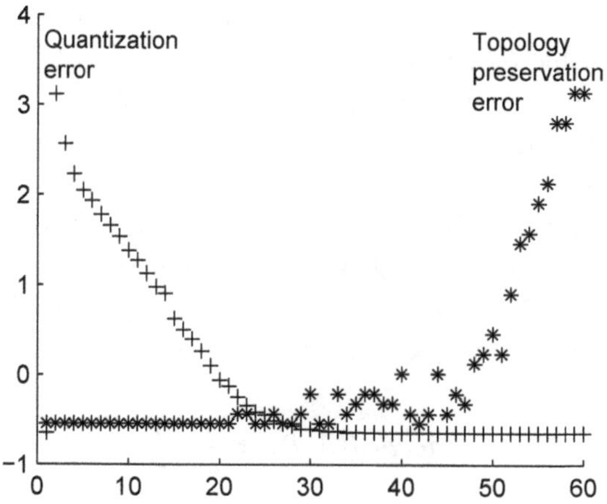

Figure 4.55: Co-plot of quantization errors and topographic errors for increasing SOM lattice size relative to MDECTE

Hennig 2007; Jain and Moreau 1987; Lange et al. 2004; Levine and Domany 2001; Pascual, Pla, and Sanchez 2010; Roth et al. 2002).

There are various ways of perturbing data, but Jain and Moreau (1987), who first proposed this approach, used bootstrapping, and this is described here. Given a set of k equal-size samples $S_1, S_2 \ldots S_k$ from a population P and a corresponding set of data matrices $D_1, D_2 \ldots D_k$ abstracted from the S_i, a clustering C is stable to the extent that the $C_1, C_2 \ldots C_k$ applied to the D_i are the same, assuming that the clustering method and its parameterizationremain constant.

In most practical applications all one has is a single sample S_0 and repeated sampling of the population from which it came is not feasible, but the k-sample sequence can be simulated using bootstrapping, which works as follows. Assume that the data matrix D_0 abstracted from S_0 contains m rows and n columns. Then D_1 is an m x n matrix each of whose rows is randomly selected from D_0; repeated selection of the same row is allowed, so D_1 may be but probably is not identical to D_0. Each successive $D_2 \ldots D_k$ in constructed in the same way. This looks at first glance like what Jain and Dubes (1988: 159) called "a classy way of cheating", but bootstrapping is a standard approach to estimating population statistics (Hennig 2007), and can respectably be used here as a substitute for actual multiple population sampling.

Once the D_i have been generated, the original data matrix D_0 is clustered to give C_0, and each of the bootstrap data matrices $D_1, D_2 \ldots D_k$ is clustered to give $C_1, C_2 \ldots C_k$. The similarity of each of the C_i to C_0 is then calculated to give an index value $I(C_i, C_0)$, the k index values are summed, and the mean of the sum is taken to be a measure of the stability of C: the larger the mean, the greater the similarity of the bootstrapped clusterings to C_0, and therefore the greater the stability of C_0 across multiple bootstrapped samples. There are many ways of measuring the similarity between clusters (Gower and Legendre 1986), but the present discussion follows Ben-Hur, Elisseeff, and Guyon (2002) and Hennig (2007) in adopting the Jaccard coefficient, which measures the similarity of two sets A and B as in Equation (4.17):

$$Jaccard(A,B) = \frac{A \cap B}{A \cup B} \tag{4.17}$$

where $\cap$ is the set-theoretic intersection and $\cup$ the union of A and B. In the present application, the Jaccard coefficient is used to compare two clusters, A from C_0 and B from C_i, and is the ratio of the number of objects which are both in A and B to the number of objects in either A or B or both. For two clusterings C_0 and C_i, each cluster in C_0 is paired with the corresponding cluster in C_i and the Jaccard index is calculated for each pair; the index values are summed, and the mean of the sum is the Jaccard index for C_0 and C_i. The stability index can be used just like the other types of validation index already discussed to look for an optimal clustering scheme by applying the above procedure to each such scheme and then selecting the scheme that generates the most stable clustering result.

Validation was until fairly recently the Cinderella of cluster analysis. Some early work on the topic has been cited; surveys of it are available in Dubes and Jain (1979), Milligan and Cooper (1985) and Milligan (1996), and the discussion in Jain and Dubes (1988) remains fundamental. It is, however, only since the fairly recent surge in availability and size of digital data and the consequent growth in the use of clustering as an analytical tool across the range of sciences that the importance of cluster validation has come to be appreciated, and this has in turn generated an increasingly copious research literature which evaluates existing methods and either proposes modifications which address perceived shortcomings or entirely new ones. Examples are: Halkidi, Batistakis, and Vazirgiannis (2001, 2002a,b), Halkidi and Vazirgiannis (2008), and Halkidi, Vazirgiannis, and Batistakis (2000), Maulik and Bandyopadhyay (2002), Bolshakova and Azuaje (2003), Stein, Eissen, and Wissbrock (2003), Kim and Ramakrishna (2005), Kovacs, Legany, and Babos (2006), Brun et al. (2007), Temizel et al. (2007), Dalton, Ballarin, and

Brun (2009), Deborah, Baskaran, and Kannan (2010), Kryszczuk and Hurley (2010), Liu et al. (2010), Rendon et al. (2011), Baarsch and Celebi (2012), Mary and Kumar (2012). Overviews of the state of work on cluster validation are given in Halkidi, Batistakis, and Vazirgiannis (2002a,b), Handl, Knowles, and Kell (2005), Tan, Steinbach, and Kumar (2006: Ch. 8.5), Gan, Ma, and Wu (2007: Ch. 17), Xu and Wunsch (2009: Ch. 10), Everitt et al. (2011: Ch. 9.4), Mirkin (2013: Ch. 6).

What emerges from these discussions and others like them is that, at present, cluster validation methods are far from foolproof. Much like the clustering methods they are intended to validate, they have biases which arise on account of the assumptions they make about the nature of clusters and about the shapes of the manifolds which clustering purports to describe (Handl, Knowles, and Kell 2005; Jain and Dubes 1988; Jain, Murty, and Flynn 1999: Ch. 3.3). Because of these biases the various methods can be unevenly effective and even misleading, as empirical studies of their application to different data and cluster shapes demonstrate. This is especially true of traditional validation methods based on cluster compactness and separation such as Dunn and Davies-Bouldin, which work well on data whose density structure is linearly separable but not otherwise, though progress in this area is being made via incorporation of density measurement into indices like *CDbw* (Halkidi and Vazirgiannis 2008). Brun et al. (2007) are pessimistic about prospects, but the current consensus appears to be that, despite their problems, cluster validation methods provide information which is not otherwise available, and as such they both reduce the risk of misinterpreting results and increase confidence in them. In any clustering application, therefore, a range of results should be generated using different clustering methods and parameter values, and these results should be assessed using appropriate validation indices. This can be onerous, but it is essential. As Jain and Dubes (1988) noted, “the validation of clustering structures is the most difficult and frustrating part of cluster analysis” but, without it, “cluster analysis will remain a black art accessible only to those true believers who have experience and great courage”.

5 Hypothesis generation

The preceding chapters have

- identified a research domain: the speech community of Tyneside in north-east England;
- asked a research question about the domain: is there systematic phonetic variation in the Tyneside speech community, and, if so, does that variation correlate systematically with social variables?
- abstracted phonetic frequency data from the DECTE sample of speakers and represented them as a matrix MDECTE;
- normalized MDECTE to compensate for variation in interview length and reduced its dimensionality;
- described and exemplified application of a selection of cluster analytical methods to the transformed MDECTE data.

The present chapter develops a hypothesis in answer to the research question based on cluster analysis of MDECTE. The discussion is in two main parts: the first part reviews and extends the clustering results presented so far, and the second uses these extended results to formulate the hypothesis.

5.1 Cluster analysis of MDECTE

Apart from Dbscan, where variation in data density was a problem, all the cluster analyses in the preceding chapter agreed that there are two highly distinctive clusters in MDECTE: a larger one containing speakers g01 to g56, and a smaller one containing speakers n57 to n63. These are not exhaustively repeated here; the MDS result in Figure 5.1 is given as representative. The agreement of all the methods on the two-cluster structure obviates the need for validation of individual results.

These results also show subsidiary structure in both main clusters. The structure in the smaller one is of less interest for present purposes than that of the larger, for reasons that will emerge, so the focus in what follows will be on the latter. To examine the structure of the g01 to g56 cluster more closely, the rows from n57 to n63 were removed from MDECTE. The row-reduced MDECTE56 was then re-analyzed using the methods applied to the full MDECTEin the preceding chapter.

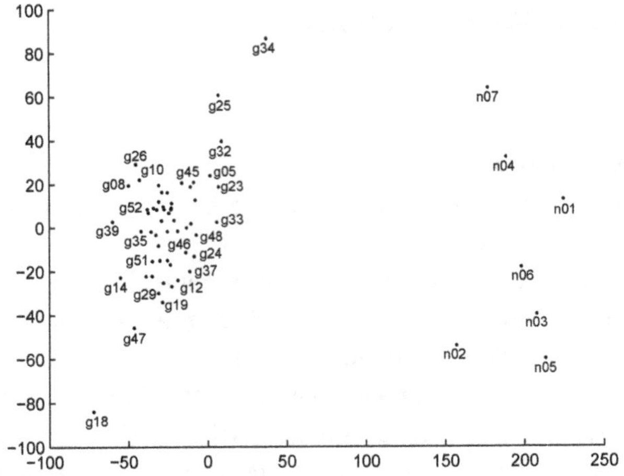

Figure 5.1: Multidimensional scaling analysis of MDECTE

5.1.1 Projection methods

The first step in the re-analysis was to apply the projection clustering methods, the results of which are given in Figures 5.2–5.6; the inset in each is intended to show the shape of the distribution unobscured by the speaker labels.

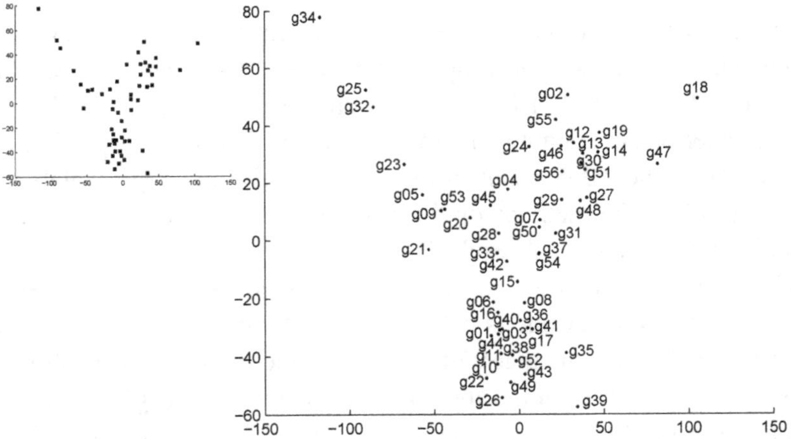

Figure 5.2: PCA clustering of MDECTE56

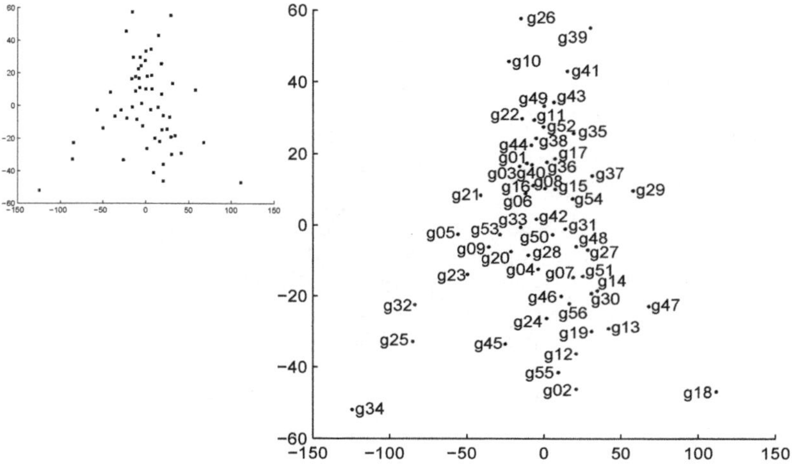

Figure 5.3: Nonmetric MDS clustering of MDECTE56

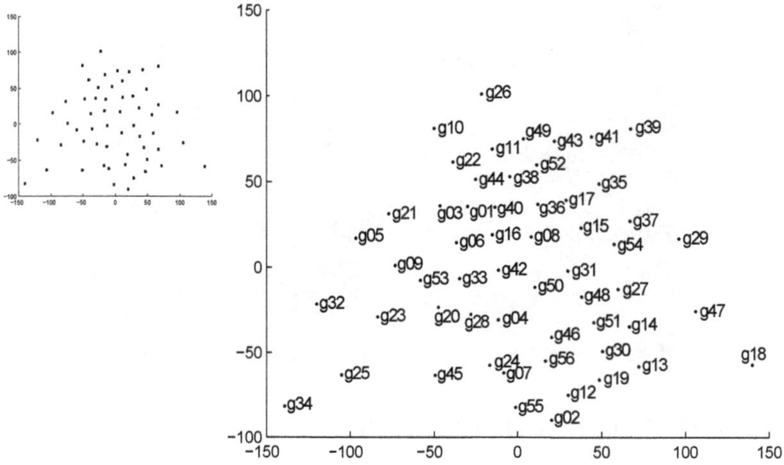

Figure 5.4: Sammon's Method clustering of MDECTE56

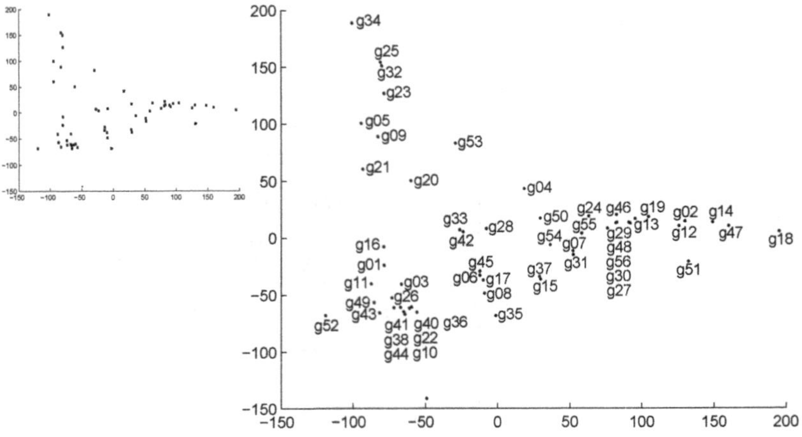

Figure 5.5: Isomap clustering of MDECTE56, $k = 7$

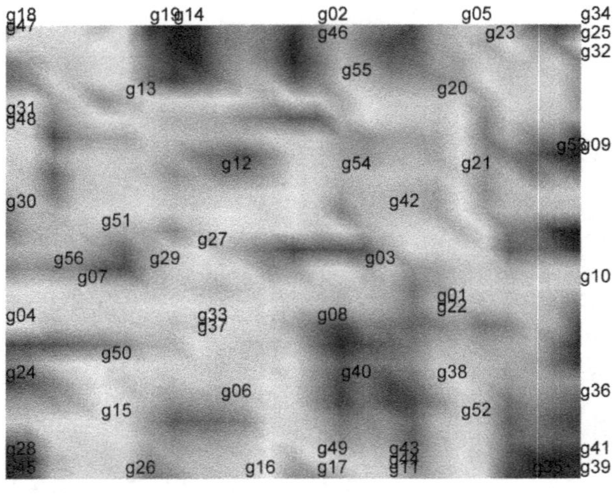

Figure 5.6: SOM clustering of MDECTE56

The SOM topology preservation indices given in the validation section of the previous chapter indicate that the 25×25 lattice given in Figure 5.6 is optimal, but apart from that none of these projections is particularly reliable.

For PCA the first two dimensions represented in Figure 5.2 capture only 34.2 percent of the cumulative variance, for MDS and Sammon's Mapping the stress associated with dimensionality 2 is relatively high, and so is the residual variance for embedding dimensionality 2 for Isomap. None of the projections show any visually very clear cluster structure, moreover: PCA, MDS, and Sammon, and Isomap all show a single cluster with a relatively few outlying points, and the U-matrix representation of the distribution of data points in the SOM lattice shows no obvious cluster demarcations.

5.1.2 Nonhierarchical partitional methods

Application of Dbscan to MDECTE56 required experimentation with numerous combinations of $MinPts$ and Eps parameters to determine the optimal ranges of both, where optimality was judged as lying between combinations which designated all data points as noise points on the one hand, and those which included all data points in a single cluster. The number of clusters is not prespecified for Dbscan but is inferred from the density structure of the data, and for no combination of parameters was anything other than a two-cluster structure found. The optimal range for $MinPts$ was found to be $2 \ldots 5$, and for Eps $59 \ldots 61$; the partition given for $MinPts = 4$ and $Eps = 60$ in Table 5.1 is representative of clustering results for these ranges.

Table 5.1: Dbscan clustering of MDECTE56

Cluster 1:	g04	g14	g19	g24	g27	g31	g46	g48	g50	g51
Cluster 2:	g01	g03	g06	g08	g11	g16	g22	g36	g38	g40
	g43	g49	g52							
Noise:	g02	g05	g07	g09	g10	g12	g13	g15	g17	g18
	g20	g21	g23	g25	g26	g28	g29	g30	g32	g33
	g34	g35	g37	g39	g41	g42	g44	g45	g47	g53
	g54	g55	g56							

To determine the optimal number of k-means clusters, k-means was applied to MDECTE56 for $k = 1 \ldots 10$, and for each value of k ten different initializations of cluster centers were used. The Dunn and Davies-Bouldin indices for this range of k indicated that $k = 2$ and $k = 3$ were roughly equally optimal, the results for which are shown in Tables 5.2 and 5.3.
The $k = 2$ and $k = 3$ clusterings relate to one another as follows:

Table 5.2: k-means partition of MDECTE56 for $k = 2$

Cluster 1:	g02	g07	g12	g13	g14	g18	g19	g24	g27	g29
	g30	g31	g37	g46	g47	g48	g50	g51	g55	g56
Cluster 2:	g01	g03	g04	g05	g06	g08	g09	g10	g11	g15
	g16	g17	g20	g21	g22	g23	g25	g26	g28	g32
	g33	g34	g35	g36	g38	g39	g40	g41	g42	g43
	g44	g45	g49	g52	g53	g54				

Table 5.3: k-means partition of MDECTE56 for $k = 3$

Cluster 1:	g02	g04	g07	g12	g13	g14	g18	g19	g24	g27
	g29	g30	g31	g46	g47	g48	g50	g51	g55	g56
Cluster 2:	g05	g09	g21	g23	g25	g32	g34	g53		
Cluster 3:	g01	g03	g06	g08	g10	g11	g15	g16	g17	g20
	g22	g26	g28	g33	g35	g36	g37	g38	g39	g40
	g41	g42	g43	g44	g45	g49	g52	g54		

- With two exceptions, cluster 1 for $k = 2$ is the same as cluster 1 for $k = 3$. The exceptions are that $k = 3$ cluster 2 adds g04 and deletes g37.
- With two exceptions, cluster 2 for $k = 2$ has been split into clusters 2 and 3 for $k = 3$. The exceptions are that $k = 3$ cluster 3 loses g04 to $k = 3$ cluster 1 and receives g37 from $k = 2$ cluster 1.

In other words, cluster 1 remains is the same for both, $k = 2$ cluster 2 splits into $k = 3$ clusters 2 and 3, and two speakers g04 and g37 are exceptions.

The relationship between the k-means and Dbscan results is straightforward:

- For $k = 2$ Dbscan cluster 1 is a subset of k-means cluster 1, and Dbscan cluster 2 is a subset of k-means cluster 2, with the exception of g04 which is in different clusters in Dbscan and k-means.
- For $k = 3$ the result is the same as for $k = 2$ in that none of the k-means cluster 2 data points are in either Dbscan cluster.

5.1.3 Hierarchical partitional methods

Trees for Single, Complete, Average, and Ward Linkage analyses of MDECTE56 are given in Figures 5.7–5.10 using both Euclidean and geodesic distance.

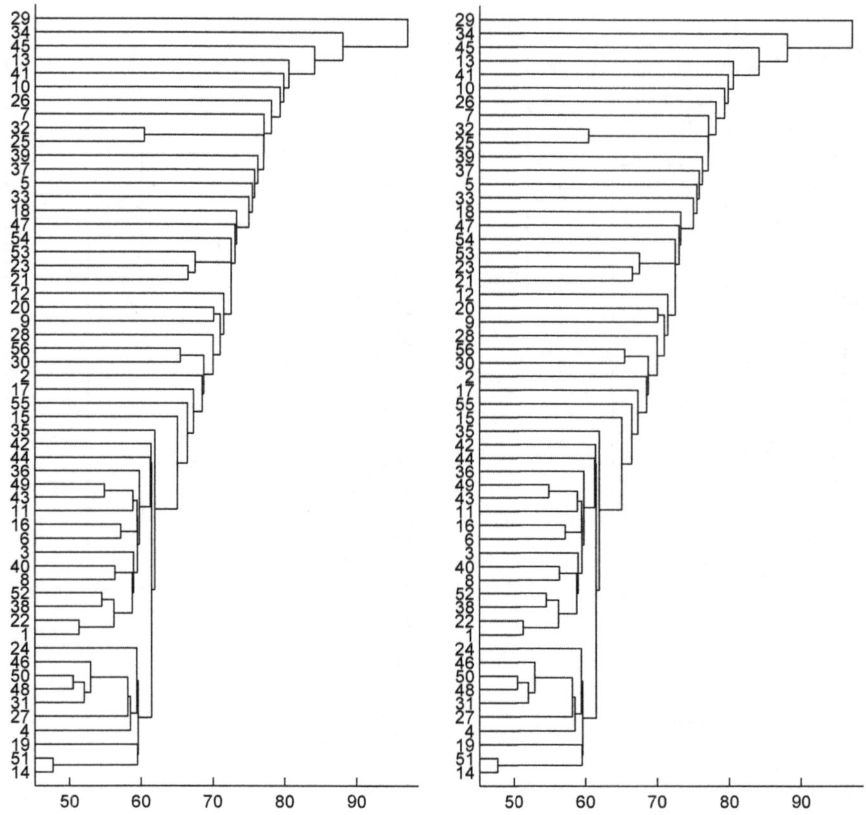

(a) Euclidean distance. Cophenetic: 0.5521 *(b)* Geodesic distance. Cophenetic: 0.3372

Figure 5.7: Single Linkage analysis of MDECTE56

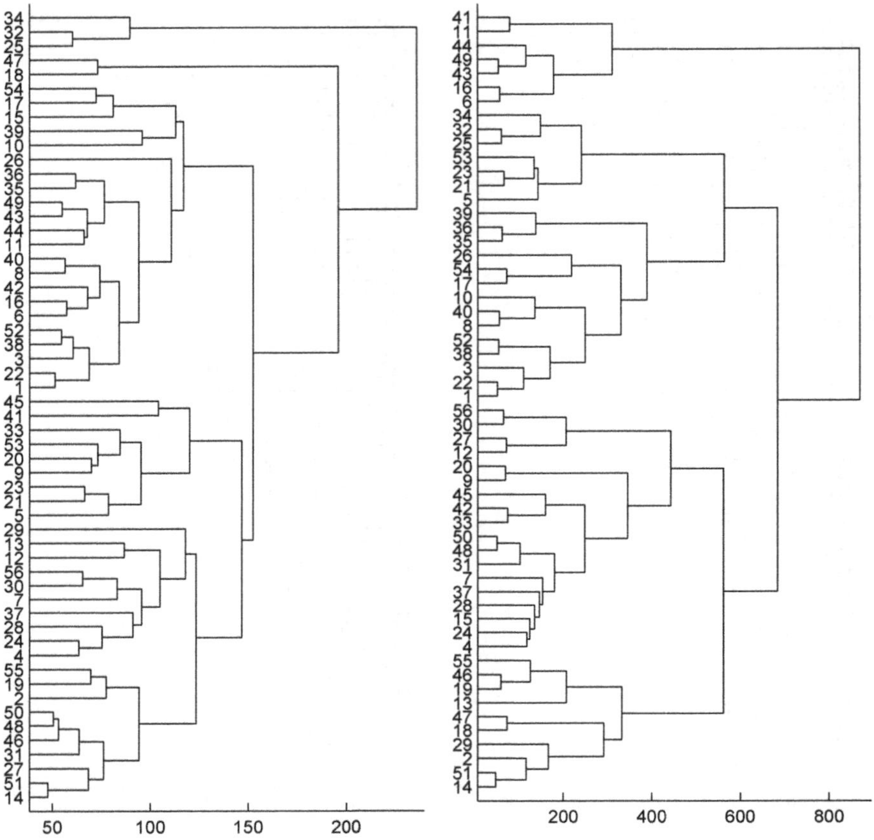

(a) Euclidean distance. Cophenetic: 0.7569

(b) Geodesic distance. Cophenetic: 0.6777

Figure 5.8: Complete Linkage analysis of MDECTE56

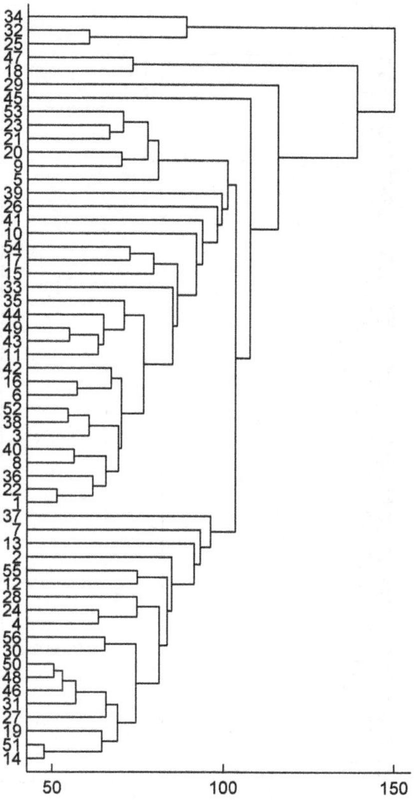

(a) Euclidean distance. Cophenetic: 0.7935

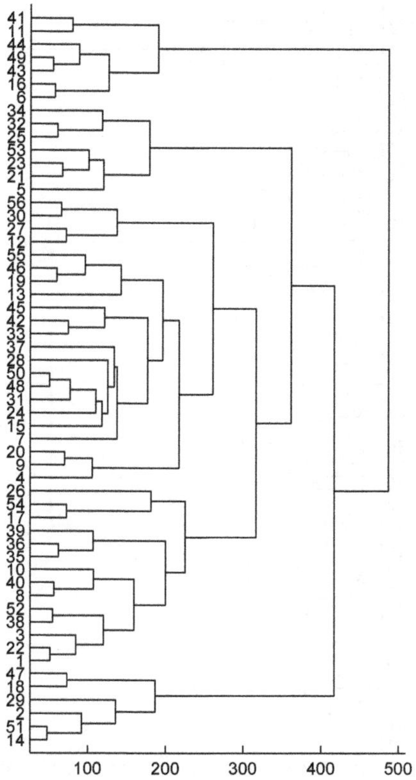

(b) Geodesic distance. Cophenetic: 0.7212

Figure 5.9: Average Linkage analysis of MDECTE56

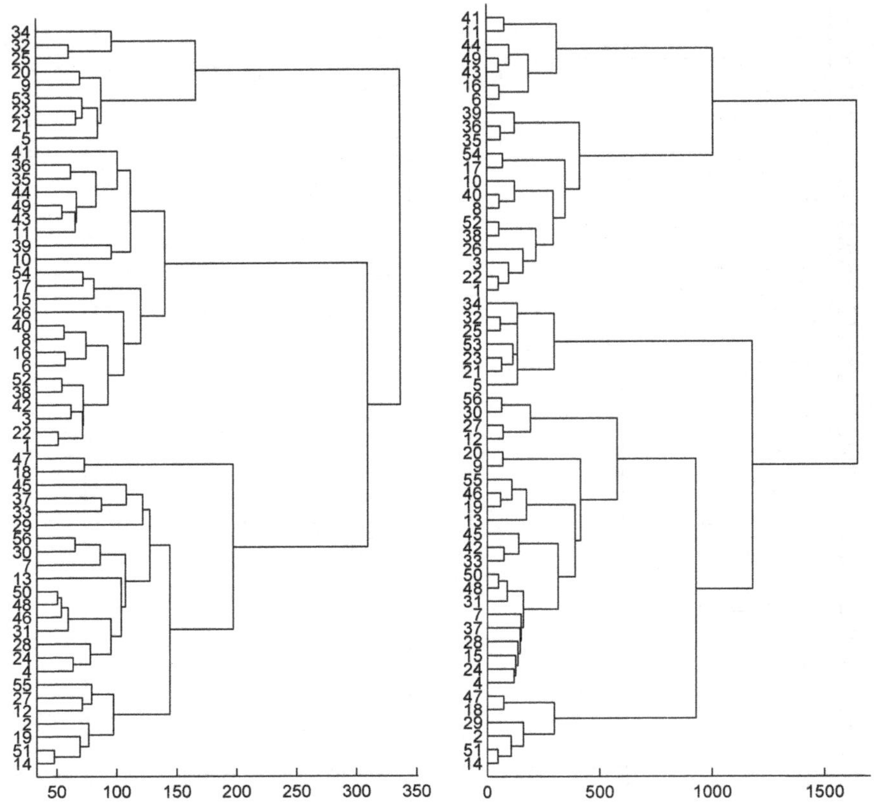

(a) Euclidean distance. Cophenetic: 0.5227   *(b)* Geodesic distance. Cophenetic: 0.6079

Figure 5.10: Ward Linkage analysis of MDECTE56

The initial visual impression is that these trees are all very different except for the Euclidean and geodesic Single Linkage ones, which are identical, but closer examination of them reveals regularities common to them all: among others, for example, the sets of data points (4 14 19 24 27 31 46 48 50 51) and (1 3 6 8 11 16 22 36 38 40 43 49 52) are in different subtrees in all cases apart from the Average Linkage tree based on geodesic distance. The detailed structural information offered by the trees makes it difficult to gain an overall impression of the ways in which they coincide with and differ from one another, however; this will emerge more clearly when the hierarchical results are compared with the nonhierarchical partitional and projection results in the section that follows.

5.1.4 Comparison of results

Having applied the different clustering method categories to MDECTE56 individually, the next step is to correlate the results in order to determine the degree to which they are mutually supporting with respect to the cluster structure of the data.

5.1.4.1 Comparison of projection and nonhierarchical partition results

Comparison of the k-means $k = 2$ and $k = 3$ results with the projection ones are dealt with separately, beginning with the former; because the Dbscan clustering is a subset of the k-means $k = 2$ one, the k-means $k = 2$ comparison subsumes it and there is consequently no need to deal with it separately. Figures 5.11–5.15 show the k-means partition for $k = 2$ on each of the projections, using the '*' symbol to represent data points belonging to k-means cluster 1, and '+' for k-means cluster 2.

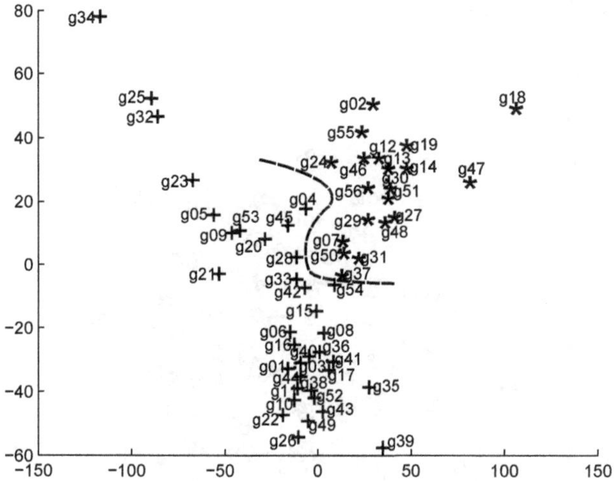

Figure 5.11: k-means $k = 2$ partition correlated with PCA projection. k-means cluster 1 = *, cluster 2 = +

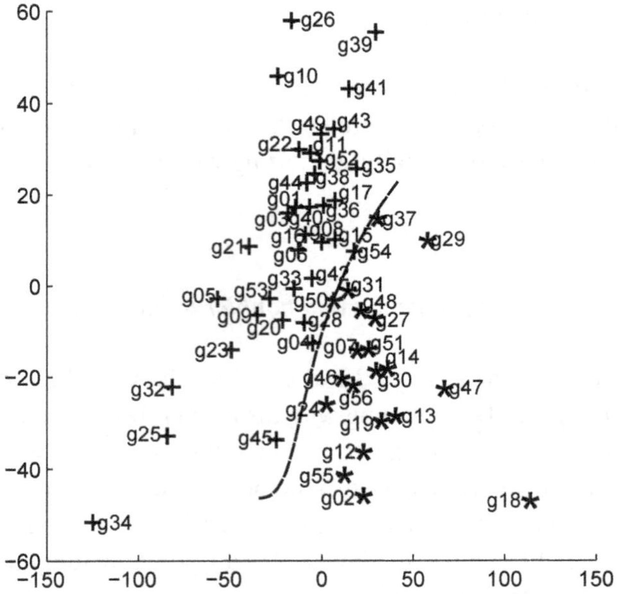

Figure 5.12: k-means $k = 2$ partition correlated with MDS projection. k-means cluster 1 = *, cluster 2 = +

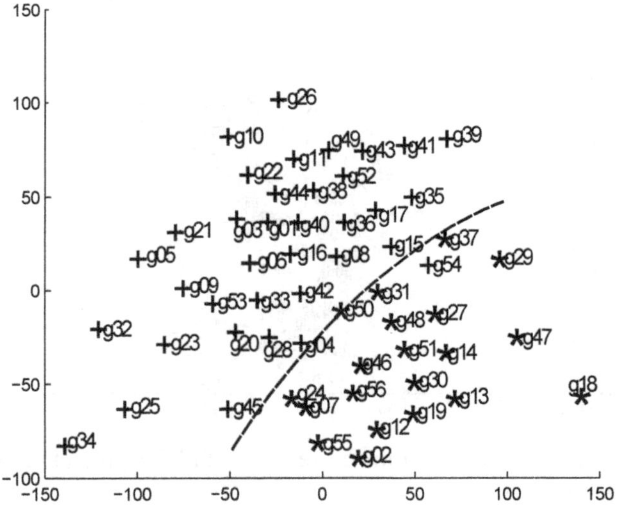

Figure 5.13: k-means $k = 2$ partition correlated with Sammon's Mapping. k-means cluster 1 = *, cluster 2 = +

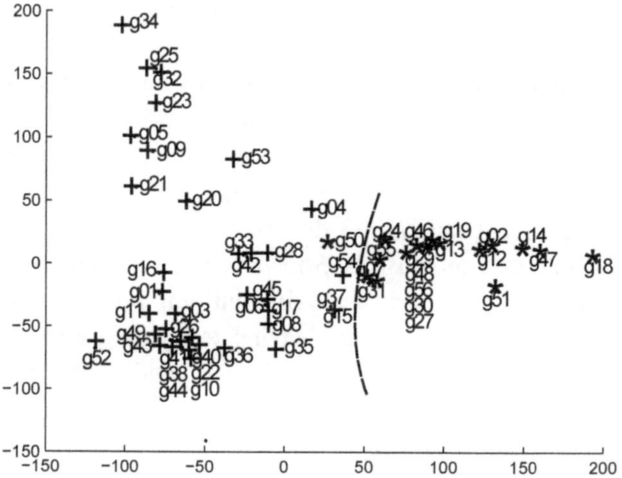

Figure 5.14: k-means $k = 2$ partition correlated with Isomap. k-means cluster 1 = *, cluster 2 = +

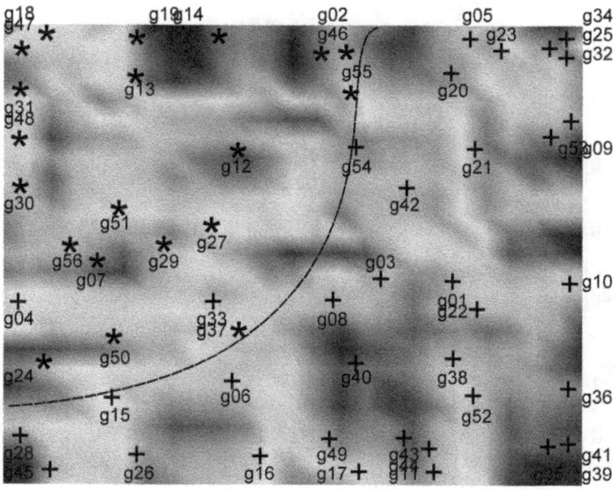

Figure 5.15: k-means $k = 2$ partition correlated with the SOM; k-means cluster 1 = *, cluster 2 = +

In each case k-means for $k = 2$ partitions the projections into two regions, which are demarcated by dashed lines for convenience of reference. This is unsurprising because, as we have seen, k-means clusters linearly separable regions of the data space, and the regions of the projections in Figures 5.11–5.15 can be regarded as linearly separable when the distortions introduced by the projection methods are taken into account.

As can be seen, k-means $k = 3$ differs from $k = 2$ only in dividing the $k = 2$ cluster 2 into two subclusters. Shown in relation to the PCA projection in Figure 5.16, the set of linearly separable outlying points noted earlier have been assigned to a separate cluster, leaving everything else the same. The situation for the other projections is analogous, and so there is no need to show these explicitly.

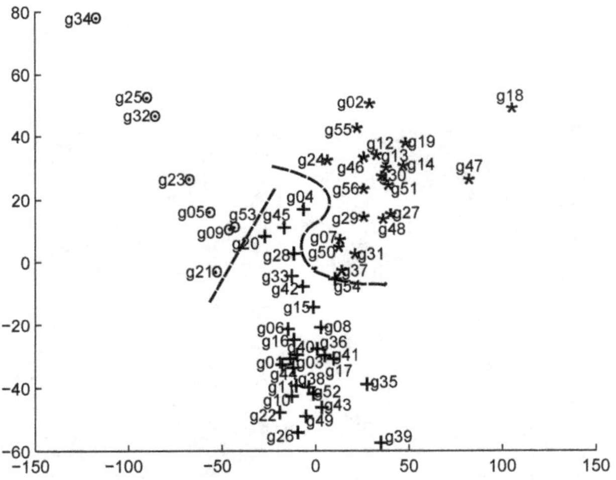

Figure 5.16: k-means $k = 3$ partition correlated with the PCA projection; k-means cluster 1 = *, cluster 2 = o, cluster 3 = +

In summary, the k-means partitions for $k = 2$ and $k = 3$ are compatible with all the projections in the sense that both assign the data points to disjoint linearly separable regions of the data space. Based on projection and non-hierarchical partitional clustering methods, therefore, the indication is that MDECTE56 has a two- or three-cluster structure, where the three-cluster one simply sub-partitions cluster 2 of the two-cluster one. It remains to see if this is compatible with results from hierarchical clustering.

5.1.4.2 Comparison of projection and nonhierarchical partition with hierarchical clustering results

In this section the hierarchical analyses are correlated with the Dbscan and *k*-means partitions of MDECTE56. Figures 5.17–5.20 represent Dbscan / *k*-means cluster 1 with the symbol '*' and Dbscan / *k*-means cluster 2 with '+', as above.

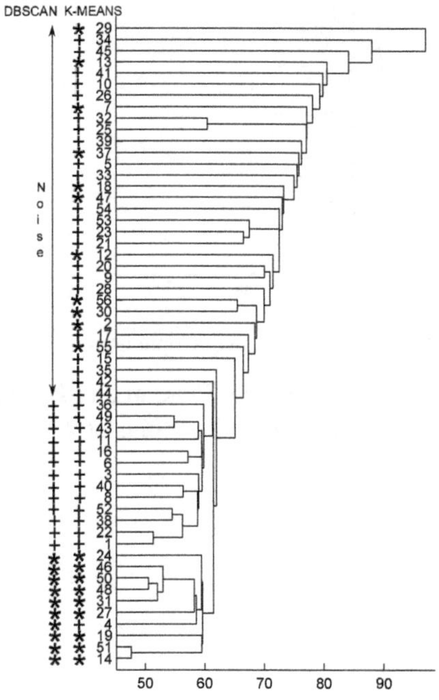

(a) Euclidean distance. Cophenetic: 0.5521

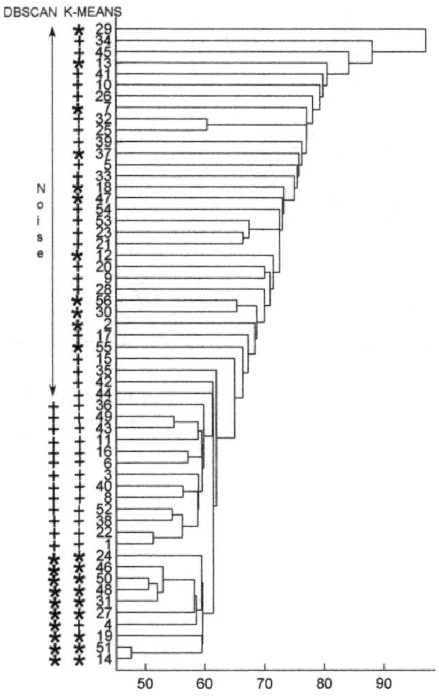

(b) Geodesic distance. Cophenetic: 0.3372

Figure 5.17: Single Linkage analysis of MDECTE56 with Dbscan and *k*-means ($k = 2$), where * = Dbscan/*k*-means cluster 1, and + = cluster 2

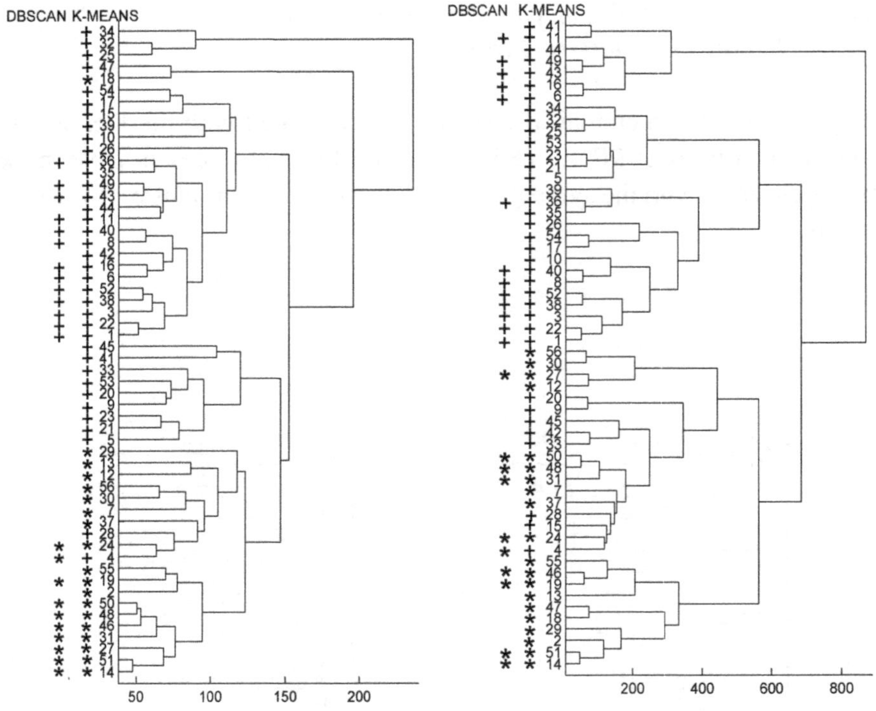

(a) Euclidean distance. Cophenetic: 0.7569 *(b)* Geodesic distance. Cophenetic: 0.6777

Figure 5.18: Complete Linkage analysis of MDECTE56 with Dbscan and k-means ($k = 2$), where * = Dbscan/k-means cluster 1, and + = cluster 2

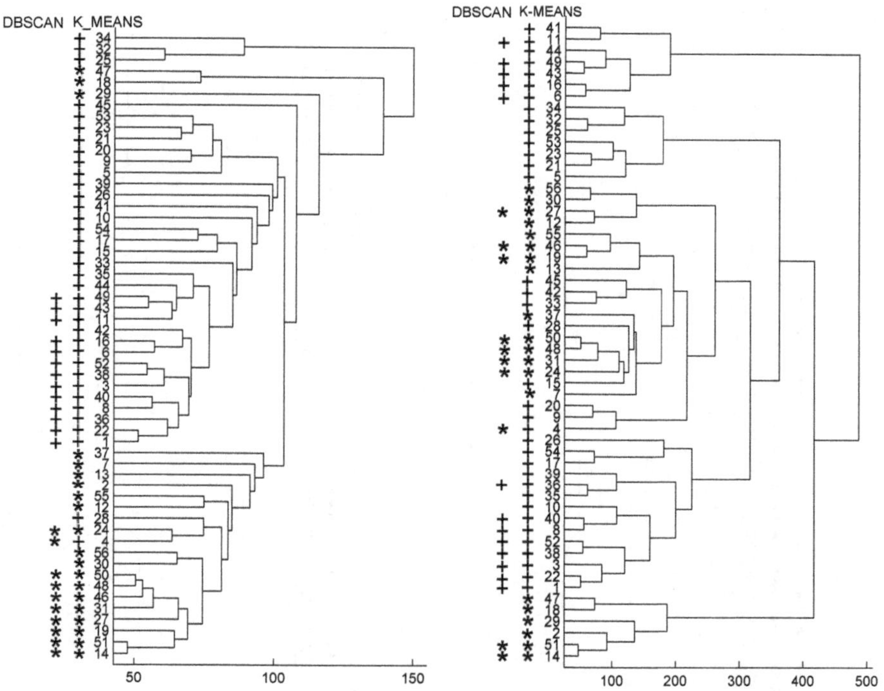

(a) Euclidean distance. Cophenetic: 0.7935 *(b)* Geodesic distance. Cophenetic: 0.7212

Figure 5.19: Average Linkage analysis of MDECTE56 with Dbscan and k-means ($k = 2$), where * = Dbscan/k-means cluster 1, and + = cluster 2

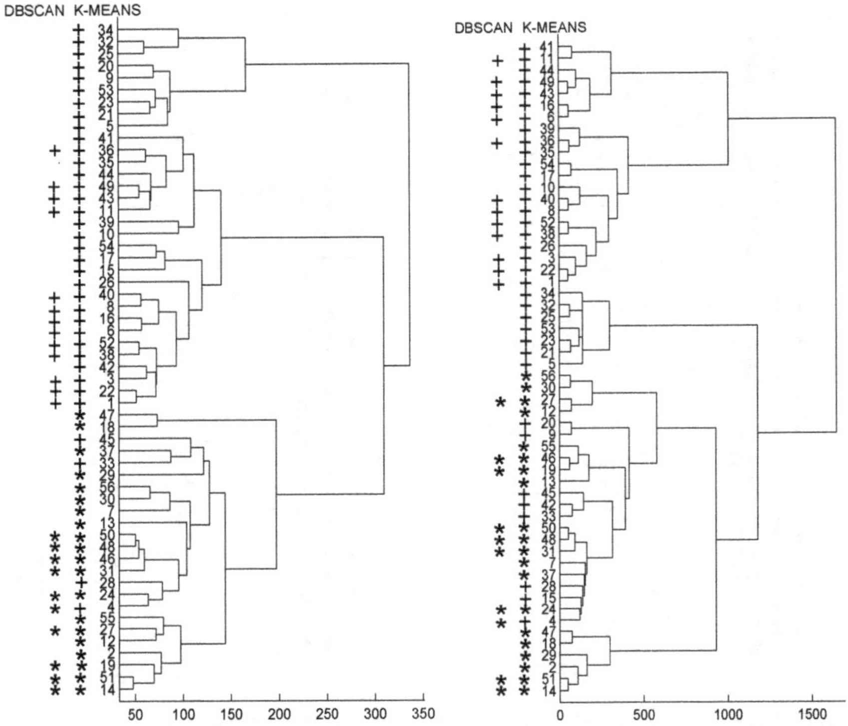

(a) Euclidean distance. Cophenetic: 0.5227

(b) Geodesic distance. Cophenetic: 0.6079

Figure 5.20: Ward Linkage analysis of MDECTE56 with Dbscan and *k*-means ($k = 2$), where * = Dbscan/*k*-means cluster 1, and + = cluster 2

Using the additional information that the correlation with the Dbscan and k-means provides, the relationships among the various trees becomes clearer.

- The Euclidean and geodesic distance based Single Linkage trees are identical, and differ from all the other trees in their characteristic chained structure. That chaining keeps all the data points corresponding to those in Dbscan cluster 1 and all the points corresponding to those in Dbscan cluster 2 together, and the rest of the structure is an apparently random mix of remaining points from the k-means clusters. Single Linkage has, in other words, identified the same clusters as Dbscan, which is unsurprising because, as noted earlier, Single Linkage is a density-based clustering method unlike the other varieties of hierarchical analysis, which are distance-based.
- Complete, Average, and Ward Linkage trees based on Euclidean distance are very similar though not identical, and correspond quite closely to the k-means partition for $k = 3$. In each case there are two main clusters corresponding to k-means clusters 1 and 3 which are, respectively, supersets of Dbscan clusters 1 and 2. There is also a small number of data points clustered separately from the main ones. These latter points are observable as outliers or at the periphery of the main data cloud in the projection plots, and though the selection of points differs somewhat from tree to tree, they correspond essentially to k-means cluster 2; the distributions are shown in Table 5.4.
- The Complete, Average, and Ward Linkage trees based on geodesic distance differ more substantially from the k-means partitions than the corresponding Euclidean distance based ones. Like the latter, the Complete and Ward Linkage trees consist of two large clusters corresponding largely to the k-means $k = 3$ clusters 1 and 3, and a single smaller one. The components of the smaller one are, however, not the outliers and data-peripheral points clustered in the Euclidean-based trees. Instead, the cluster consists of a subset of k-means cluster 3, and the outliers and data-peripheral points are included in the main clusters. The Average Linkage tree, moreover, is much more complex than the other two geodesic distance based ones: like them, it has the small cluster, but it fragments the neat two-cluster structure of the remaining data points.

Table 5.4: The distribution of data points corresponding to k-means cluster 2 in the Euclidean distance based Complete, Average, and Ward Linkage trees

Complete	Average	Ward	k-means
		g05	g05
		g09	g09
g18	g18		
		g20	
		g21	g21
		g23	g23
g25	g25	g25	g25
	g29		
g32	g32	g32	g32
g34	g34	g34	g34
	g45		
g47	g47		
		g53	g53

In summary, all the hierarchical trees apart from the geodesic distance based Average Linkage one are compatible with k-means for $k = 3$ to the extent that they show two main, relatively large clusters and a small one. They differ, however, in the constituents of the small cluster: the constituents in the Euclidean distance based trees correspond essentially to those of k-means cluster 2, while those in the geodesic distance based trees are a subset of k-means cluster 3. The exception, the geodesic distance based Average Linkage tree, shares the small cluster with the other geodesic trees, but fragments the two-cluster structure of the other trees for the remaining data points.

5.2 Hypothesis formulation

The first part of the research question, *Is there systematic phonetic variation in the Tyneside speech community?*, can now be answered affirmatively on the basis of the foregoing results. The DECTE speakers fall into two main clusters, a larger one containing speakers $g01 \ldots g56$ and a smaller one containing speakers $n57 \ldots n63$. The $g01 \ldots g56$ cluster itself has a subcluster structure, for which the foregoing analyses have revealed two alternatives. All the projection and nonhierarchicalpartition results as well as all the hierarchical results apart from Average Linkage based on geodesic distance agree that there are two main subclusters, but that these do not include all the speakers:

1. The Single Linkage tree and Dbscan partition a minority of the data points into two clusters and regard the rest of the data as noise.
2. The Euclidean distance based Complete, Average, and Ward Linkage trees group a small number of speakers separately from the two main clusters in slightly different ways; these speakers are observable at the periphery of the main data cloud in the projection plots or as outliers to it, and correspond substantially to the smallest of the clusters in the k-means result for $k = 3$.
3. The geodesic distance based Complete and Ward Linkage trees are partially compatible with (2) in partitioning most of the speakers into two main clusters and the remaining speakers into a small one, but differ from (2) in the constituents of the small cluster.

The other alternative is that offered by the geodesic distance based Average Linkage tree, which partitions the data into fairly numerous small clusters and is compatible with none of the other analyses, hierarchical or otherwise.

It remains to consider the second part of the question: *Does that variation correlate systematically with associated social variables?*. To answer it, the cluster results are supplemented with social data associated with the speakers in the DECTE corpus.

The unanimous partition of the 63 speakers into two subclusters $g01 \ldots g56$ and $n57 \ldots n63$ corresponds to speakers from Gateshead on the south bank of the river Tyne, betokened by the 'g' prefix, and those from Newcastle on the north bank betokened by the 'n' prefix. DECTE does not include any social data for the Newcastle speakers, though surviving members of the original *Tyneside Linguistic Survey* team have confirmed that the $n57 \ldots n63$ speakers were in fact the academics who comprised the team. The Gateshead speakers, on the other hand, were with a few exceptions working class with the minimum legal level of education and in manual skilled and unskilled employment. The primary clustering of the DECTE speakers therefore has a clear sociolinguistic interpretation based on educational level and employment type.

For the Gateshead subcluster the two alternatives identified above are available for interpretation. Since the aim here is methodological, that is, to exemplify the application of cluster analysis to hypothesis generation, only the first of them is addressed, though clearly the second would also have to be considered if the aim were an exhaustive investigation of the Geordie dialect as represented by DECTE. For clarity of presentation, the hierarchical result with the best cophenetic index, the Average Linkage tree based on Euclidean distance, is used as the basis for discussion.

Initially the full range of social variables provided by DECTE was correlated with the Average Linkage tree, and variables with no discernible systematic correlation with the tree structure were eliminated, The surviving variables are shown in Figure 5.21.

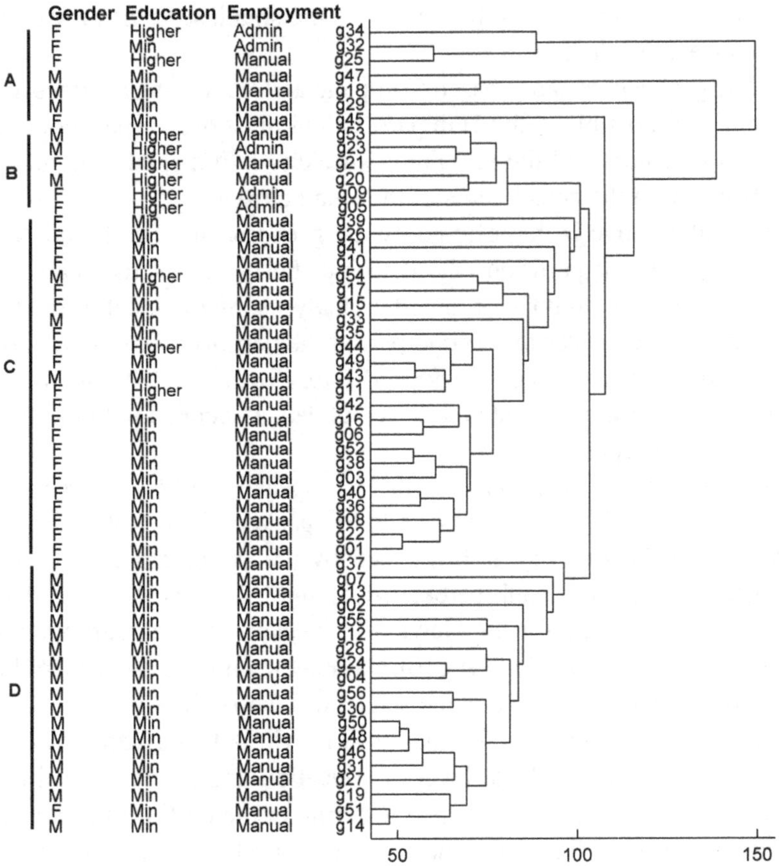

Figure 5.21: Euclidean distance based Average Linkage tree with DECTE social data

Note that the value 'Higher' in the *Education* column does not necessarily or even usually mean university-level education, but simply a level higher than the legal minimum, designated 'Min', at the time when the TLS project worked.

The two main clusters, labelled *C* and *D* in Figure 5.21, show a gender split, with *D* predominantly male and *C* predominantly female, and both consisting almost exclusively of speakers with minimal education and in manual employment. *B* contains a mix of male and female speakers, but these have higher-than-minimal education and, in a few cases, non-manual employment. A, finally, consists of two small composite and two singleton clusters containing male and female speakers with a mix of educational and employment levels; all but g29 and g45 in the small grouping are either outliers to or lie at the periphery of the two main clusters in the projection plots and are treated as noise by Dbscan, and this together with their separation from the two main clusters in Figure 5.21 indicates that these speakers are anomalous relative to the others in the sample in some way, and require separate investigation.

The second part of the research question can now also be answered affirmatively on the above evidence. The primary distinction in phonetic usage in DECTE separates a small group of highly-educated, middle-class Newcastle professionals from a much larger Gateshead group whose members were predominantly speakers in manual employment and with a minimal or sometimes slightly higher level of education. The indication is that the primary differentiating factor among the Gateshead speakers is gender, though the existence of cluster B suggests that educational level and type of employment are also factors.

The hypothesis developed thus far can be augmented by identification of the phonetic segments which are chiefly responsible for differentiating the DECTE speakers into clusters. This is done using cluster centroids, as introduced in Chapter 2. The centroid of any given cluster represents the average usage of its constituent speakers; variable-by-variable comparison of any two centroids reveals the degree to which the clusters differ, on average, for every variable, and allows those with the largest differences to be identified. Centroids for successive pairs of clusters were calculated and compared using bar plots to represent the comparisons graphically. Figure 5.22 shows the comparison for the 12 largest differences between the centroids of the Newcastle and Gateshead clusters, with the white bars representing Newcastle and the grey bars Gateshead; the choice of a dozen is arbitrary, and the selection could of course be increased or decreased in size according to need. The differences are arranged in descending order of magnitude from the left, and, for each, the symbolic representation of the phonetic segment in question and a lexical example are given, together with the corresponding DECTE phonetic code to facilitate comparison with the transcriptions in the DECTE corpus.

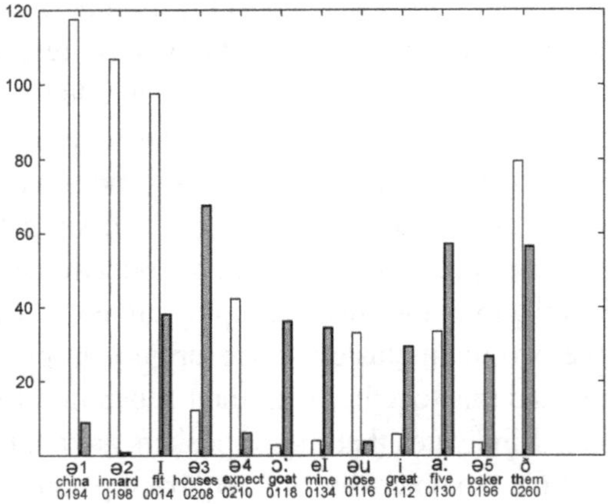

Figure 5.22: Comparison of centroids for Newcastle and Gateshead clusters

The phonetic segments primarily responsible for differentiating the Newcastle and Gateshead speakers can be read from the plot, with the most important on the left and decreasing in importance as one moves to the right. The primary phonetic determinants of subclusters can be found in the same way to whatever tree depth one finds useful. Figure 5.23, for example, compares the centroids for the two main Gateshead clusters, with the white bars representing the predominantly female cluster and the grey bars the predominantly male one.

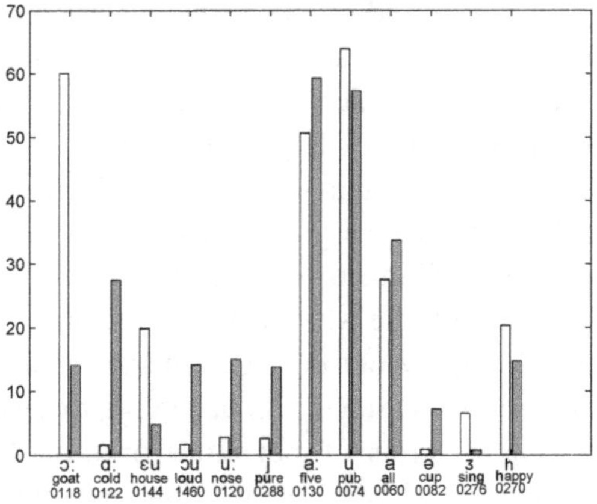

Figure 5.23: Comparison of centroids for the two main Gateshead clusters

Based on the foregoing analysis and interpretation of the DECTE sample, the following hypothesis about the Tyneside speech community from which the sample was taken can be stated:

> *There is systematic phonetic variation in the Tyneside speech community, and this variation correlates in a sociolinguistically significant way with educational level, employment type, and gender. The phonetic segments shown in Figures 5.22 and 5.23 account of most of the variation, and their distribution across social factors is shown in Figure 5.21.*

As already noted, a full rather than the present expository study of DECTE would have to go into greater detail, investigating such matters as the variation in the cluster structuring and allocation of speakers to clusters found in the foregoing discussion, the possible relevance of additional social factors, and centroid analysis of additional subcluster differences. The next step is to test this hypothesis with respect to data drawn from speakers not included in DECTE.

In principle, this discussion of hypothesis generation based on cluster analysis of DECTE should stop here, but it happens that existing work on the phonetics of Tyneside English provides results based on samples apart from the TLS component of DECTE (Moisl and Maguire 2008), and these results confirm the usefulness of cluster analysis as a tool for hypothesis generation. This work relates in part to the segments [ɔ:] (DECTE 0118), [ɑ:] (DECTE 0122), and [əu] (DECTE 0116), which are included by Wells (1982) in the GOAT lexical set. These are briefly discussed in terms of how well they support the results in Figures 5.22 and 5.23.

Variation in the GOAT vowel is a prominent feature of Tyneside English. Basing their results on the somewhat more recent PVC component of DECTE, Watt and Milroy (1999) identified four chief variants of the GOAT vowel in the corpus, three of which are included in the foregoing centroid comparisons:

- PVC [o:] / DECTE [ɔ:] is the unmarked variant preferred by all speakers apart from working-class males in the PVC corpus.
- PVC [ɵ:] / DECTE [ɑ:] is the almost exclusive preserve of working class males, and is described by Watt and Allen (2003) as 'archaic' and characteristic of 'older speakers' in the PVC sample.
- PVC [ʊə] / DECTE [əu] is almost completely restricted to the speech of middle-class females, old and young, and of young middle-class males. This variant is described as characteristic of 'high prestige supra-local speech patterns' (Watt and Milroy 1999: 37f.).

All of these are consistent with the relevant column pairs in Figures 5.22 and 5.23: PVC [ʊə] / DECTE [əu] is used mainly by the middle class academic Newcastle speakers, PVC [o:] / DECTE [ɔ:] is used by the predominantly female cluster C in Figure 5.21, and PVC [ɵ:] / DECTE [ɑ:] by the predominantly male cluster D.

These confirmations indicate that cluster analysis has generated an empirically supportable hypothesis in the present case, and suggest that it can do so both for other phonetic segments in DECTE and in corpus linguistic applications more generally.

6 Literature Review

This chapter reviews the use of cluster analysis in corpus linguistics to date.

6.1 Scope

(Larsen and Ins 2010) have investigated the growth rate of worldwide science publication from 1907 to 2007 based on information derived from databases such as the Science Citation Index and on existing growth data recorded in the relevant research literature. Their conclusion was that "old, well established disciplines like mathematics and physics have had slower growth rates than the new disciplines including computer science and engineering sciences, but that the overall growth rate for science still has been at least 4.7 percent per year". This corresponds to an approximate doubling in the number of publications every 15 years; there are almost twice as many publications this year than there were in 2000. In the same year a group of five professors of English, mechanical engineering, medicine, management, and geography respectively published a Commentary entitled *We must stop the avalanche of low-quality research* (Bauerlein et al. 2010) in the *Chronicle of Higher Education*, a respected Washington D.C.-based news service publication for the U.S. academic community. It argued that the 'astounding growth' of academic publication in recent years threatens the integrity of academic endeavour as a whole both because it has generated increasing amounts of poor-quality and/or unnecessary published output and because the sheer volume of publication renders the normal research procedures of reading, assimilating, and peer reviewing the discipline-specific literature increasingly unworkable. What proportion of the research output is of poor quality and/or unnecessary is open to debate, but the perception of the crushing effect of volume strikes a chord: conscientious attempts to bring the relevant literature to bear on any given research topic usually open up seemingly unending vistas of recently-published books, articles, chapters, and conference proceedings.

As such, the following review sets a tractable limit on the literature it surveys, and that limit is the literature specific to corpus linguistics. The Introduction defined corpus linguistics for the purposes of the present discussion as a methodology in the service of the science of language. This implicitly excludes a range of language technologies such as information retrieval, document classification, data mining, and speech processing, as well as areas of artificial intelligence like natural language generation / understanding and machine learning. These technologies work with natural language text and speech and thereby have much in common methodologically with corpus linguistics, including application to cluster analysis to text corpora; indeed, many of the concepts and techniques presented in the foregoing chapters come from their literatures. Their aims are, however, not language science but language engineering, and they therefore fall outside the self-imposed remit of this review. Also excluded on these grounds are quantitative stylometry and author attribution, whose aims are literary rather than linguistic.

The review is in two main parts: the first part outlines work on quantitative methods in corpus linguistics to serve as a context for the second, which deals with the cluster analytic work specifically.

6.2 Context

The hypothesis generation methodology described in the foregoing chapters is intended as a contribution to corpus linguistics, whose remit the Introduction described as development of methodologies for creating collections of natural language speech and text, abstracting data from them, and analysing those data with the aim of generating or testing hypotheses about the structure of language and its use in the world. This is a not-uncontroversial position. The literature draws a distinction between corpus-driven and corpus-based linguistics, where the former is taken to be a scientific paradigm in the sense that behaviourist and generative linguistics are paradigms, that is, ontologies in terms of which linguistic theories can be stated, and the latter to be a methodology, that is, a collection of techniques for the formulation and testing of hypotheses within some already-existing paradigm. The present discussion does not engage with this debate, and simply adopts the corpus-based view of corpus linguistics; for recent discussions see Taylor (2008), Gries (2009c, 2011a, 2012), McEnery and Hardie (2012) and the papers in volume 15(3) of the *International Journal of Corpus Linguistics* for 2010. For the development and current state of corpus linguistics see Stubbs (1996), McEnery and Wilson (1996), Biber, Conrad, and Reppen (1998), Kennedy (1998), McEnery and Wilson (2001), Meyer (2002), Sampson and McCarthy (2004), Facchinetti (2007), Johansson (2008), Lüdeling and Kytö (2008, 2009), O'Keefe, and McCarthy (2010), McEnery and Hardie (2012).

Despite its relatively well-developed state, corpus linguistic methodology has historically been unevenly distributed across the linguistics research community as a whole. This is especially the case for application of quantitative methods: simple quantifications such as frequencies, means, and percentages together with uni- and bivariate graphs are fairly extensively represented in the literature, but application of anything more complex than that is much rarer.

In grammatical linguistics, here understood as the study of the architecture of human language, statistical analysis of data derived from empirical observation was part of the standard methodology in the heyday of behaviourism in the first half of the twentieth century. Linguists at this time studied ways of inferring grammatical structure from linguistic performance using distributional information extracted from natural language corpora. A major figure in this approach to linguistics was Zellig Harris, who in a series of publications – for example Harris (1954, 1962, 1968) – brought mathematical concepts from areas like linear algebra and information theory to bear on the study of data abstracted from corpora. For a review of his legacy see Nevin (2002) and Nevin and Johnson (2002); other examples are Miller and Nicely (1955) and Stolz (1965). The demise of behaviourism and the rise of generative linguistics in the mid-1950s ushered in a research paradigm which relied and continues to rely on native speaker grammaticality judgment rather than on corpus-derived data as the empirical basis for hypothesis testing – cf. Gilquin and Gries (2009), Gries (2010b). This methodological commitment to native speaker judgment is reflected in the paucity of reference to corpora and to quantitative analysis of corpus-derived data in the generative linguistics literature. That literature is very extensive, and no claim to have searched it comprehensively is made here. Recent textbooks, handbooks together with the contents of a few of the main generative linguistics journals were reviewed, however, and the snapshot of the current research culture which these provided made it abundantly clear that corpus-based methodology in general and quantitative methods specifically are not a prominent part of it. One might conclude from this that corpus-based methodology has no obvious role in the present and future study of the architecture of human language. That conclusion would, however, be mistaken.

The eclipse of corpus-based methodology is characteristic of linguistics as practised in the United States and Britain. In continental Europe the application of mathematical and statistical concepts and methods to analysis of corpus data for derivation of linguistic laws has continued to be developed. An exhaustive listing of the relevant earlier European literature, most of which is not in English and therefore little known in the English-speaking linguistics world, is given in Köhler (1995). More recently, this European paradigm in linguistics has been championed by the *International Quantitative Linguistics Association* via its journal, the *Journal of Quantitative Linguistics*, and the monograph series *Quantitative Linguistics*, to which the present monograph belongs. Examples of recent work in this paradigm are Baayen (2001, 2008), Köhler (2005), Köhler, Altmann, and Piotrowski (2005), Best (2006), Grzybek and Köhler (2007), Kornai (2008), Köhler (2011, 2012), Köhler and Altmann (2011).

Quantitative methods have, moreover, always been central in some linguistics subfields concerned with the architecture of language: quantitative analysis of empirical and, increasingly, corpus-derived data is fundamental in psycholinguistics – cf. Gilquin and Gries (2009), McEnery and Hardie (2012: Chs. 3–6) –, and phonetics has a long tradition of quantifying empirical observations of pronunciation and then using statistical techniques like hypothesis testing, regression and factor extraction methods like PCA in analyzing the data (Johnson 2008: Ch. 3). And, finally, cognitive linguistics (Geeraerts and Cuykens 2010) has emerged strongly as a competitor paradigm to generative linguistics in recent years, and corpus-based methods are increasingly being proposed as a methodology well suited to it – cf. Gries (2003, 2012) and Gries and Stefanowitsch (2006), Stefanowitsch (2010), Arppe et al. (2010), McEnery and Hardie (2012: Chs. 6–8).

Variationist linguistics, here understood as the study of how language use varies in chronological, social, and geographical domains, is fundamentally empirical in that it uses data abstracted from observation of language use either to infer hypotheses about patterns of linguistic variation in or to test hypotheses about a language community. These three domains have historically been the preserves of historical linguistics, sociolinguistics, and dialectology respectively, though there is substantial overlap among them. Because they are based on analysis of data abstracted from language use, all three are naturally suited to quantitative and more specifically statistical methods, and, as corpus linguistics has developed in recent years, the corresponding research communities have increasingly adopted it, albeit unevenly. This unevenness is particularly evident with respect to quantitative methods, as sampling of the recent literatures in a way analogous to that for grammatical linguistics above testifies.

- Social variation research has a long-established quantitative tradition (Bayley 2013): Labov's studies of the social motivation for sound change (Labov 1963, 1966), Ladefoged, Glick, and Criper's (1971) study of language in Uganda, Fasold's (1972) study of Black American English, and Trudgill's (1974) study of social differentiation of English in Norwich are early examples. In a series of publications in the 1970s Sankoff and various collaborators carried out quantitative analyses of Canadian language varieties and more specifically of Montreal French; examples are: Sankoff and Cedergren (1971, 1976), Sankoff and Lessard (1975), Sankoff, Lessard, and Truong (1977), and Sankoff and Rousseau (1974), Cedergren and Sankoff (1974). This work generated the VARBRUL software extensively used for regression analysis in sociolinguistics to the present day Paolillo (2002), Bayley (2013), Tagliamonte (2006). Further examples of the use of quantitative methods in social variation research are Horvath (1985), Horvath and Sankoff (1987), Girard and Larmouth (1988), Sankoff (1988), Kroch (1989), Labov (1994); for recent reviews of the use of corpus methodology in sociolinguistics generally and quantitative methods specifically see Kretzschmar and Schneider (1996: Ch. 1), Milroy and Gordon (2003: Ch. 6), Baker (2010), Kendall and Herk (2011), McEnery and Hardie (2012: Ch. 5).
- Geographical variation research also has a long-established quantitative tradition. As early as 1952, Reed and Spicer (1952) used statistical analysis of covariance to analyze data taken from a corpus of Midwest American English. Examples of more recent though still relatively early work are proposals of statistical and computational methodologies for linguistic geography by Houck (1967), Rubin (1970), and Wood (1972), and application of statistical methods to data from Fijian by Schütz and Wenker (1966), from Piedmontese by Levine and Crockett (1966) and from Bantu by Coupez, Evrard, and Vansina (1975). In 1973 Séguy (1973a,b,c) proposed a numerical measure of dialect difference on the basis of which the distribution of a mix of phonetic, phonological, morphological, morphosyntactic, and lexical variables was plotted for numerous geographical locations on maps of the Gascony region of France. Séguy called his approach to dialectology 'dialectometry', and this has since been used as a general term for the quantitative study of geographical variation. Séguy's work inspired further dialectometric research; examples, in roughly chronological order,

are: Guiter (1974, 1987), Fossat and Philps (1976), Naumann (1976, 1977, 1982), Putschke (1977), Thomas (1977, 1980, 1988), Berdan (1978), Goebl (1982, 1983, 1993a,b, 1997, 2005, 2006, 2010), Viereck (1984, 1988), Linn and Regal (1985, 1988, 1993), Cichocki (1988, 1989, 2006), Girard and Larmouth (1988), Miller (1988), Kretzschmar, Schneider, and Johnson (1989), Guy (1993), Johnson (1994), Kretzschmar (1996), Kretzschmar and Schneider (1996), and the Groningen Group from 1996; the work of the last-mentioned makes extensive use of cluster analysis and is referenced later in this chapter. For snapshots of the current state of dialectometry see Händler, Hummel, and Putschke (1989), Nerbonne (2003), Nerbonne and Kretzschmar (2003, 2006), Gries, Wulff, and Davies (2010), Szmrecsanyi (2011), and the special issues of the journal *Literary and Linguistic Computing* vol. 26(4) [2006] and vol. 28(1) [2013].

- The use of quantitative methods in chronological variation research has been concentrated in linguistic phylogeny, the study of relatedness among languages. This study has historically been associated primarily with the Indo-European group of languages, and was throughout the nineteenth and most of the twentieth centuries dominated by the Neogrammarian paradigm. This paradigm assumed that the Indo-European language interrelationships and, latterly, language interrelationships in general could be modelled by acyclic directed graphs, or trees, and used the Comparative Method to construct such trees. The Comparative Method has been and remains largely non-quantitative; quantitative comparative methods were first introduced in the mid-twentieth century via lexicostatistics, which involves calculation of proportions of cognates among related languages, and glottochronology, which posits a constant rate of linguistic change analogous to that of biological genetic mutation – cf. Embleton (1986, 2000), Rexová, Frynta, and Zrzavy (2003), McMahon and McMahon (2005: Ch. 2), McMahon and Maguire (2011). As early as 1937 Kroeber and Chrétien (1937) had urged the use of statistical methods for linguistic phylogeny, but they were swimming against the non-quantitative Neogrammarian tide, and significant adoption of such methods had, here as in other areas of linguistics, to wait until the advent of computational IT in the second half of the twentieth century made their application feasible. Since then a synthesis of language evolution and relatedness, archaeology, population dynamics, and genetics together with associated quantitative methods has emerged as cladistic language phylogeny; for the principles of phylogeny see Felsenstein (2004), and for its application in cladistics Ridley (1986), Renfrew (1987, 1999), Warnow (1997), Sims-Williams (1998), Renfrew, McMahon, and Trask (2000), Cavalli-Sforza (2000), Ringe, Warnow, and Taylor (2002), Gray and Jordan (2000), Holden (2002), McMahon and McMahon (2003, 2005), Rexová, Bastin, and Frynta (2006) and Rexová, Frynta, and Zrzavy (2003), Nichols and Warnow (2008), McMahon and Maguire (2011).

The overall impression of the literature review underlying the foregoing summaries is that some researchers in some areas of linguistics have made substantial use of quantitative methods to analyze empirically-derived data, but that most researchers in most areas have not. This is corroborated by various subject specialists. In 1996 Kretzschmar and Schneider (1996) wrote that "Labov's breakthrough study of New York's Lower East Side (1966) secured the central role of quantitative techniques in sociolinguistics and so provided a model for dialectologists, but traditional dialectology has been slow to institutionalize the benefits of good counting…"; Nichols and Warnow (2008) note that "over the last 10 or more years, there has been a tremendous increase in the use of computational techniques…for estimating evolutionary histories of languages…it is probably fair to say that much of the community of historical linguists has been skeptical of the claims made by these studies, and perhaps even dubious of the potential for such approaches to be of use"; Baker (2010) notes that "corpus linguistics has made only a relatively small impact on sociolinguistics"; Kendall and Herk (2011) observe that while "much work in sociolinguistics is firmly empirical and based on the analysis, whether quantitative or qualitative, of data of actual language use" derived from corpora, "the uptake for this kind of work appears to be greater among researchers coming from corpus linguistic perspectives than among those coming from sociolinguistic backgrounds. Sociolinguists have been slower to adopt conventional corpora for research…"; Szmrecsanyi (2011) notes that "while corpus-linguistic methodologies have increasingly found their way into the dialectological toolbox and while more and more dialect corpora are coming on-line…it is fair to say that the corpus-linguistic community is not exactly drowning in research that marries the qualitative-philological jeweller's-eye perspective inherent in the analysis of naturalistic corpus data with the quantitative-aggregational bird's-eye perspective that is the hallmark of dialectometrical research".

Why should this be so? It is not for a lack of literature describing and applying quantitative corpus linguistic methodology. There are:

- Textbooks, monographs, and tutorial papers: Butler (1985), Woods, Fletcher, and Hughes (1986), Davis (1990), Rietveld and Hout (1993), Young and Bloothooft (1997), Oakes (1998), Lebart, Salem, and Berry (1998), McMahon and Smith (1998), Manning and Schütze (1999), Hubey (1999), Reppen, Fitzmaurice, and Biber (2002), Gries (2003), Kepser and Reis (2005), Rietveld and Hout (2005), Baayen (2008), Raisinger (2008), Johnson (2008), Gries (2009a, 2010c), Baroni and Evert (2009), Gries and Stefanowitsch (2010), Moisl (2009), Maguire and McMahon (2011).
- Collections of research papers and conference proceedings: Garside, Leech, and Sampson (1987), Sampson and McCarthy (2004), Gries and Stefanowitsch (2006), Stefanowitsch and Gries (2006), Baker (2009), Lindquist (2009), Renouf and Kehoe (2009), Gries and Stefanowitsch (2010).
- Numerous quantitatively-oriented articles in the main corpus linguistics and quantitative linguistics journals, which are, in no particular order, *International Journal of Corpus Linguistics, ICAME Journal, Corpus Linguistics and Linguistic Theory, Journal of Quantitative Linguistics, Corpora, Literary and Linguistic Computing, Empirical Language Research, Computer Speech and Language*, and *Language Resources and Evaluation.*
- Well developed quantitatively-oriented research programmes which have been available for decades, such as Séguy's dialectometry and its developments mentioned above, and Biber's MD in pragmatics and register research, on which see for example Biber and Finegan (1986), Biber (1992, 1996, 2006, 2009), together with detailed discussion of the programme in McEnery and Hardie (2012: Ch. 5) and the papers in the special issue of the journal *Corpora*, vol. 8 (2013) on *Twenty-five years of Biber's Multidimensional Analysis.*

The Introduction suggested that the reason for the reluctant adoption of quantitative methods by the linguistic research community has to do with a long-established and persistent arts-science divide. This is not an isolated view. Händler, Hummel, and Putschke (1989) thought that the cause in dialectology was the "traditionell geisteswissenschaftliche" orientation of the discipline, and in his introduction to English corpus linguistics Meyer (2002) wrote:

> Because many modern-day corpus linguists have been trained as linguists, not statisticians, it is not surprising that they have been reluctant to use statistics in their studies. Many corpus linguists come from a tradition that has provided them with ample background in linguistic theory and the techniques of linguistic description but little experience of statistics. As they begin doing analyses of corpora they find themselves practising their linguistic tradition in the realm of numbers, the discipline of statistics, which many corpus linguists find foreign and intimidating. As a consequence, many corpus linguists have chosen not to do any statistical analysis, and work instead with frequency counts...

Rexová, Frynta, and Zrzavy (2003) identified "the barrier between the humanities and the sciences" as "the main reason cladistic methodology has not until very recently been introduced into comparative linguistics for the evaluation of lexical data". At least some corpus linguists feel that this situation needs to change, particularly with respect to an enhanced use of quantitative methods; see for example Gries (2007, 2011a) and McEnery and Hardie (2012: Ch. 2). The hope is that corpus-based linguistics and the associated quantitative methods will increasingly penetrate mainstream linguistics, and the present book is offered as a contribution to that.

6.3 Cluster analysis in corpus linguistics

Several of the above-cited textbooks and monographs on quantitative methods in linguistics include accounts of cluster analysis and its application. The following discussion of specific application areas adopts the foregoing categorization of subdisciplines into grammatical and variationist linguistics, and the latter into social, geographical, and chronological variation.

6.3.1 Cluster analysis in grammatical research

In the aftermath of Chomsky's critique of empirical approaches to the study of language in his 1959 review of Skinner's *Verbal Behaviour* Chomsky (1959), interest in distributional linguistics research waned, as noted above. Empirical work continued to be done, however, and some of it used cluster analysis. Miller (1971) and Kiss (1973) used hierarchical analysis of word distribution data in text to derive syntactic and semantic lexical categories.

Shepard (1972) applied multidimensional scaling to analysis of Miller and Nicely's (1955) English consonant phonetic data and found a reduced two-dimensional representation of it based on nasality and voicing; a similar approach was used by Berdan (1978). Baker and Derwing (1982) applied hierarchical cluster analysis to study the acquisition of English plural inflections by children.

It was not until the later 1980s, however, that there was a resurgence of interest in empirical and more specifically statistical approaches among linguists as a consequence partly of the advent of 'connectionist' cognitive science with its emphasis on the empirical learning of cognitive functions by artificial neural networks, partly of the success of stochastic methods like Hidden Markov Models in the natural language and speech processing communities, and partly of the increasing availability of large-scale digital electronic natural language corpora from which to extract distributional information reliably, including corpora augmented with grammatical information such as the Penn Treebank and WordNet. Since ca. 1990 the volume of empirical corpus-based linguistic research has increased quite rapidly, as sketched above, and with it the use of cluster analysis. Most of the work has concentrated on lexis, but there is also some on phonology, syntax, and language acquisition (Stefanowitsch and Gries 2009).

Over the past two decades or so a large amount of work has been done on inference of grammatical and semantic lexical categories from text (Korhonen 2010), driven mainly by the requirements of natural language processing tasks like computational lexicography, parsing, word sense disambiguation, and semantic role labelling, as well as by language technologies like information extraction, question answering, and machine translation systems. This work is based on the intuition that there are restrictions on which words can co-occur within some degree of contiguity in natural language strings, and that the distribution of words in text can therefore be used to infer grammatical and semantic categories and category relatedness for them. This intuition underlies the "distributional hypothesis" central to Harris' work on empirically-based inference of grammatical objects Harris (1954, 1962, 1968), according to which "the meaning of entities and the meaning of grammatical relations among them is related to the restriction of combinations of these entities relative to other entities" Harris (1968), or, in Firth (1957)'s more economical formulation, "a word is characterized by the company it keeps". This intuition also accords with psychological evidence that humans judge the similarity of words by the similarity of lexical contexts in which they are used; Miller and Charles (1991) hypothesize that "two words are semantically similar to the extent that their contextual representations are similar". The volume of published work in this area precludes an exhaustive review of individual contributions; for overviews see for example Patwardhan, Banerjee, and Pedersen (2006), Pedersen (2006, 2010), Peirsman, Geeraerts, and Speelman (2010), Turney and Pantel (2010). Instead, a selection of representative selection of work is described and related work cited.

Schütze (1992, 1995, 1998) epitomizes earlier work on corpus-based category induction. He aims "to induce senses from contextual similarity" using a sense discrimination algorithm for ambiguous words in which word senses are represented as clusters of similar contexts in a multidimensional vector space. Given an ambiguous word type w, the data on which the algorithm operates is an $m \times n$ matrix M, where m is the number of tokens of w in whatever corpus is being used and n is the number of distinct lexical types t which occur in that corpus: each row M_i represents a different token w_i, each column M_j represents a different lexical type t_j, and the value at $M_{i,j}$ is the number of times t_j is a close neighbour of w_i, where close neighbours of w_i are lexical types t which co-occur with w_i "in a sentence or a larger context". The context vectors M_i are then clustered using a combination of expectation-minimization (EM) and hierarchical methods, which results in a set of sense clusters; the representation of the sense of any given cluster is the centroid of that cluster. The EM method is used because it is guaranteed to converge on a locally optimal clustering solution. It is, however, recognized that this does not guarantee a globally optimal solution, and that good initial cluster centres need to be selected to avoid local minima. The rows of M are therefore clustered using the average-link hierarchical method to find acceptable cluster centres. Moreover, because the context vectors of M are very high-dimensional and therefore typically sparse, singlular value decomposition is used to reduce the dimensionality of M, and the reduced version of the data is clustered. The validity of the derived clusters for word sense discrimination is externally validated using a test set with known word senses. The main advantage of this approach to word sense discrimination over others proposed in the literature is said to be that, unlike the others, it requires no information additional to the corpus text itself. Other examples of earlier work on lexical category induction via clustering before are Hindle (1990), Zernik (1991), Brown et al. (1992), Kneser and Ney (1993), Hatzivassiloglou and McKeown (1993), Jardino and Adda (1993), Pereira, Tishby, and Lee (1993), Tokunaga, Iwayama, and Tanaka (1995), Ueberla (1995), Yang, Lafferty, and Waibel (1996), Basili, Pazienza, and Velardi (1996), McMahon and Smith (1996), Ushioda (1996), Andreev (1997), Honkela (1997), Hogenhout and Matsumoto (1997), Chen and Chang (1998), Li and Abe (1998), Martin, Liermann, and Ney (1998), Rooth et al. (1998, 1999), Lee and Pereira (1999).

Representative examples of more recent work in this area are:

- Sun and Korhonen (2009) investigate the potential of verb selectional preferences acquired from corpus data using cluster analysis for automatic induction of verb classes. They introduce a new approach to verb clustering that involves a subcategorization frame acquisition system, syntactic-semantic feature sets extracted from the data generated by this system, and a variant of spectral clustering suitable for high-dimensional feature space. Using this approach, they show on two well established test sets that automatically acquired verb selectional preferences can be highly effective for verb clustering, particularly when used in combination with syntactic features. An advantage of this approach is that, like Schütze's, it does not require resources additional to the corpus text itself.
- Devereux et al. (2009) review and assess work in the computational linguistics and psycholinguistics communities on measuring the semantic relatedness between words and concepts using the idea that semantically similar or related words occur in similar textual contexts, and propose a way of using this idea for automatic acquisition of feature-based mental representation models from text corpora using cluster analysis. Several similarity metrics are considered, all using the WordNet ontology for calculating similarity on the grounds that concepts which fall close together under specific superordinate classes in WordNet will tend to be highly similar. The similarity data is analyzed using agglomerative hierarchical clustering, and the quality of the clusters is evaluated relative to the corresponding WordNet ontology.
- Shutova, Sun, and Korhonen (2010) propose a method for automatic metaphor identification in unrestricted text. Starting from a small seed set of metaphorical expressions, the method uses associations that underlie their production and comprehension to generalize over the associations by means of verb and noun clustering. The obtained clusters then represent potential source and target concepts between which metaphorical associations hold. The knowledge of such associations is then used to annotate metaphoricity in a large corpus. The motivation for the use of clustering is that of the foregoing studies: that the lin-

guistic environment in which a lexical item occurs can shed light on its meaning. Like Devereux et al. (2009), similarity measurement is based on the WordNet ontology, and spectral clustering is used.

- Gries and Stefanowitsch (2010) address the relationship between words and grammatical constructions, and more specifically the problem of determining the number and nature of semantic classes that are instantiated by a given set of words without recourse to prior assumptions or intuitions. They investigate the effectiveness of cluster analytic techniques for this purpose by determining how well these techniques identify on the most prototypical sense(s) of a construction as well as subsenses instantiated by coherent semantic classes of words occurring in it. The usual data creation approach of abstracting and counting the collocates of target words from a corpus within some user-defined span is modified by including only the covarying collexemes of target words, that is, words which occur in a well-defined slot of the same construction as the target word. Hierarchical cluster analysis is applied to the data, and it yields "a relatively coherent classification of verbs into semantic groups", and one superior to a classification based solely on collocate-counting. The conclusion is that clustering techniques can be a useful step towards making the semantic analysis of constructions more objective and more precise. For related work see also Gries (2007), Divjak and Gries (2006, 2008, 2009), Berez and Gries (2009), Divjak (2010).

Other recent work on lexical category and meaning induction includes: Walde (2000, 2006), Walde and Brew (2002), Brew and Walde (2002), Allegrini, Montemagni, and Pirrelli (2000), Andersen (2001), Lin and Pantel (2001), Gildea and Jurafsky (2002), Pantel and Lin (2002), Watters (2002), Almuhareb and Poesio (2004), Bekkerman et al. (2003), Linden (2004), Gamallo, Agustini, and Lopes (2005), Jickels and Kondrak (2006), Joanis, Stevenson, and James (2008), Bergsma, Lin, and Goebel (2008), Gries (2010a), Gries and Otani (2010), Hauer and Kondrak (2011), Schütze and Walsh (2011), Sun and Korhonen (2011). For a recent review of work on this topic see Korhonen (2010).

Work in other areas of grammar involving cluster analysis is less extensive, as noted. Croft and Poole (2008) proposed multidimensional scaling as a general methodology for inference of language universals. Waterman (1995) categorized syntactic structures of lexical strings using hierarchical clustering applied to Levenshtein distance measurement of the relative proximity of strings, and Gamallo, Agustini, and Lopes (2005) use hierarchical clustering to identify syntactic and semantic features of nouns, verbs, and adjectives from partially parsed text corpora. For phonology and morphology see Shepard (1972), Berdan (1978), Baker and Derwing (1982), Jassem and Lobacz (1995), Müller, Möbius, and Prescher (2000), Calderone (2009), Moisl (2012), Li, Zhang, and Wayland (2012). Gries and Stoll (2009) use variability-based neighbour clustering, a modified form of hierarchical clustering, to identify groups in language acquisition data.

6.3.2 Cluster analysis in chronological variation research

As noted, the use of quantitative methods in chronological variation research has been concentrated in linguistic phylogeny, and application of cluster analysis in this subfield shares that orientation.

Examples of earlier work are as follows. Black (1976) used multidimensional scaling to cluster lexicostatistical data abstracted from Slavic dialects. Beginning in the mid-1980s, Embleton and her collaborators Embleton (1986, 1987, 1993, 2000), Embleton, Uritescu, and Wheeler (2004, 2013), and Embleton and Wheeler (1997a,b) championed the use of multidimensional scaling in historical linguistics in general and for linguistic phylogeny edging over into dialectology in particular. Cortina-Borja, Stuart-Smith, and Valinas-Coalla (2002) and Cortina-Borja and Valinas-Coalla (1989) analyzed phonological, morphological, and lexical data using a range of clustering methods – hierarchical, multidimensional scaling, *k*-means, Sammon's mapping, and partitioning around medoids – to classify Uto-Aztecan languages. Batagelj, Pisanski, and Kerzic (1992) is a methodological study of how cluster analysis can be applied to language classification. Several string similarity metrics, including Levenshtein distance, are used to measure the proximities of 16 common words in 65 languages, and these languages are clustered on the basis of these measures using hierarchical analysis. Dyen, Kruskal, and Black (1992) is the summation of an extensive series of publications on lexicostatistical language classification in general and of Indo-European language classification in particular beginning with Carroll and Dyen (1962). Work by Black (1976), one of Dyen's collaborators, was noted above, and further references are given in Dyen, Kruskal, and Black (1992); lexicostatistical data is used to classify the languages "by some method that is systematic and reasonable but is not specified in detail", but in practice this is hierarchical analysis. Kita (1999) proposed a method for automatic clustering of languages whereby a probabilistic model was empirically developed for each language, distances between languages were computed according to a distance measure defined on the language models, and the distances were used to classify the languages using hierarchical clustering.

Since about 2000, language classification has increasingly been associated with cladistic language phylogeny, and cluster analysis is one of a variety of quantitative methods used in cladistic research. The general focus in this work is not so much achievement of a definitive result for some language group, but methodological discussion of data creation and clustering issues. McMahon and McMahon (2005), Nichols and Warnow (2008), Johnson (2008), Kessler (2008), and Delmestri and Cristianini (2012) provide introductions to and reviews of quantitative methods in linguistic phylogeny, including distance measurement and clustering methods; for a detailed discussion of proximity metrics in linguistic phylogeny see Kessler (2005, 2007). Some examples are of recent work in this field are:

- Holm (2000) uses hierarchical analysis to classify the Indo-European languages on the basis of data derived from Pokorny's etymological dictionary, but was methodologically dissatisfied with the result, and Holm (2007) assessed the reliability of such clustering with respect to Indo-European phylogeny, with largely negative results. Cysouw, Wichmann, and Kamholz (2006) provide a critique of Holm's methodology; McMahon and McMahon (2003) urged greater attention by the cladistics community to understanding of how the various tree-building methods, including hierarchical clustering, work, and to assessment of their validity relative to the data on which they are based.
- Wichmann and Saunders (2007) look at ways of using databases that document the distribution of typological features across the world's languages to make inferences about genealogical relationships with phylogenetic algorithms developed in biology. The focus is on methodology not only for enhancement of procedures for identifying relationships but also for assessment of the stability of results. One of the proposed classification procedures is neighbour-joining, a clustering method introduced by Saitou and Nei (1987), which is used on combination with bootstrapping for result validation. The discussion includes a review of work in quantitative linguistic phylogeny to 2007.
- Petroni and Serva (2008) base their work on the proposition that languages evolve over time in a way similar to the evolution of mitochon-

> drial DNA in complex biological organisms, and argue that is possible in principle to verify hypotheses about the relationships among languages on this basis. Central to this is the definition of distances among pairs of languages by analogy with the genetic distances among pairs of organisms. The distance between two languages is defined as the Levenshtein distance among words with the same meaning, using language vocabulary as the analogue for DNA among organisms. This approach is applied to Indo-European and Austronesian language groups, looking at 50 languages in each. Two 50×50 Levenshtein language distance matrices are derived. These are clustered using the average linkage hierarchical method and the stability of the result is validated using bootstrapping. The language relationships so obtained are similar to those obtained for these groups by glottochronologists, with some important differences.

For further similar work on phylogeny see Kessler and Lehtonen (2006), Downey et al. (2008), Johnson (2008: Ch. 6), Blanchard et al. (2011), Mehler, Pustylnikov, and Diewald (2011), Abramov and Mehler (2011).

As diachronic corpora have begun to appear, so have applications of cluster analysis in chronological variation research apart from language phylogeny Gries (2011). Gries and Hilpert (2008) address a problem in application of standard clustering methods to grouping of features in diachronic corpus data: that standard clustering methods are blind to temporal sequencing in data in that they cluster all data points in accordance with their relative proximities in the data space. The authors propose variability-based neighbour clustering (VNC), a modification of standard hierarchical agglomerative clustering which provides takes account of chronological stages in diachronic corpus data; for VNC see also Hilpert and Gries (2009) and Gries and Hilpert (2010). Ji (2010) proclaims itself "the first systematic large-scale investigation of the various morpho-syntactic patterns underpinning the evolution of Chinese lexis". Occurrence data for recurrent morpho-syntactic patterns are abstracted from the *Sheffield Corpus of Chinese* and analyzed using PCA and hierarchical clustering.

6.3.3 Cluster analysis in geographical variation research

Applications of cluster analysis in geographical variation research are associated primarily with the universities of Salzburg and Groningen. These are reviewed first, and work by other individuals and groups is covered subsequently.

In a series of articles from 1971 onwards, Goebl developed a dialectometric methodology which culminated in three books: *Dialektometrie. Prinzipien und Methoden des Einsatzes der Numerischen Taxonomie im Bereich der Dialektgeographie* (Goebl 1982), *Dialektometrische Studien. Anhand italoromanischer, rätoromanischer und galloromanischer Sprachmaterialien aus AIS und ALF* (Goebl 1984b), and *Dialectology* (Goebl 1984a). Being in German, the first two are not as well known in the English-speaking research community as they ought to be, but the third established Goebl as a pioneer in dialectometry. In extensive publications since then – for example Goebl (1993a,b, 1997, 2005, 2006, 2010) –, he and his collaborators have refined Séguy's approach to quantitative measurement of dialect distance and developed or adapted a range of graphically-oriented methods for interpretation of dialect distance data, one of which is hierarchical cluster analysis. This has included work on validation of hierarchical clustering results Mucha and Haimerl (2004), Mucha (2006), Haimerl and Mucha (2006) and development of the *Visual DialectoMetry* (VDM) software application which covers the steps of the Salzburg dialectometric evaluation procedure: management of preclassified atlas data, calculation of matrices, and visualizations using different types of maps Haimerl (2006).

Since the late 1990s Nerbonne and his collaborators have developed a methodology based on the dialectometry pioneered by Séguy and Goebl, whose defining characteristics are creation of data based on quantification of observed language use and multivariate analysis of that data, and which incorporates the most sophisticated application of cluster analysis in current linguistic research. To gain an understanding of that methodology a good place to begin is with Nerbonne et al. (1996), which sets out its essentials. In that publication the domain of interest is the Dutch dialect area. Twenty geographical locations are selected to represent the distribution of phonetic usage in that area. The pronunciations of a set of 100 words in each of the twenty locations were recorded and phonetically transcribed. For each pair of locations the transcriptions of the 100 words were compared and the differences quantified; the mean quantitative difference across all 100 words was taken to be the dialect distance between the locations. The dialect distances between all possible pairs of locations were recorded in a 20×20 distance matrix, which was hierarchically cluster analyzed to identify any regularities in the dialect distance data, and the clustering results were found to reconstruct the traditional division of Dutch dialects into Lower Saxon, Frisian, Franconian, and Flemish. Subsequent publications have developed aspects of this methodology; for convenience, the researchers involved in these publications will be referred to as the 'Groningen group' because Nerbonne is the central figure among them, and he is based at the University of Groningen in the Netherlands. A good snapshot of the current state of the Groningen group's methodology is Wieling (2012).

Quantification of observed language use

The discussion in the foregoing chapters of this book has been based on quantification of observed language use in terms of the vector space model: a set of n variables was defined to describe a set of m objects – in the case of DECTE, speakers – in the research domain, and the frequency of occurrence of each of the variables in the domain for each of the objects or speakers was recorded. This approach to quantification has been used in dialectometry by, for example, Hoppenbrouwers and Hoppenbrouwers (1988, 2001) and Babitsch and Lebrun (1989), but not by the Groningen group, which instead adopts the Levenshtein string edit distance – cf. Levenshtein (1966), Sankoff and Kruskal (1999) –, first applied in dialectometry by Kessler (1995). The Levenshtein distance measures the difference between two strings s_1 and s_2, and its value is the minimum number of editing operations required to change s_1 into s_2 or vice versa, where the defined editing operations are single-character insertion, deletion, and substitution. In Table 6.1, for example, the Levenshtein distance between s_1 and s_2 is 2, because substitution of p for c and of u for a transforms s_1 into s_2.

Table 6.1: Edit transformation of s_1 into s_2

s_1	cat
s_2	put

The advantage of string edit over vector space distance measurement for dialectometry stems from a fundamental characteristic of natural language strings: that the occurrence of any given linguistic feature at any given level of analysis is not random but is rather dependent on its context. Selection of a set of words like the 100 in Nerbonne et al. (1996) defines a canonical set of contexts in which the linguistic features of interest occur, and the use of Levenshtein distance between and among corresponding words therefore captures the contextual distribution of the features in those words across the different geographical locations. The vector space model, on the other hand, simply counts features without regard to their contexts.

Given its centrality in the Groningen group's methodology, the Levenshtein distance is described in greater or lesser degrees of detail in many of its publications; see for example Nerbonne and Heeringa (1997, 2001, 2007), Nerbonne, Heeringa, and Kleiweg (1999), Nerbonne and Hinrichs (2006), and Nerbonne and Kleiweg (2007), Heeringa et al. (2006), Wieling (2012). There have also been various developments of the Levenshtein distance since its introduction in the mid-1960s, and a selection of these has been adapted for dialectometric application; evaluation with respect to other edit distances are found in Prokic, Wieling, and Nerbonne (2009), Wieling, Margaretha, and Nerbonne (2011, 2012) and Wieling, Prokic, and Nerbonne (2009), Nerbonne and Heeringa (2010).

Multivariate analysis

Multivariate analysis in dialectometric terms is the simultaneous use of multiple linguistic features to infer dialect distributions from data. The Groningen group describes this simultaneous use of linguistic features as 'aggregation', and contrasts it with the traditional dialectological procedure of analyzing the distribution of a single or at most a very few features. The reasons for the group's preference for aggregation-based analysis are made explicit in Nerbonne (2006, 2008, 2009, 2010), and are, in essence, that it is both scientifically and methodologically superior: on the one hand, inferring dialect distributions from the variabilities of numerous linguistic features both allows for more general hypotheses about these distributions in the population and also endows the hypotheses with greater empirical support, and on the other the arbitrariness inherent in selection of individual features for analysis is mitigated.

Cluster analysis is an important class of multivariate method, and has been fundamental to the Groningen methodology from the outset; see Nerbonne (2010) for a statement of its role. The main emphasis have been on hierarchical clustering Nerbonne and Heeringa (2001), Nerbonne and Kretzschmar (2003), and Nerbonne et al. (1996), Heeringa and Nerbonne (2001), Shackleton (2007), Prokic and Nerbonne (2008) and Prokic, Wieling, and Nerbonne

(2009), Osenova, Heeringa, and Nerbonne (2009), Houtzagers, Nerbonne, and Prokic (2010), Wieling, Nerbonne, and Baayen (2011), Valls, Wieling, and Nerbonne (2013) and Valls et al. (2012) and multimensional scaling Heeringa and Nerbonne (2001), Spruit (2006), Gooskens (2006), Vriend et al. (2008), Prokic and Nerbonne (2008) and Prokic et al. (2009), Spruit, Heeringa, and Nerbonne (2009), Osenova, Heeringa, and Nerbonne (2009), Houtzagers, Nerbonne, and Prokic (2010), Valls, Wieling, and Nerbonne (2013) and Valls et al. (2012), though principal component analysis Shackleton (2007), neighbour-joining clustering Prokic and Nerbonne (2008), *k*-means Prokic and Nerbonne (ibid.), and self-organizing maps Nerbonne and Heeringa (2001) have also been applied. In recent years a graph-based clustering method, bipartite spectral clustering, has increasingly been used Wieling (2012), Wieling, Margaretha, and Nerbonne (2011, 2012) and Wieling and Nerbonne (2009, 2010a,b), Montemagni et al. (2013), and there has been an emphasis on stability and validation of clustering results Nerbonne et al. (2008), Prokic and Nerbonne (2008) and Prokic et al. (2009), Osenova, Heeringa, and Nerbonne (2009), Valls, Wieling, and Nerbonne (2013) and Valls et al. (2012).

Geographical scope

The Groningen group's original application domain was the Dutch dialect area (Nerbonne, Heeringa, and Kleiweg 1999), and this has continued to be cultivated (Heeringa and Nerbonne 2001; Spruit 2006; Spruit, Heeringa, and Nerbonne 2009; Vriend et al. 2008; Wieling 2012; Wieling and Nerbonne 2009, 2010a,b), but its methodology has also been extended to Norwegian (Gooskens 2006), American and British English (Shackleton 2007; Wieling 2012; Wieling, Shackleton, and Nerbonne 2013), Bulgarian (Houtzagers, Nerbonne, and Prokic 2010; Osenova, Heeringa, and Nerbonne 2009; Prokic, Wieling, and Nerbonne 2009; Prokic et al. 2009), Catalan (Valls, Wieling, and Nerbonne 2013; Valls et al. 2012; Wieling 2012), Estonian (Uiboaed et al. 2013), and Tuscan (Montemagni et al. 2013; Wieling 2012).

Software

Gabmap is "a web application aimed especially to facilitate explorations in quantitative dialectology" Nerbonne et al. (2011), and implements the essence of the Groningen Group's methodology.

Applications of cluster analysis in dialectometry apart from those referenced so far are cited briefly and in roughly chronological order in what follows, and more recent ones are then described in greater detail.

For the period before 2000, the earliest example of the use of cluster analysis for geographical variation research appears to be that of Houck (1967), who included factor analysis and hierarchical analysis among other quantitative methods as part of the statistical methodology he proposed for variationist linguistics. Shaw (1974) derived lexical occurrence data from the *Survey of English Dialects* and found three distinct clusters of East Midlands villages using hierarchical cluster analysis. He also proposed but did not implement application of multidimensional scaling and principal component analysis to the data. Black (1976) introduced multidimensional scaling to variationist research by applying it to dialect distance data for Philippine, African, and North American language groups. Embleton (1987, 1993) developed a methodology for application of multidimensional analysis to dialectometry, Embleton and Wheeler (1997a) applied it to Finnish dialect data, Embleton and Wheeler (1997b) to English dialect data, and Embleton, Uritescu, and Wheeler (2004, 2013) to Romanian. Chambers (1997) applied multidimensional scaling to American dialect data; this work is also presented in Chambers and Trudgill (1998). Hoppenbrouwers and Hoppenbrouwers (1988, 2001) used phonetic feature frequency as the basis for measuring dialect distance for Dutch dialect data and applied hierarchical analysis to identification of speaker clusters. Cichocki (1988, 1989) applied dual scaling, a variant of multidimensional scaling, to cluster Canadian French dialect data, and Labov (1994: 485ff.) used MDS on English dialect data. Kessler (1995) grouped dialects according to data taken from various Gaelic-speaking regions using Levenshtein distance and hierarchical cluster analysis, and Schiltz (1996) used hierarchical clustering for German dialectometry.

Examples of work since 2000 are:

- Palander, Opas-Hänninen, and Tweedie (2003) studied the transition zone between two Finnish dialect areas. For each speaker, the number of times each of 198 speakers used each of 10 phonological and morphological variables was recorded and expressed as a percentage of the number of times the variable occurred across the whole corpus. A covariance matrix was calculated from this data and then hierarchically cluster analyzed.
- Speelman, Grondelaers, and Geeraerts (2003) proposed "profile-based linguistic uniformity", described as "a method designed to compare

language varieties on the basis of a wide range or potentially heterogeneous linguistic variables", where "a profile for a particular concept or linguistic function in a particular language variety is the set of alternative linguistic means used to designate that concept or linguistic function in that language variety together with their frequencies (expressed as relative frequencies, absolute frequencies or both)". Distances between pairs language varieties are calculated from these frequencies, and the varieties are clustered using multidimensional scaling and hierarchical analysis.

- Hyvönen, Leino, and Salmenkivi (2007) analyzed lexical variation in Finnish dialects using principal component analysis, hierarchical cluster analysis, and multidimensional scaling. Leino and Hyvönen (2008) empirically compared the usefulness of 'component models' such as principal component analysis, factor analysis, and independent component analysis for identifying structure in the spatial distribution of dialect features relative to two Finnish data sets, one phonetic and one lexical.
- Leinonen (2008) extracted two principal components from phonetic data for 1014 speakers in 91 Swedish dialects, which captured more than 75% of the data variance. A two-dimensional plot of these components showed a strong resemblance to vowels in a formant plane.
- Vriend et al. (2008) integrated linguistic, geographic and social distance data to study the impact the Dutch-German state border has had on the linguistic characteristics of a sub-area of the Kleverlandish dialect area. Multidimensional scaling was applied to a vector space model of the data to extract speaker-clusters.
- Mukherjee and Gries (2009) investigate verb transitivity across various New Englishes using hierarchical clustering to analyze data derived from a range of international English corpora. Gries and Mukherjee (2010) applied hierarchical clustering to lexical co-occurrence preferences and investigated if and to what degree *n*-grams distinguish between different modes and varieties in the same components of the *International Corpus of English.*
- Szmrecsanyi (2008) and Szmrecsanyi and Kortmann (2009) used multidimensional scaling, hierarchical clustering, and PCA to analyze morphosyntactic variability in British English dialects. Szmrecsanyi (2011) is a sketch of methodologies "to tap corpora for exploring aggregate linguistic distances between dialects or varieties as a function of prop-

erties of geographic space". The paper describes the creation of distance matrices and visualization techniques for dialectal distributions, including cluster maps, relative to the *Freiburg Corpus of English Dialects* and focussing on regional variation in morphosyntax. Szmrecsanyi (2013) is a book-length study of grammatical variation in British English dialects using corpus-based dialectometry. It gives an overview of the development of dialectometry and includes chapters on quantitative data creation, proximity measurement, and multivariate analysis, including multidimensional scaling and hierarchical clustering.

- Grieve, Speelman, and Geeraerts (2011) propose a methodology for analysis of regional linguistic variation which "identifies individual and common patterns of spatial clustering in a set of linguistic variables measured over a set of locations based on a combination of three statistical techniques: spatial autocorrelation, factor analysis, and cluster analysis". The methodology is exemplified using hierarchical clustering applied to lexical variation data for 206 American cities.
- Meschenmoser and Pröll (2012) and Pröll (2013) cluster data from dialect maps using the empirical covariance and fuzzy clustering.

6.3.4 Cluster analysis in social variation research

Several researchers have used cluster analysis for sociolinguistic interpretation of data relating to variation in phonetic usage among speakers. In chronological order:

- Sankoff and Cedergren (1976) used multidimensional scaling to measure the dimensionality of sociolinguistic variation in Montreal French.
- Jones-Sargent (1983) used hierarchical clustering to analyze phonetic data abstracted from the Tyneside Linguistic Survey (TLS) and interpreted the result in relation to social data associated with the TLS speakers to see if any sociolinguistically interesting hypotheses could be derived. This work is the direct ancestor of the present book.
- Horvath (1985) and Horvath and Sankoff (1987) used principal component and principal coordinate analysis on vowel variation data taken from samples of Australian speech to group speakers on the basis of their linguistic behaviour and to determine the major linguistic and social dimensions of variation in the data.

- Labov (1994: 485ff.) used multidimensional scaling to study lexical diffusion, and Labov (2001) studied the role of social factors in language change using PCA.
- Moisl and Jones (2005) validated clustering of the DECTE data used for exemplification in the foregoing chapters by comparing results derived by a linear method, hierarchical analysis, and a nonlinear one, the self-organizing map.
- Moisl, Maguire, and Allen (2006) and Moisl and Maguire (2008) interpreted hierarchical analysis of the DECTE data in terms of the social data associated with the DECTE speakers to derive sociologically relevant hypotheses.
- Moisl (2012) used PCA to map the distribution of phonetic usage across the DECTE speakers.

Another area of sociolinguistics where cluster analysis has been used is in the study of linguistic register, that is, how different language varieties are used for particular purposes or in particular social settings. This is primarily associated with the work of Biber, who in an extensive series of publications developed his Multi-Dimensional (MD) approach to linguistic register and textual genre variation using factor analysis and hierarchical clustering (for example Biber and Finegan (1986), Biber (1992, 1996, 2006, 2009), Biber, Conrad, and Reppen (1998); for a recent overview of this work see Baker (2010) and the special issue of the journal *Corpora*, vol. 8 [2013]. Gries, Newman, and Shaoul (2011) explored the use of different-length *n*-grams as a basis for identifying relationships between registers using *BNC Baby* and *ICE-GB* corpora and hierarchical clustering
indexsubcluster!linguistic register.

Recent research has seen the development of sociolectometry as a parallel to dialectometry, which studies lexical variation across language varieties in relation to social factors using vector space models based on the distributional hypothesis referred to earlier, and a variety of clustering methods such as multidimensional scaling, principal component analysis, hierarchical clustering, and clustering by committee (Pantel and Lin 2002); see for example Peirsman, Geeraerts, and Speelman (2010), Ruette, Speelman, and Geeraerts (2013), Grieve, Speelman, and Geeraerts (2011), Heylen, Speelman, and Geeraerts (2012), Heylen and Ruette (2013).

7 Conclusion

The foregoing discussion has proposed cluster analysis as a tool for generating linguistic hypotheses from natural language corpora. The motivation for doing so was practical: as the size and complexity of corpora and of data abstracted from them have grown, so the traditional paper-based approach to discerning structure in them has become increasingly intractable, and cluster analysis offers a solution. Hypothesis generation based on cluster analysis has two further advantages in terms of scientific methodology, however. These are objectivity and replicability (Audi 2010; Chalmers 1999; Daston and Galison 2010; Gauch 2003).

- **Objectivity.** The question of whether humans can have objective knowledge of reality has been central in philosophical metaphysics and epistemology since Antiquity and, in recent centuries, in the philosophy of science. The issues are complex, controversy abounds, and the associated academic literatures are vast – saying what an objective statement about the world might be is anything but straightforward, as Chapter 2 has already noted. The position adopted here is that objective knowledge is ultimately impossible simply because no observation of the natural world and no interpretation of such observation can be independent of the constraints which the human cognitive system and the physiological structures which implement it impose on these things. On this assumption, objectivity in science becomes a matter of attempting to identify sources of subjectivity and to eliminate them as factors in formulating our species-specific understanding of nature. An important way of doing this in science is to use generic methods grounded in mathematics and statistics, since such methods minimize the chance of incorporating subjective assumptions into the analysis, whether by accident or design.
- **Replicability** is a foundational principle of scientific method. Given results based on a scientific experiment, replicability requires that the data creation and analytical methods used to generate those results are sufficiently well specified to allow the experiment to be reproduced and for that reproduction to yield results identical to or at least compatible with the originals. The obvious benefit is elimination of fraud.

> There have been and presumably always will be occasional cases of scientific fraud, but this is not the main motivation for the replicability requirement. The motivation is, rather, avoidance of error: everyone makes mistakes, and by precise and comprehensive specification of procedures the researcher enables subject colleagues suspicious of the validity of results to check them.

Cluster analysis and the associated data representation and transformation concepts are objective in the above sense in that they are mathematically grounded, and analyses based on them are replicable as a consequence in that experimental procedures can be precisely and comprehensively specified. In the foregoing search for structure in the DECTE corpus the initial selection of phonetic variables was subjective, but the data representation and the transformation methods used to refine the selection were generic, as were the clustering methods used to analyze the result. The data and clustering methodology was, moreover, specified precisely enough for anyone with access to the DECTE corpus to check the results of the analysis.

Given the methodological advantages of cluster analysis for hypothesis generation, the hope is that this book will foster its adoption for that purpose in the corpus linguistics community.

8 Appendix

This Appendix lists software implementations of the clustering methods presented earlier. The coverage is not exhaustive: only software known to the author to be useful either via direct experience or online reviews is included.

8.1 Cluster analysis facilities in general-purpose statistical packages

Most general-purpose statistics / data analysis packages provide some subset of the standard dimensionality reduction and cluster analysis methods: principal component analysis, factor analysis, multidimensional scaling, *k*-means clustering, hierarchical clustering, and sometimes others not covered in this book. In addition, they typically provide an extensive range of extremely useful data creation and transformation facilities. A selection of them is listed in alphabetical order below; URLs are given for each and are valid at the time of writing.

8.1.1 Commercial

- GENSTAT
  `http://www.vsni.co.uk/software/genstat`
- MINITAB
  `http://www.minitab.com/en-US/products/minitab/`
- NCSS
  `http://www.ncss.com/`
- SAS
  `http://www.sas.com/`
- SPSS
  `http://www-01.ibm.com/software/uk/analytics/spss/`
- STATA
  `http://www.stata.com/`
- STATGRAPHICS
  `http://www.statgraphics.com/`

- STATISTICA
  `http://www.statsoft.com/`
- SYSTAT
  `http://www.systat.com/`

8.1.2 Freeware

- CHAMELEON STATISTICS
  `http://www.seventh-sense-software.com/chameleon.htm`
- MICROSIRIS
  `http://www.microsiris.com/`
- ORIGINLAB
  `http://www.originlab.com/`
- PAST
  `http://folk.uio.no/ohammer/past/`
- PSPP
  `http://www.gnu.org/software/pspp/`
- TANAGRA
  `http://eric.univ-lyon2.fr/~ricco/tanagra/en/tanagra.html`
  - This is unusual in including the self-organizing map in addition to the standard methods.
- WINIDAMS
  `www.unesco.org/idams/`
- WINSTAT
  `http://www.winstat.com/`

8.2 Cluster analysis-specific software

The following software is designed specifically for cluster analysis.

8.2.1 Commercial

- ANTHROPAC
  `http://www.analytictech.com/anthropac/apacdesc.htm`
  - Principal component analysis, factor analysis, hierarchical, multidimensional scaling

- BMDP
  http://www.statistical-solutions-software.com
  /bmdp-statistical-software/cluster-analysis/
  - Hierarchical, k-means
- CLUSTAN
  http://www.clustan.com/
  - Hierarchical, k-means
- GELCOMPAR
  http://www.applied-maths.com/gelcompar-ii
  - Principal component analysis, hierarchical, multidimensional scaling
- KCS
  http://www.kovcomp.co.uk/mvsp/.
  - Principal component analysis, hierarchical
- STATISTIXL
  http://statistixl.software.informer.com/
  - Principal component analysis, factor analysis, hierarchical
- VISCOVER
  http://www.viscovery.net/
  - Self-organizing map
- VISIPOINT
  http://www.visipoint.fi/
  - Self-organizing map, Sammon's mapping

8.2.2 Freeware

- CLUSTER 3.0
  http://bonsai.hgc.jp/~mdehoon/software/cluster/
  - Hierarchical, k-means, self-organizing map, principal component analysis
- DATABIONIC ESOM
  http://databionic-esom.sourceforge.net/
  - Emergent self-organizing map, and extension of the SOM described earlier

- GENESIS
  `http://genome.tugraz.at/genesisclient/genesisclient-description.shtml`
  - Principal component analysis, hierarchical, *k*-means, self-organizing map
- MICROARRAYS CLUSTER
  `http://derisilab.ucsf.edu/microarray/software.html` or `http://rana.lbl.gov/EisenSoftware.htm`
  - Principal component analysis, hierarchical, *k*-means, self-organizing map
- MULTIBASE
  `http://www.numericaldynamics.com/`
  - Principal component analysis, hierarchical
- OC
  `http://www.compbio.dundee.ac.uk/Software/OC/oc.html`
  - Hierarchical
- PERMUTMATRIX
  `http://www.lirmm.fr/~caraux/PermutMatrix/`
  - Hierarchical
- SERF CLUSTERS
  `http://www.bram.org/serf/Clusters.php`
  - Hierarchical

8.3 Programming languages

All the foregoing packages are good, most are excellent, and any corpus linguist who is seriously interested in applying cluster analysis to his or her research can use them with confidence. That corpus linguist should, however, consider learning how to use at least one programming language for this purpose. The packages listed above offer a small subset of the dimensionality reduction and cluster analysis methods currently available in the research literature, and users of them are restricted to this subset; developments of and alternatives to these methods, such as DBSCAN and the many others that were not even mentioned, remain inapplicable. These developments and alternatives have appeared and continue to appear for a reason: to refine cluster analytic methodology. In principle, researchers should be in a position to use the best methodology available in their field, and programming makes the current state of clustering methodology accessible to corpus linguists because it renders implementation of any current or future clustering method feasible. A similar case for programming is made by Gries (2011a).

There are numerous programming languages, and in principle any of them can be used for corpus linguistic applications. In practice, however, two have emerged as the languages of choice for quantitative natural language processing generally: Matlab and *R*. Both are high-level programming languages in the sense that they provide many of the functions relevant to statistical and mathematical computation as language-native primitives and offer a wide range of excellent graphics facilities for display of results. For any given algorithm, this allows programs to be shorter and less complex than they would be for lower-level, less domain-specific languages like, say, Java or C++, and makes the languages themselves easier to learn.

Matlab (`http://www.mathworks.co.uk/`) is described by its website as "a high-level language and interactive environment for numerical computation, visualization, and programming". It provides numerous and extensive libraries of functions specific to different types of quantitative computation such as signal and image processing, control system design and analysis, and computational finance. One of these libraries is called "Math, Statistics, and Optimization", and it contains a larger range of dimensionality reduction and cluster analysis functions than any of the above software packages: principal component analysis, canonical correlation, factor analysis, singular value decomposition, multidimensional scaling, Sammon's mapping, hierarchical clustering, *k*-means, self-organizing map, and Gaussian mixture models. This is a useful gain in coverage, but the real advantage of Matlab over the packages is twofold. On the one hand, Matlab makes it possible for users to contribute application-specific libraries to the collection of language-native ones. Several such contributed libraries exist for cluster analysis, and these substantially expand the range of available methods. Some examples are:

- D. Corney: Clustering with Matlab
  `http://www.dcorney.com/ClusteringMatlab.html`
- J. Abonyi: Clustering and Data Analysis Toolbox
  `http://www.mathworks.co.uk/matlabcentral/fileexchange/7473-clustering-and-data-analysis-toolbox`
- Non-linearity and Complexity Research Group, Aston University
  `http://www1.aston.ac.uk/ncrg/`

- A. Strehl: Cluster Analysis and Cluster Ensemble Software
  `http://www.ideal.ece.utexas.edu/~strehl/soft.html`
- SAGE Research Methods: Cluster Analysis. A Toolbox for Matlab
  `http://srmo.sagepub.com/view/sage-hdbk-quantitative-methods-in-psychology/n20.xml`
- Matlab Toolbox for Dimensionality Reduction
  `http://homepage.tudelft.nl/19j49/Matlab-Toolbox-for-Dimensionality-Reduction.html`
- SOM Toolbox
  `http://www.cis.hut.fi/projects/somtoolbox/`

On the other hand, because the user has access to the program code both for the native Matlab and the contributed libraries, functions can be modified according to need in line with current research requirements, or, as a last resort, the required functions can be written *ab initio* using the rich collection of already-existing mathematical and statistical ones. Finally, there is a plethora of online tutorials and Matlab textbooks ranging from introductory to advanced, so accessibility is not a problem.

R (`http://www.r-project.org/`) is described by its website as "a free software environment for statistical computing and graphics", and it has the same advantages over the clustering software packages as Matlab. R provides an extensive range of dimensionality reduction and cluster analysis functions, which are listed at the following websites:

- Dimensionality reduction:
  `http://cran.r-project.org/web/views/Multivariate.html`
- Cluster analysis:
  `http://cran.r-project.org/web/views/Cluster.html`

Examples of user-contributed libraries are:

- H. Fritz et al: TCLUST
  `http://cran.r-project.org/web/packages/tclust/vignettes/tclust.pdf`
- CRAN: cluster
  `http://cran.r-project.org/web/packages/cluster/index.html`
- R. Suzuki and H. Shimodaira: PVCLUST
  `http://www.is.titech.ac.jp/~shimo/prog/pvclust/`
- V. Gopal et al: BAYESCLUST
  `http://www.stat.ufl.edu/~viknesh/publications/bayesclustPlain.pdf`

There are numerous online tutorials for *R* generally and for cluster analysis using *R* specifically, as well as a range of textbooks. Of particular interest to corpus linguists are: Baayen (2008), Gries (2009a,b), and Oksanen (2010).

In recent years *R* has been emerging as the preferred language for quantitative natural language processing and corpus linguistics, not least because, unlike Matlab, it is available online free of charge.

8.4 Corpus linguistic-specific applications using clustering

The dialectometric methodology of the Groningen group, described in the foregoing literature review, has been implemented in *Gabmap*, described as "a Web application aimed especially to facilitate explorations in quantitative dialectology – or dialectometry – by enabling researchers in dialectology to conduct computer-supported explorations and calculations even if they have relatively little computational expertise" (Nerbonne et al. 2011). As the literature review showed, cluster analysis is an important aspect of that methodology, and it is duly included among the *Gabmap* facilites. The application is available online free of charge at `http://www.gabmap.nl/`.

References

Abramov, O. and A. Mehler (2011): "Automatic Language Classification by means of Syntactic Dependency Networks." In: *Journal of Quantitative Linguistics* 18 (1); 291–336.

Aggarwal, C., A. Hinneburg, and D. Keim (2001): "On the surprising behaviour of distance metrics in high dimensional space." In: *Proceedings of the 8th International Conference on Database Theory*; 420–34.

Agrawal, R. et al. (1998): "Automatic subspace clustering of high-dimensional data for data mining applications." In: *Proceedings of the ACM SIGMOD '98 International Conference on Management of Data*; 94–105.

— (2005): "Automatic Subspace Clustering of High Dimensional Data." In: *Data Mining and Knowledge Discovery* 11; 5–33.

Allegrini, P., S. Montemagni, and V. Pirrelli (2000): "Learning word clusters from data types." In: *COLING-2000*; 8–14.

Allen, W. et al. (2007): "A Linguistic Time-Capsule: The Newcastle Electronic Corpus of Tyneside English." In: *Creating and Digitising Language Corpora, Vol. 2: Diachronic Databases*. Ed. by J. Beal, K. Corrigan, and H. Moisl. Basingstoke: Palgrave Macmillan; 16–48.

Allinson, N. et al. (2001): *Advances in Self-Organising Maps*. Berlin: Springer.

Almuhareb, A. and M. Poesio (2004): "Attribute-based and value-based clustering: An evaluation." In: *Proceedings of the conference on empirical methods in natural language processing (EMNLP 2004*; 158–165.

Anderberg, M. (1973): *Cluster Analysis for Applications*. Waltham, Massachusetts: Academic Press.

Andersen, S. (2001): "The emergence of meaning: generating symbols from random sounds – a factor analytic model." In: *Journal of Quantitative Linguistics* 8; 101–36.

Andreev, S. (1997): "Classification of Verbal Characteristics Based on their Implication Force by Means of Cluster- and Factor Analysis." In: *Journal of Quantitative Linguistics* 4; 23–29.

Ankerst, M. et al. (1999): "OPTICS: Ordering points to identify the clustering structure." In: *Proceedings of the ACM SIGMOD International Conference on Management of Data*; 49–60.

Anscombe, F. (1973): "Graphs in Statistical Analysis." In: *American Statistician* 27; 17–21.

Arabie, P. and L. Hubert (1996): "An overview of combinatorial data analysis." In: *Clustering and Classification*. Ed. by P. Arabie, L. Hubert, and G. De Soete. Singapore: World Scientific; 5–63.

Arora, S. and B. Barak (2009): *Computational Complexity. A Modern Approach*. Cambridge, UK: Cambridge University Press.

Arppe, A. et al. (2010): "Cognitive Corpus Linguistics: five points of debate on current theory and methodology." In: *Corpora* 5; 1–27.

Audi, R. (2010): *Epistemology: A Contemporary Introduction to the Theory of Knowledge*. London: Routledge.

Baarsch, J. and M. Celebi (2012): "Investigation of internal validity measures for k-means clustering." In: *Proceedings of the 2012 IAENG International Conference on Data Mining and Applications (ICDMA 2012)*; 471–476.

Baayen, H. (2001): *Word Frequency Distributions*. Berlin: Springer.

— (2008): *Analyzing Linguistic Data. A practical Introduction to Statistics using R*. Cambridge, UK: Cambridge University Press.

Babitsch, R. and E. Lebrun (1989): "Dialectometry as computerized hierarchical classification analysis." In: *Journal of English Linguistics* 22; 83–87.

Baker, F. and L. Hubert (1975): "Measuring the power of hierarchical cluster analysis." In: *Journal of the American Statistical Association* 70; 31–38.

Baker, P. (2009): *Contemporary Corpus Linguistics*. London: Continuum.

— (2010): *Sociolinguistics and Corpus Linguistics*. Edinburgh: Edinburgh University Press.

Baker, W. and B. Derwing (1982): "Response coincidence analysis for language acquisition strategies." In: *Applied Psycholinguistics* 3; 193–221.

Balasumbramanian, M. and E. Schwartz (2002): "The Isomap algorithm and topological stability." In: *Science* 295; 7.

Ball, G. and D. Hall (1967): "A clustering technique for summarixing multivariate data." In: *Behavioural Science* 12; 153–155.

Banerjee, A. and R. Dave (2004): "Validating clusters using the Hopkins statistic." In: *Proceedings of the 2004 IEEE International Conference on Fuzzy Systems*; 149–153.

Baroni, M. and S. Evert (2009): "Statistical methods for corpus exploitation." In: *Corpus Linguistics. An International Handbook. Vol. 2*. Ed. by A. Lüdeling and M. Kytö. Berlin: Walter de Gruyter; 777–803.

Basili, R., M. Pazienza, and P. Velardi (1996): "Integrating general-purpose and corpus-based verb classification." In: *Computational Linguistics* 22; 559–568.

Batagelj, V., T. Pisanski, and D. Kerzic (1992): "Automatic clustering of languages." In: *Computational Linguistics* 18; 339–352.

Bauer, H. and K. Pawelzik (1992): "Quantifying the neighbourhood preservation of self-organizing feature maps." In: *IEEE Transactions on Neural Networks* 3; 570–579.

Bauer, H. and T. Villmann (1997): "Growing a hypercubical output space in a self-organizing feature map." In: *IEEE Transactions on Neural Network Learning Systems* 8; 218–226.

Bauerlein, M. et al. (2010): "We must stop the avalanche of low-quality research." In: *The Chronicle of Higher Education* June 13.

Bayley, R. (2013): "The quantitative paradigm." In: *The Handbook of Language Variation and Change, 2nd edition*. Ed. by P. Chambers J. Trudgill and N. Schilling-Estes. Hoboken, NJ: Blackwell; 85–107.

Beal, J., L. Burbano-Elzondo, and C. Llamas (2012): *Urban North-Eastern English: Tyneside to Teesside*. Edinburgh: Edinburgh University Press.

Bekkerman, R. et al. (2003): "Distributional word clusters vs. words for text categorization." In: *Journal of Machine Learning Research* 3; 1183–1208.

Belkin, M. and P. Niyogi (2003): "Laplacian Eigenmaps for Dimensionality Reduction and Data Representation." In: *Neural Computation* 15; 1373–1396.

Ben-Hur, A., A. Elisseeff, and I. Guyon (2002): "A stability based method for discovering structure in clustered data." In: *Pacific Symposium on Biocomputing* 7; 6–17.

Berdan, R. (1978): "Multidimensional analysis of vowel variation." In: *Linguistic Variation: Models and Methods*. Ed. by D. Sankoff. Waltham, Massachusetts: Academic Press; 149–160.

Berez, A. and S. Gries (2009): "In defense of corpus-based methods: a behavioral profile analysis of polysemous get in English." In: *Proceedings of the 24th Northwest Linguistics Conference, University of Washington Working Papers in Linguistics Vol. 27*. Ed. by S. Moran, D. Tanner, and M. Scanlon; 157–166.

Bergsma, S., D. Lin, and R. Goebel (2008): "Discriminative learning of selectional preference from unlabeled text." In: *Proceedings of the 2008 Conference on Empirical Methods in Natural Language Processing*; 59–68.

Berkhin, P. (2006): "Survey of Clustering Data Mining Techniques." In: *Grouping Multidimensional Data – Recent Advances in Clustering*. Ed. by J. Kogan, C. Nicholas, and M. Teboulle. Berlin: Springer; 25–71.

Berkhin, P. and I. Dhillon (2009): "Knowledge discovery: clustering." In: *Encyclopedia of Complexity and Systems Science*. Ed. by R. Meyers. Berlin: Springer; 5051–64.

Bertuglia, C. and F. Vaio (2005): *Nonlinearity, Chaos, and Complexity: The Dynamics of Natural and Social Systems*. Oxford: Oxford University Press.

Best, K. (2006): *Quantitative Linguistik – Eine Annäherung*. 3rd ed. Göttingen: Peust and Gutschmidt.

Beyer, K. et al. (1999): "When Is 'Nearest Neighbor' Meaningful?" In: *ICDT '99 Proceedings of the 7th International Conference on Database Theory*; 217–235.

Bezdek, J. and R. Hathaway (2002): "VAT: a tool for visual assessment of (cluster) tendency." In: *Proceedings of the 2002 International Joint Conference on Neural Networks, Honolulu, IEEE*; 2225–2230.

Bezdek, J., R. Hathaway, and J. Huband (2006): "Scalable visual assessment of cluster tendency for large data sets." In: *Pattern Recognition* 39; 1315–1324.

Bezdek, J. and N. Pal (1998): "Some new indexes of cluster validity." In: *IEEE Transaction on Systems,Man, and Cybernetics(Part B)* 28; 301–315.

Biber, D. (1992): "The multi-dimensional approach to linguistic analyses of genre variation: An overview of methodology and findings." In: *Language Resources and Evaluation* 26; 331–345.

— (1996): "Investigating Language Use through Corpus-based Analyses of Association Patterns." In: *International Journal of Corpus Linguistics* 1; 171–197.

— (2006): *University Language. A Corpus-Based Study of Spoken and Written Registers.* Amsterdam: John Benjamins.

— (2009): "Multi-dimensional approaches." In: *Corpus Linguistics. An International Handbook. Vol 2*. Ed. by A. Lüdeling and M. Kytö. Berlin: Walter de Gruyter; 822–855.

Biber, D., S. Conrad, and R. Reppen (1998): *Corpus Linguistics. Investigating Language Structure and Use*. Cambridge: Cambridge University Press.

Biber, D. and E. Finegan (1986): "An Initial Typology of English Text Types." In: *Corpus Linguistics II*. Ed. by J. Aarts and W. Meijis. Amsterdam: Rodopi; 19–46.

Bishop, C. (1995): *Neural Networks for Pattern Recognition.* Oxford: Clarendon Press.

— (2006): *Pattern Recognition and Machine Learning*. Berlin: Springer.

Bishop, C., M. Svensen, and C. Williams (1998): "The Generative Topographic Mapping." In: *Neural Computation* 10; 215–234.

Black, P. (1976): "Multidimensional scaling applied to linguistic relationships." In: *Cahiers de l'Institut de linguistique de Louvain* 3; 43–92.

Blackmore, J. and R. Miikkulainen (1995): "Visualizing high-dimensional structure with the incremental grid growing neural network." In: *Machine Learning: Proceedings of the 12th International Conference*. Ed. by A. Prieditis and S. Russell; 55–63.

Blanchard, P. et al. (2011): "Geometric representation of language taxonomies." In: *Computer Speech and Language* 25; 679–99.

Bolshakova, N. and F. Azuaje (2003): "Cluster validation techniques for genome expression data." In: *Signal Processing* 83; 825–833.

Borg, I. and P. Groenen (2005): *Modern Multidimensional Scaling*. 2nd ed. Berlin: Springer.

Boslaugh, S. and P. Watters (2008): *Statistics in a Nutshell.* Cambridge MA: O'Reilly.

Bradley, P. and U. Fayyad (1998): "Refining initial points for *k*-means clustering." In: *Proceedings of the Fifteenth international Conference on Machine Learning*; 91–99.

Brew, C. and S. Schulte im Walde (2002): "Spectral clustering for German verbs." In: *Proceedings of the 7th Conference on Empirical Methods in Natural Language Processing, Philadelphia (EMNL)*; 117–124.

Brown, P. et al. (1992): "Class-based *n*-gram models of natural language." In: *Computational Linguistics* 18; 467–479.

Brun, M. et al. (2007): "Mode-based evaluation of clustering validation measures." In: *Pattern Recognition* 40; 807–24.

Burnham, K. and D. Anderson (2002): *Model Selection and Multi-Model Inference: A Practical Information-Theoretic Approach.* 2nd ed. Berlin: Springer.

Butler, C. (1985): *Statistics in Linguistics.* Oxford: Basil Blackwell.

Calderone, B. (2009): "Learning Phonological Categories by Independent Component Analysis." In: *Journal of Quantitative Linguistics* 16; 132–156.

Camastra, F. (2003): "Data dimensionality estimation methods: a survey." In: *Pattern Recognition* 36; 2945–54.

Carreira-Perpinan, M. (1997): *A Review of Dimension Reduction Techniques.* Tech. rep. CS-96-09. Deptartment of Computer Science, University of Sheffield.

— (2011): *Dimensionality Reduction.* London: Chapman and Hall CRC.

Carroll, J. and I. Dyen (1962): "High speed computation of lexicostatistical indices." In: *Language* 38; 274–8.

Cavalli-Sforza, L. (2000): *Genes, Peoples, and Languages.* London: Allen Lane.

Cedergren, H. and D. Sankoff (1974): "Variable rules: performance as a statistical reflection of competence." In: *Language* 50; 333–55.

Chaithin, G. (2001): *Exploring Randomness.* Berlin: Springer.

Chalmers, A. (1999): *What is this thing called science?* 3rd ed. Burr Ridge IL: Open University Press.

Chambers, J. (1997): "Mapping variability." In: *Issues and Methods in Dialectology.* Ed. by A. Thomas. Cardiff: University of Wales Press; 284–93.

Chambers, J. and P. Trudgill (1998): *Dialectology.* Cambridge: Cambridge University Press.

Chen, J. and J. Chang (1998): "Topical clustering of MRD senses based on information retrieval techniques." In: *Computational Linguistics* 24; 61–95.

Chomsky, N. (1959): "Review of Skinner's Verbal Behavior." In: *Language* 35; 26–58.

Chu, C., J. Holliday, and P. Willett (2009): "Effect of data standardization on chemical clustering and similarity searching." In: *Journal of Chemical Information and Modeling* 49; 155–161.

Cichocki, W. (1988): "Uses of dual scaling in social dialectology: multidimensional analysis of vowel variation." In: *Methods in Dialectology: Proceedings of the Sixth International Conference at University College, North Wales, August 1987*. Ed. by M. Thomas; 187–99.

— (1989): "An application of dual scaling in dialectometry." In: *Journal or English Linguistics* 22; 91–95.

— (2006): "Geographic variation in Acadian French /r/: what can correspondence analysis contribute toward and explanation?" In: *Literary and Linguistic Computing* 21; 529–41.

Cormen, T. et al. (2009): *Introduction to Algorithms*. 3rd ed. Cambridge MA: MIT Press.

Corrigan, K., A. Mearns, and H. Moisl (2012): *The Diachronic Electronic Corpus of Tyneside English*. URL: `http://research.ncl.ac.uk/decte/`.

Cortina-Borja, M., J. Stuart-Smith, and L. Valinas-Coalla (2002): "Multivariate classification methods for lexical and phonological dissimilarities and their application to the Uto-Aztecan family." In: *Journal of Quantitative Linguistics* 9; 97–124.

Cortina-Borja, M. and L. Valinas-Coalla (1989): "Some remarks on Uto-Aztecan classification." In: *International Journal of American Linguistics* 55; 214–39.

Cottrell, M., E. de Bodt, and M. Verleysen (2001): "A statistical tool to assess the reliability of self-organizing maps." In: *Advances in Self-Organising Maps*. Ed. by N. Allinson et al. Berlin: Springer; 7–14.

Cottrell, M., J. Fort, and G. Pages (1998): "Theoretical aspects of the SOM algorithm." In: *Neurocomputing* 21; 119–38.

Coupez, A., E. Evrard, and J. Vansina (1975): "Classification d'un échantillon de langues bantoues d'apres la lexicostatistique." In: *Africana Linguistica* 6; 131–58.

Cover, T. (1965): "Geometrical and statistical properties of systems of linear inequalities with applications in pattern recognition." In: *IEEE Transactions on Electronic Computers* EC–14; 326–334.

Croft, W. and K. Poole (2008): "Inferring universals from grammatical variation: Multidimensional scaling for typological analysis." In: *Theoretical Linguistics* 34; 1–37.

Cysouw, M., S. Wichmann, and D. Kamholz (2006): "A critique of the separation base method for genealogical subgrouping with data from Mixezoquean." In: *Journal of Quantitative Linguistics* 13; 225–264.

Dalton, L., V. Ballarin, and M. Brun (2009): "Clustering algorithms: on learning, validation, performance, and applications to genomics." In: *Current Genomics* 10; 430–45.

Daston, L. and P. Galison (2010): *Objectivity*. Cambridge MA: MIT Press.

Daszykowski, M., B. Walczak, and D. Massart (2001): "Looking for natural patterns in data, part 1: density-based approach." In: *Chemometrics and Intelligent Laboratory Systems* 56; 83–92.

Davies, D. and D. Bouldin (1979): "Cluster separation measure." In: *IEEE Transactions on Pattern Analysis and Machine Intelligence* 1; 224–7.

Davis, L. (1990): *Statistics in Dialectology*. Tuscaloosa AL: University of Alabama Press.

Dawoud, H. and W. Ashour (2012): "Modified DBSCAN Clustering Algorithm for Data with Different Densities." In: *Computing and Information Systems* 16(3); 16–21.

De Bodt, E., M. Cottrell, and M. Verleysen (2002a): "Are they really neighbors? A statistical analysis of the SOM algorithm output." In: *AISTATS 2001. Proceedings of the International workshop on Artificial Intelligence and Statistics*; 35–40.

— (2002b): "Statistical tools to assess the reliability of self-organizing maps." In: *Neural Networks* 15; 967–78.

Deborah, L., R. Baskaran, and A. Kannan (2010): "A survey on internal validity measure for cluster validation." In: *International Journal of Computer Science and Engineering Survey* 1; 85–101.

Delmestri, A. and N. Cristianini (2012): "Linguistic phylogenetic inference by PAM-like matrices." In: *Journal of Quantitative Linguistics* 19; 95–120.

Demartines, P. and J. Hérault (1997): "Curvilinear component analysis. A self-organizing neural network for nonlinear mapping of data sets." In: *IEEE Transactions on Neural Networks* 8; 148–54.

DeSilva, V. and J. Tenenbaum (2003): "Unsupervised learning of curved manifolds." In: *Nonlinear Estimation and Classification*. Ed. by D. Denison et al. Berlin: Springer; 453–466.

Devaney, M. and A. Ram (1997): "Efficient Feature Selection in Conceptual Clustering." In: *Proceedings of the Fourth International Conference on Machine Learning*; 92–7.

Devereux, B. et al. (2009): "Towards unrestricted, large-scale acquisition of feature-based conceptual representations from corpus data." In: *Research on Language and Computation* 7; 137–170.

Deza, M. and E. Deza (2009): *Encyclopedia of Distances*. Berlin: Springer.

Dhillon, I., Y. Guan, and B. Kulis (2004): "Kernel k-means, spectral clustering and normalized cuts." In: *Proceedings of the Tenth ACM SIGKDD International Conference on Knowledge Discovery and Data Mining*; 551–556.

— (2005): *A unified view of kernel k-means, spectral clustering and graph cuts*. Tech. rep. UTCS Technical Report TR 04 25. Department of Computer Sciences, University of Texas at Austin.

Diamantaras, K. and S. Kung (1996): *Principal Component Neural Networks. Theory and Applications*. Hoboken NJ: John Wiley and Sons.

Divjak, D. (2010): *Structuring the Lexicon: a Clustered Model for Near-Synonymy*. Berlin: Mouton de Gruyter.

Divjak, D. and S. Gries (2006): "Ways of trying in Russian. Clustering behavioral profiles." In: *Corpus Linguistics and Linguistic Theory* 2; 23–60.

— (2008): "Clusters in the mind? Converging evidence from near synonymy in Russian." In: *The Mental Lexicon* 3; 188–213.

— (2009): "Corpus-based cognitive semantics: a contrastive study of phrasal verbs in English and Russian." In: *Studies in cognitive corpus linguistics*. Ed. by K. Dziwirek and B. Lewandowska-Tomaszczyk. New York: Peter Lang; 273–96.

Docherty, G. and P. Foulkes (1999): "Sociophonetic variation in glottals in Newcastle English." In: *Proceedings of the 14th International Congress of Phonetic Sciences*.

Donoho, D. (2000): *High-dimensional data analysis: the curses and blessings of dimensionality.* URL: http://www-stat.stanford.edu/~donoho/Lectures/AMS2000/Curses.pdf.

Downey, S. et al. (2008): "Computational Feature-Sensitive Reconstruction of Language Relationships: Developing the ALINE Distance for Comparative Historical Linguistic Reconstruction." In: *Journal of Quantitative Linguistics* 15; 340–369.

Draper, N. and H. Smith (1998): *Applied Regression Analysis.* 3rd. Hoboken NJ: John Wiley and Sons.

Dubes, R. and A. Jain (1979): "Validity studies in clustering methodologies." In: *Pattern Recognition* 11; 235–54.

Dunn, J. (1974): "Well separated clusters and optimal fuzzy partitions." In: *Journal of Cybernetics* 4; 95–104.

Dy, J. (2008): "Unsupervised feature selection." In: *Computational Methods of Feature Selection.* Ed. by H. Liu and H. Motada. London: Chapman and Hall CRC; 19–39.

Dy, J. and C. Bodley (2004): "Feature selection for unsupervised learning." In: *Journal of Machine Learning Research* 5; 845–89.

Dyen, I., J. Kruskal, and P. Black (1992): "An indoeuropean classification: a lexicostatistical experiment." In: *Transactions of the American Philosophical Society* 82; 1–132.

Elizondo, D. (2006): "The linear separability problem: some testing methods." In: *IEEE Transactions on Neural Networks* 17; 330–44.

Embleton, S. (1986): *Statistics in historical linguistics.* Bochum: Studienverlag Dr. N. Brockmeyer.

— (1987): "Multidimensional scaling as a dialectometrical technique." In: *Papers from the 11th Annual Meeting of the Atlantic Provinces Linguistic Association.* Ed. by R. Babitch; 33–49.

— (1993): "Multidimensional scaling as a dialectometrical technique: Outline of a research project." In: *Contributions to Quantitative Linguistics, Proceedings of the First Quantitative Linguistics Conference, September 23–27.* Ed. by R. Köhler and B. Rieger; 277–286.

— (2000): "Lexicostatistics / glottochronology: from Swadesh to Sankoff to Starostin to future horizons." In: *Time Depth in Historical Linguistics.* Ed. by C. Renfrew, A. McMahon, and L. Trask. Cambridge: McDonald Institute for Archaeological Research; 143–66.

Embleton, S., D. Uritescu, and E. Wheeler (2004): "An Exploration into the Management of High Volumes of Complex Knowledge in the Social Sciences and Humanities." In: *Journal of Quantitative Linguistics* 11; 183–192.

— (2013): "Defining dialect regions with interpretations: advancing the multidimensional scaling approach." In: *Literary and Linguistic Computing* 28; 13–22.

Embleton, S. and E. Wheeler (1997a): "Finnish dialect atlas for quantitative studies." In: *Journal of Quantitative Linguistics* 4; 99–102.

— (1997b): "Multidimensional scaling and the SED data." In: *The computer developed linguistic atlas of England 2*. Ed. by W. Viereck and H. Ramisch. Tübingen: Max Niemeyer; 5–11.

Ertöz, L., M. Steinbach, and V. Kumar (2003): "Finding clusters of different sizes, shapes, and densities in noisy, high dimensional data." In: *Proceedings of the 2003 SIAM International Conference on Data Mining (SDM 2003)*; 47–58.

Erwin, E., K. Obermayer, and K. Schulten (1992): "Self-organizing maps: ordering, convergence properties and energy function." In: *Biological Cybernetics* 67; 47–65.

Ester, M. et al. (1996): "A density-based algorithm for discovering clusters in large spatial databases with noise." In: *Proceedings of the Second International Conference on Knowledge Discovery and Data Mining*; 226–31.

Everitt, B. and G. Dunn (2001): *Applied Multivariate Data Analysis*. 2nd ed. London: Arnold.

Everitt, B. et al. (2011): *Cluster Analysis*. 5th ed. Hoboken NJ: Wiley.

Facchinetti, R. (2007): *Corpus Linguistics 25 Years On*. Amsterdam: Rodopi.

Fasold, R. (1972): *Tense Marking in Black English*. Washington DC: Washington DC: Centre for Applied Linguistics.

Felsenstein, J. (2004): *Inferring Phyologenies*. Sunderland MA: Sinauer Associates.

Firth, J. (1957): "A synopsis of linguistic theory 1930-1955." In: *Studies in Linguistic Analysis*. Ed. by J. F. et al. Blackwell; Hoboken, NJ.

Fodor, I. (2002): *A survey of dimensional reduction techniques*. URL: `https://e-reports-ext.llnl.gov/pdf/240921.pdf`.

Forgy, E. (1965): "Cluster analysis of multivariate data: efficiency versus interpretability of classifications." In: *Biometrics* 21; 768–769.

Fossat, J. and D. Philps (1976): *Microdialectologie et Dialectométrie des Pyrénées Gasconnes*. Toulouse: Toulouse II de Miral.

Frakes, W. and R. Baeza-Yates (1992): *Information Retrieval: Data Structures and Algorithms*. London: Prentice–Hall.

Fraley, C. and A. Raferty (2002): "Model-based clustering, discriminant analysis, and density estimation." In: *Journal of the American Statistical Association* 97; 611–31.

François, D., V. Wertz, and M. Verleysen (2007): "The concentration of fractional distances." In: *IEEE Transaction on Knowledge and Data Engineering* 19; 873–86.

Fritzke, B. (1993): "Growing Cell Structures. A self-organizing neural network for unsupervised and supervised learning." In: *Neural Networks* 7; 1441–60.

— (1995): "Growing Grid. A self-organizing network with constant neighbourhood range and adaptation strength." In: *Neural Processing Letters* 2; 9–13.

Fritzke, P. (1999): "Growing self-organizing networks – history, status quo, and perspectives." In: *Kohonen Maps*. Ed. by E. Oja and S. Kaski. London: Elsevier; 131–44.

Gamallo, P., A. Agustini, and G. Lopes (2005): "Clustering syntactic positions with similar semantic requirements." In: *Computational Linguistics* 31; 107–45.

Gan, G., C. Ma, and J. Wu (2007): *Data Clustering. Theory, Algorithms, and Applications*. Alexandria VA: American Statistical Association.

Garside, R., G. Leech, and G. Sampson (1987): *The Computational Analysis of English. A corpus-Based Approach*. London: Longman.

Gauch, H. (2003): *Scientific Method in Practice*. Cambridge: Cambridge University Press.

Geeraerts, D. and H. Cuykens (2010): *The Oxford Handbook of Cognitive Linguistics*. Oxford: Oxford University Press.

Gildea, D. and D. Jurafsky (2002): "Automatic labelling of semantic roles." In: *Computational Linguistics* 28; 245–88.

Gilquin, G. and S. Gries (2009): "Corpora and experimental methods: a state-of-the-art review." In: *Corpus Linguistics and Linguistic Theory* 5; 1–26.

Girard, D. and D. Larmouth (1988): "Log-linear statistical models: explaining the dynamics of dialect diffusion." In: *Methods in Dialectology: Proceedings of the Sixth International Conference at University College, North Wales*. Ed. by M. Thomas; 251–77.

Gnanadesikan, R. (1997): *Methods for Statistical Data Analysis of Multivariate Observations*. 2nd ed. Hoboken, NJ: Wiley–Interscience.

Gnanandesikan, R., S. Tsao, and J. Kettenring (1995): "Weighting and selection of variables for cluster analysis." In: *Journal of Classification* 12; 113–136.

Goebl, H. (1982): *Dialektometrie: Prinzipien und Methoden des Einsatzes der Numerischen Taxonomie im Bereich der Dialektgeographie*. Wien: Österreichische Akademie der Wissenschaften.

— (1983): " 'Stammbaum' and 'Welle'. Vergleichende Beitragen aus numerisch-taxonomischer Sicht." In: *Zeitschrift für Sprachwissenschaft* 2; 3–44.

— (1984a): *Dialectology*. Tübingen: Niemeyer.

— (1984b): *Dialektometrische Studien. Anhand italoromanischer, rätoromanischer und galloromanischer Sprachmaterialien aus AIS und ALF*. Tübingen: Niemeyer.

— (1993a): "Dialectometry: A Short Overview of the Principles and Practice of Quantitative Classification of Linguistic Atlas Data." In: *Contributions to Quantitative Linguistics*. Ed. by R. Köhler and B. Rieger. Dordrecht: Kluwer; 277–315.

— (1993b): "Probleme und Methoden der Dialektometrie: Geolinguistik in globaler Perspektive." In: *Proceedings of the International Congress of Dialectologists I*. Ed. by W. Viereck; 27–81.

— (1997): "Some Dendrographic Classification of the Data of CLAE 1 and CLAE 2." In: *The Computer Developed Linguistic Atlas of England 1*. Ed. by W. Viereck and H. Ramisch. Tübingen: Max Niemeyer; 23–32.

— (2005): "La dialectométrie corrélative: un nouvel outil pour l'etude de l'aménagement dialectal de l'espace par l'homme." In: *Revue de Linguistique Romane* 69; 321–67.

— (2006): "Recent Advances in Salzburg dialectometry." In: *Literary and Linguistic Computing* 21; 411–35.

Goebl, H. (2010): "Dialectometry: theoretical prerequisites, practical problems, and concrete applications (mainly with examples drawn from the "Atlas linguistique de la France" 1902-1910." In: *Dialectologia* 1; 63–77.

Goldreich, O. (2008): *Computational Complexity: A Conceptual Perspective*. Cambridge: Cambridge University Press.

— (2010): *P, NP, and NP-Completeness: The Basics of Computational Complexity*. Cambridge: Cambridge University Press.

Goodman, L. and W. Kruskal (1954): "Measures of association for cross classifications, part I." In: *Journal of the American Statistical Association* 49; 732–764.

Gooskens, C. (2006): "The relative contribution of pronunciational, lexical, and prosodic differences to the perceived distances between Norwegian dialects." In: *Literary and Linguistic Computing* 21; 477–492.

Gorban, A. et al. (2007): *Principal Manifolds for Data Visualization and Dimension Reduction*. Berlin: Springer.

Gordon, A. (1996): "Hierarchical classification." In: *Clustering and Classification*. Ed. by P. Arabie, L. Hubert, and G. D. Soete. London: World Scientific; 65–121.

— (1998): "Cluster validation." In: *Data Science, Classification, and Related Methods*. Ed. by C. Hayashi et al. Berlin: Springer; 22–39.

— (1999): *Classification*. 2nd ed. London: Chapman and Hall.

Gower, J. and P. Legendre (1986): "Metric and Euclidean properties of dissimilarity coefficients." In: *Journal of Classification* 3; 5–48.

Gray, R. and F. Jordan (2000): "Language tree supports the express train sequence of Austronesian expansion." In: *Nature* 405; 1052–1055.

Greengrass, E. (2001): *Information retrieval: A survey*. URL: `http://www.freetechbooks.com/information-retrieval-a-survey-t595.html`.

Gries, S. (2003): *Multifactorial analysis in corpus linguistics: a study of particle placement*. London: Continuum.

— (2007): "Corpus-based methods and cognitive semantics: the many meanings of to run." In: *Corpora in cognitive linguistics: corpus-based approaches to syntax and lexis*. Ed. by S. Gries and A. Stefanowitsch. Berlin: Mouton de Gruyter; 57–99.

— (2009a): *Quantitative Corpus Linguistics with R*. London: Routledge.

— (2009b): *Statistics for Linguistics with R*. Berlin: Mouton de Gruyter.

— (2009c): "What is corpus linguistics?" In: *Language and Linguistics Compass* 3; 1–17.

— (2010a): "Behavioral Profiles: a fine-grained and quantitative approach in corpus-based lexical semantics." In: *The Mental Lexicon* 5; 323–346.

— (2010b): "Corpus linguistics and theoretical linguistics: a love-hate relationship? Not necessarily. . . " In: *International Journal of Corpus Linguistics* 15; 327–343.

— (2010c): "Useful statistics for corpus linguistics." In: *A mosaic of corpus linguistics: selected approaches*. Ed. by A. Sanchez and M. Almela. New York: Peter Lang; 269–291.

— (2011): "Commentary." In: *Current Methods in Historical Semantics*. Ed. by K. Allan and J. Robinson. Berlin: Mouton de Gruyter; 184–195.

— (2011a): "Methodological and interdisciplinary stance in corpus linguistics." In: *Perspectives on corpus linguistics: connections and controversies*. Ed. by G. Barnbrook, V. Viana, and S. Zyngier. Amsterdam: John Benjamins; 81–98.

— (2012): "Corpus linguistics, theoretical linguistics, and cognitive / psycholinguistics: towards more and more fruitful exchanges." In: *Corpus linguistics and variation in English: Theory and description*. Ed. by J. Mukherjee and M. Huber. Amsterdam: Rodopi; 41–63.

Gries, S. and M. Hilpert (2008): "The identification of stages in diachronic data: variability-based neighbour clustering." In: *Corpora* 1; 59–81.

— (2010): "Variability-based neighbor clustering: a bottom-up approach to periodization in historical linguistics." In: *The Oxford Handbook on the History of English*. Ed. by T. Nevalainen and E. Traugott. Oxford: Oxford University Press; 134–144.

Gries, S. and J. Mukherjee (2010): "Lexical gravity across varieties of English: an ICE-based study of *n*-grams in Asian Englishes." In: *International Journal of Corpus Linguistics* 15; 520–548.

Gries, S., J. Newman, and C. Shaoul (2011): "N-grams and the clustering of registers." In: *Empirical Language Research* 5; nr. 1.

Gries, S. and N. Otani (2010): "Behavioral profiles. A corpus-based perspective on synonymy and antonymy." In: *ICAME Journal* 34; 121–50.

Gries, S. and A. Stefanowitsch (2006): *Corpora in Cognitive Linguistics. Corpus Based Approaches to Syntax and Lexis*. Berlin: Mouton de Gruyter.

Gries, S. and A. Stefanowitsch (2010): "Cluster analysis and the identification of collexeme classes." In: *Empirical and experimental methods in cognitive/functional research*. Ed. by S. Rice and J. Newman. Stanford CA: CSLI; 73–90.

Gries, S. and S. Stoll (2009): "Finding Developmental Groups in Acquisition Data: Variability-based Neighbour Clustering." In: *Journal of Quantitative Linguistics* 16; 217–242.

Gries, S., S. Wulff, and M. Davies (2010): *Corpus linguistic applications: current studies, new directions*. Amsterdam: Rodopi.

Grieve, J., D. Speelman, and D. Geeraerts (2011): "A statistical method for the identification and aggregation of regional linguistic variation." In: *Language Variation and Change* 23; 193–221.

Groenen, P. and M. Van de Velden (2005): "Multidimensional Scaling." In: *Encyclopedia of Statistics in Behavioral Science, Volume II*. Ed. by B. Everitt, and D. Howell. Hoboken NJ: Wiley; 1280–1289.

Gross, J. and J. Yellen (2006): *Graph Theory and its Applications*. 2nd ed. London: Chapman and Hall.

Grzybek, P. and R. Köhler (2007): *Exact Methods in the Study of Language and Text. Dedicated to Gabriel Altmann on the Occasion of his 75th Birthday*. Berlin: Mouton de Gruyter.

Guha, S., R. Rastogi, and K. Shim (1998): "CURE: an efficient clustering algorithm for large databases." In: *Proceedings of the ACM SIGMOD International Conference in Management of Data*; 73–84.

Guiter, H. (1974): "Une vérification de loi linguistique par corrélation." In: *Revue de Linguistique Romane* 38; 253–64.

— (1987): "Etalongage d'une méthode géolinguistique." In: *Revue de Linguistique Romane* 51; 55–62.

Guy, G. (1993): "The quantitative analysis of linguistic variation." In: *American Dialect Research*. Ed. by D. Preston. Amsterdam: John Benjamins; 223–49.

Guyon, I. and A. Elisseeff (2003): "An introduction to variable and feature selection." In: *Journal of Machine Learning Research* 3; 1157–82.

Haimerl, E. (2006): "Database Design and Technical Solutions for the Management, Calculation, and Visualization of Dialect Mass Data." In: *Literary and Linguistic Computing* 21; 437–44.

Haimerl, E. and H. Mucha (2006): "Comparing the stability of different clustering results of dialect data." In: *Advances in Data Analysis. Proceedings of the 30th Meeting of the Gesellschaft für Klassifikation, Dortmund, Mar. 8-10*. Ed. by R. Decker and H. Lenz; 619–26.

Hair, J. et al. (2010): *Multivariate Data Analysis. A Global Perspective*. London: Pearson.

Halkidi, M., Y. Batistakis, and M. Vazirgiannis (2001): "On clustering validation techniques." In: *Journal of Intelligent Information Systems* 17; 107–45.

— (2002a): "Cluster validity methods: part I." In: *SIGMOD Record* 31; 40–45.

— (2002b): "Cluster validity methods: part II." In: *SIGMOD Record* 31; 19–27.

Halkidi, M. and M. Vazirgiannis (2008): "A density-based cluster validity approach using multi-representatives." In: *Pattern Recognition Letters* 29; 773–86.

Halkidi, M., M. Vazirgiannis, and I. Batistakis (2000): "Quality Scheme Assessment in the Clustering Process." In: *Proceedings of PKDD, Lyon, France*.

Ham, J. et al. (2004): "A kernel view of the dimensionality reduction of manifolds." In: *Proceedings of the Twenty-first International Conference in Machine Learning (ICML 2004)*; 47–55.

Hämäläinen, T. (2002): "Parallel implementations of self-organizing maps." In: *Self-organizing Neural Networks: Recent Advances and Applications*. Ed. by U. Seiffert and L. Jain. Berlin: Springer; 245–76.

Han, J. and M. Kamber (2001): *Data Mining*. London: Morgan Kaufmann Publishers.

Handl, J., J. Knowles, and D. Kell (2005): "Computational cluster validation in post-genomic data analysis." In: *Bioinformatics* 21; 3201–12.

Händler, H., L. Hummel, and W. Putschke (1989): "Computergestützte Dialektologie." In: *Computational Linguistics. An International Handbook on Computer Oriented Language Research and Applications*. Ed. by I. Batori, W. Lenders, and W. Putschke. Walter De Gruyter; Berlin.

Harris, Z. (1954): "Distributional structure." In: *Word* 10; 146–162.

— (1962): *String Analysis of Language Structure*. Berlin: Mouton.

Harris, Z. (1968): *Mathematical Structures of Language*. Hoboken NJ: Interscience Publishers.

Hastie, T. and W. Stuetzle (1989): "Principal Curves." In: *Journal of the American Statistical Association* 84; 502–516.

Hatzivassiloglou, V. and K. McKeown (1993): "Towards an automatic identification of adjectival scales: clustering adjectives according to meaning." In: *Proceedings of the 31st Annual Meeting of the Association for Computational Linguistics*; 172–82.

Hauer, B. and G. Kondrak (2011): "Clustering Semantically Equivalent Words into Cognate Sets in Multilingual Lists." In: *The 5th International Joint Conference on Natural Language Processing (IJCNLP 2011)*; 865–73.

Havens, T. and J. Bezdek (2012): "An efficient formulation of the improved Visual Assessment of Cluster Tendency (iVAT) algorithm." In: *IEEE Transaction on Knowledge and Data Engineering* 24; 813–22.

Haykin, S. (1999): *Neural Networks. A Comprehensive Foundation*. London: Prentice Hall International.

Heeringa, W. and J. Nerbonne (2001): "Dialect areas and dialect continua." In: *Language Variation and Change* 13; 375–400.

Heeringa, W. et al. (2006): "Evaluation of String Distance Algorithms for Dialectology." In: *Linguistic Distances Workshop at the joint conference of International Committee on Computational Linguistics and the Association for Computational Linguistics, The Association for Computational Linguistics (ACL)*. Ed. by J. Nerbonne and E. Hinrichs; 51–62.

Hennig, C. (2007): "Cluster-wise assessment of cluster stability." In: *Computational Statistics and Data Analysis* 52; 258–71.

Heskes, T. (1999): "Energy functions for self-organizing maps." In: *Kohonen Maps*. Ed. by E. Oja and S. Kaski. London: Elsevier; 303–16.

Heylen, K. and T. Ruette (2013): "Degrees of semantic control in measuring aggregated lexical distance." In: *Approaches to measuring linguistic differences*. Ed. by L. Borin, A. Saxena, and T. Rama. Mouton de Gruyter; Berlin.

Heylen, K., D. Speelman, and D. Geeraerts (2012): "Looking at word meaning. An interactive visualization of Semantic Vector Spaces for Dutch synsets." In: *Proceedings of the EACL 2012 Joint Workshop of LINGVIS and UNCLH, Association for Computational Linguistics*; 16–24.

Hilpert, M. and S. Gries (2009): “Assessing frequency changes in multistage diachronic corpora: Applications for historical corpus linguistics and the study of language acquisition.” In: *Literary and Linguistic Computing* 24; 385–401.

Hindle, D. (1990): “Noun classification from predicate argument structures.” In: *Proceedings of the 28th Annual Meeting of the Association for Computational Linguistics*; 268–75.

Hinneburg, A., C. Aggarwal, and D. Keim (2000): “What is the nearest neighbor in high dimensional spaces?” In: *Proceedings of the 26th International Conference on Very Large Data Bases*; 506–15.

Hinneburg, A. and H. Gabriel (2007): “DENCLUE 2.0: Fast clustering based on kernel density estimation.” In: *Proceedings of the 7th International Conference on Intelligent Data Analysis*; 70–80.

Hinneburg, A. and D. Keim (1998): “An efficient approach to clustering in large multimedia databases with noise.” In: *Proceedings of the Fourth International Conference on Knowledge Discovery and Data Mining*; 58–65.

— (1999): “Optimal Grid-Clustering: Breaking the curse of dimensionality in high-dimensional clustering.” In: *Proceedings of the 25th VLDB Conference, Edinburgh*; 506–17.

— (2003): “A general approach to clustering in large databases with noise.” In: *Knowledge and Information Systems* 5(4); 387–415.

Hogenhout, W. and Y. Matsumoto (1997): “Word clustering from syntactic behaviour.” In: *CoNLL97: Computational Natural Language Learning, ACL*. Ed. by T. Ellison; 16–24.

Holden, C. (2002): “Bantu language trees reflect the spread of farming across sub-Saharan Africa: a maximum-parsimony analysis.” In: *Proceedings of the Royal Society of London* 269; 793–799.

Holm, H. (2000): “Genealogy of the Main Indo-European Branches Applying the Separation Base Method.” In: *Journal of Quantitative Linguistics* 7; 73–95.

— (2007): “The new arboretum of Indo-European “trees”. Can new algorithms reveal the phylogeny and even prehistory of Indo-European?” In: *Journal of Quantitative Linguistics* 14; 167–214.

Honkela, T. (1997): "Comparisons of self-organized word category maps." In: *Proceedings of WSOM'97, Workshop on Self-Organizing Maps, Helsinki University of Technology, Neural Networks Research Centre*; 298–303.

Hoppenbrouwers, C. and G. Hoppenbrouwers (1988): "De feature frequentie methode en de classificatie van Nederlandse dialecten." In: *TABU, Bulletin voort Taalwetenschap* 18; 51–92.

— (2001): *De indeling van de Nederlands streektalen: dialecten van 156 steden en dorpen geklasseerd volgens de FFM*. Assen: Koninklijke Van Gorcum.

Horvath, B. (1985): *Variation in Australian English*. Cambridge: Cambridge University Press.

Horvath, B. and D. Sankoff (1987): "Delimiting the Sydney speech community." In: *Language in Society* 16; 179–204.

Houck, C. (1967): "A computerized statistical methodology for linguistic geography: a pilot study." In: *Folia Linguistica* 1; 80–95.

Houtzagers, P., J. Nerbonne, and J. Prokic (2010): "Quantitative and traditional classifications of Bulgarian dialects compared." In: *Scando-Slavica* 59; 163–188.

Hu, Y. (2012): "VATdt: Visual assessment of cluster tendency using diagonal tracing." In: *American Journal of Computational Mathematics* 2; 27–41.

Hu, Y. and R. Hathaway (2008): "An algorithm for clustering tendency assessment." In: *WSEAS Transactions on Mathematics* 7; 441–50.

Huan, L. and H. Motada (1998): *Feature Selection for Knowledge Discovery and Data Mining*. Berlin: Springer.

Hubel, D. and T. Wiesel (2005): *Brain and Visual Perception*. Oxford: Oxford University Pres.

Hubey, H. (1999): *Mathematical Foundations of Linguistics*. Munich: LINCOM Europa.

Hull, D. (1996): "Stemming algorithms: a case study for detailed evaluation." In: *Journal of the American Society for Information Science* 47; 70–84.

Hyvärinen, A., J. Karhunen, and E. Oja (2001): *Independent Component Analysis*. Hoboken NJ: Wiley Blackwell.

Hyvönen, S., A. Leino, and M. Salmenkivi (2007): "Multivariate Analysis of Finnish Dialect Data – An Overview of Lexical Variation." In: *Literary and Linguistic Computing* 22; 271–90.

Izenman, A. (2008): *Modern Multivariate Statistical Techniques. Regression, Classification, and Manifold Learning*. Berlin: Springer.

Jackson, J. (2003): *A User's Guide to Principal Components*. Hoboken NJ: Wiley Interscience.

Jain, A. (2010): "Data clustering: 50 years beyond K-means." In: *Pattern Recognition Letters* 31; 651–66.

Jain, A. and R. Dubes (1988): *Algorithms for Clustering Data*. London: Prentice Hall.

Jain, A. and J. Moreau (1987): "Bootstrap technique in cluster analysis." In: *Pattern Recognition* 20; 547–68.

Jain, A., M. Murty, and P. Flynn (1999): "Data clustering: a review." In: *ACM Computing Surveys* 31; 264–323.

Jardino, M. and G. Adda (1993): "Automatic word classification using simulated annealing." In: *Proceedings of the International Conference on Acoustics, Speech, and Signal Processing, vol. 2, Minneapolis*; 41–44.

Jassem, W. and P. Lobacz (1995): "Multidimensional scaling and its applications in a perceptual analysis of Polish consonants." In: *Journal of Quantitative Linguistics* 2; 105–24.

Ji, M. (2010): "A corpus-based study of lexical periodization in historical Chinese." In: *Literary and Linguistic Computing* 25; 199–213.

Jickels, T. and G. Kondrak (2006): "Unsupervised Labeling of Noun Clusters." In: *Proceedings of the Nineteenth Canadian Conference on Artificial Intelligence (Canadian AI 2006)*; 278–287.

Jimenez, L. and D. Landgrebe (1998): "Supervised Classification in High Dimensional Space: Geometrical, Statistical, and Asymptotical Properties of Multivariate Data." In: *IEEE Transactions on Systems, Man, and Cybernetics* 28, Part C, Number 1; 39–54.

Joanis, E., S. Stevenson, and D. James (2008): "A general feature space for automatic verb classification." In: *Natural Language Engineering* 14; 337–367.

Johansson, S. (2008): "Some aspects of the development of corpus linguistics in the 1970s and 1980s." In: *Corpus Linguistics. An International Handbook*. Ed. by A. Lüdeling and M. Kytö. Berlin: De Gruyter Mouton; 33–53.

Johnson, E. (1994): "The distribution of variants of /aI/ in the Middle and South Atlantic States." In: *Verhandlungen des Internationalen Dialektologenkongresses Bamberg 1990*. Ed. by W. Viereck; 345–67.

Johnson, K. (2008): *Quantitative Methods in Linguistics*. Hoboken NJ: Wiley-Blackwell.

Jolliffe, I. (2002): *Principal Component Analysis*. 2nd ed. Berlin: Springer.

Jones-Sargent, V. (1983): *Tyne Bytes. A computerised sociolinguistic study of Tyneside*. New York: Peter Lang.

Karypis, G., E. Han, and V. Kumar (1999): "CHAMELEON: hierarchical clustering using dynamic modelling." In: *IEEE Computer* 32; 68–75.

Kaski, S. (1997): *Data exploration using self-organizing maps*. Helsinki: Acta Polytechnica Scandinavica. Mathematics, Computing, and Management in Engineering Series 82.

Kaski, S. and C. Lagus (1996): "Comparing self-organizing maps." In: *Proceedings of the ICANN'96 International Conference on Neural Networks*. Ed. by C. von der Malsburg et al.; 809–14.

Kaski, S., J. Nikkila, and T. Kohonen (2000): "Methods for exploratory cluster analysis." In: *Proceedings of the International Conference on advances in Infrastructure for Electronic Business, Science, and Education on the Internet*.

Kaufman, L. and P. Rousseeuw (1990): *Finding Groups in Data*. Hoboken NJ: Wiley–Interscience.

Kendall, T. and G. van Herk (2011): "Corpora from a sociolinguistic perspective." In: *Corpus Linguistics and Linguistic Theory* 7; 1–6.

Kennedy, G. (1998): *An Introduction to Corpus Linguistics*. London: Longman.

Kepser, S. and M. Reis (2005): *Linguistic Evidence: Empirical, Theoretical and Computational Perspectives*. Berlin: Mouton de Gruyter.

Kessler, B. (1995): "Computational dialectology in Irish Gaelic." In: *Proceedings of the seventh conference of the European chapter of the Association for Computational Linguistics (EACL)*; 60–66.

— (2005): "Phonetic comparison algorithms." In: *Transactions of the Philological Society* 103; 243–60.

— (2007): "Word similarity metrics and multilateral comparison." In: *Proceedings of the Ninth Meeting of the ACL Special Interest Group on Computational Morphology and Phonology (SIGMORPHON),Prague, Czech Republic*. Ed. by J. Nerbonne, M. Ellison, and G. Kondrak; 6–14.

— (2008): "The mathematical assessment of long-range linguistic relationships." In: *Language and Linguistics Compass* 2; 821–839.

Kessler, B. and A. Lehtonen (2006): "Multilateral comparison and significance testing of the Indo-Uralic question." In: *Phylogenetic methods and the prehistory of languages*. Ed. by P. Forster and C. Renfrew. Cambridge: McDonald Institute for Archaeological Research; 33–42.

Kettenring, J. (2006): "The practice of cluster analysis." In: *Journal of Classification* 23; 3–30.

Kilgarriff, A. (2005): "Language is never, ever, ever, random." In: *Corpus Linguistics and Linguistic Theory* 1.2; 263–76.

Kim, M. and R. Ramakrishna (2005): "New indices for cluster validity assessment." In: *Pattern Recognition Letters* 26; 2353–63.

Kim, Y., N. Street, and F. Menczer (2003): "Feature selection in data mining." In: *Data Mining*. Ed. by J. Wang. Hershey PA: IGI Publishing; 80–105.

Kiss, G. (1973): "Grammatical Word Classes: A Learning Process and Its Simulation." In: *Psychology of Learning and Motivation* 7.

Kita, K. (1999): "Automatic clustering of languages based on probabilistic models." In: *Journal of Quantitative Linguistics* 6; 167–71.

Kiviluoto, K. (1996): "Topology preservation in self-organizing maps." In: *Proceedings of the IEEE International Conference on Neural Networks*; 294–99.

Kneser, R. and H. Ney (1993): "Improved clustering techniques for class-based statistical language modeling." In: *EUROSPEECH '93. Third European Conference on Speech Communication and Technology, Berlin, Germany*; 973–76.

Kogan, J. (2007): *Introduction to Clustering Large and High-dimensional Data*. Cambridge: Cambridge University Press.

Köhler, R. (1995): *Bibliography of Quantitative Linguistics*. Amsterdam: John Benjamins.

Köhler, R. (2005): "Synergetic Linguistics." In: *Quantitative Linguistik. Ein internationales Handbuch. Quantitative Linguistics. An International Handbook*. Ed. by R. Köhler, G. Altmann, and R. Piotrowski. Berlin: De Gruyter Mouton; 760–75.

— (2011): "Laws of Language." In: *The Cambridge Encyclopedia of the Language Sciences*. Ed. by P. Hogan. Cambridge: Cambridge University Press; 424–426.

— (2012): *Quantitative Syntax Analysis*. Berlin: Mouton de Gruyter.

Köhler, R. and G. Altmann (2011): "Quantitative Linguistics." In: *The Cambridge Encyclopedia of the Language Sciences*. Ed. by P. Hogan. Cambridge: Cambridge University Press; 695–697.

Köhler, R., G. Altmann, and R. Piotrowski (2005): *Quantitative Linguistik. Ein internationales Handbuch. Quantitative Linguistics. An International Handbook*. Berlin: Mouton de Gruyter.

Kohonen, T. (2001): *Self Organizing Maps*. 3rd ed. Berlin: Springer.

Köppen, M. (2000): "The curse of dimensionality." In: *5th Online World Conference on Soft Computing in Industrial Applications (WSC5)*; 4–8.

Korhonen, A. (2010): "Automatic lexical classification: bridging research and practice." In: *Philosophical Transaction of the Royal Society A* 368; 3621–3632.

Korn, F., B. Pagel, and C. Faloutsos (2001): "On the 'Dimensionality Curse' and the 'Self-Similarity Blessing'." In: *IEEE Transactions on Knowledge Data Enginerring* 13; 96–111.

Kornai, A. (2008): *Mathematical Linguistics*. Berlin: Springer.

Kovacs, F., C. Legany, and A. Babos (2006): "Cluster validity measurement techniques." In: *AIKED'06 Proceedings of the 5th WSEAS International Conference on Artificial Intelligence, Knowledge Engineering and Data Bases*; 388–393.

Kretzschmar, W. (1996): "Quantitative areal analysis of dialect features." In: *Language Variation and Change* 8; 13–39.

Kretzschmar, W. and E. Schneider (1996): *Introduction to Quantitative Analysis of Linguistic Survey Data: Atlas by the Numbers*. London: Sage Publications.

Kretzschmar, W., E. Schneider, and E. Johnson (1989): "Introduction to quantitative analysis of linguistic survey data: An atlas by numbers." In: *Journal of English Linguistics* 22; i–ii.

Kriegel, H., P. Kröger, and A. Zimek (2009): "Clustering high-dimensional data: a survey on subspace clustering, pattern-based clustering, and correlation clustering." In: *ACM Transactions on Knowledge Discovery from Data* 3; 1–58.

Krishna, K. and M. Murty (1999): "Genetic K-means algorithm." In: *IEEE Transactions on Systems, Man, and Cybernetics B* 29.3; 433–39.

Kroch, A. (1989): "Reflexes of grammar in patterns of linguistic change." In: *Language Variation and Change* 1; 199–244.

Kroeber, A. and C. Chrétien (1937): "Quantitative classification of Indo-European languages." In: *Language* 13; 83–103.

Kruskal, J. and M. Wish (1978): *Multidimensional Scaling*. London: Sage Publications.

Kryszczuk, K. and P. Hurley (2010): "Estimation of the number of clusters using multiple clustering validity indices." In: *Multiple Classifier Systems: 9th Internaltional Workshop on Multiple Classifier Systems*; 114–23.

Labov, W. (1963): "The social motivation of a sound change." In: *Word* 18; 1–42.

— (1966): *The Social Stratification of English in New York City*. Washington DC: Washington DC: Center for Applied Linguistics.

— (1994): *Principles of Linguistic Change. Volume 1: Internal Factors*. Hoboken, NJ: Blackwell.

— (2001): *Principles of Linguistic Change. Volume 2: Social Factors*. Hoboken, NJ: Blackwell.

Ladefoged, P., R. Glick, and C. Criper (1971): *Language in Uganda*. Oxford: Oxford University Press.

Lange, T. et al. (2004): "Stability-based validation of clustering solutions." In: *Neural Computation* 16; 1299–1323.

Larsen, P. and M. von Ins (2010): "The rate of growth in scientific publication and the decline in coverage provided by Science Citation Index." In: *Scientometrics* 84; 575–603.

Lawson, R. and P. Jurs (1990): "New index for clustering tendency and its application to chemical problems." In: *Journal of Chemical Information Compututer Science* 30; 36–41.

Lay, D. (2010): *Linear Algebra and its Applications*. 4th. London: Pearson.

Lebart, L., A. Salem, and L. Berry (1998): *Exporing Textual Data*. Dordrecht: Kluwer.

Lee, J. (2010): *Introduction to Topological Manifolds*. 2nd ed. Berlin: Springer.

Lee, J., M. Lendasse, and M. Verleysen (2004): "Nonlinear projection with curvilinear distances: Isomap versus curvilinear distance analysis." In: *Neurocomputing* 57; 49–76.

Lee, J. and M. Verleysen (2007): *Nonlinear Dimensionality Reduction*. Berlin: Springer.

Lee, L. and F. Pereira (1999): "Distributional similarity models: clustering vs. nearest neighbours." In: *Proceedings of the 37th Annual Meeting of the ACL*; 33–40.

Leino, A. and S. Hyvönen (2008): "Comparison of Component Models in Analysing the Distribution of Dialectal Features." In: *International Journal of Humanities and Arts Computing* 2; 173–87.

Leinonen, T. (2008): "Factor Analysis of Vowel Pronunciation in Swedish Dialects." In: *International Journal of Humanities and Arts Computing* 2; 189–204.

Levenshtein, V. (1966): "Binary codes capable of correcting deletions, insertions, and reversals." In: *Soviet Physics Doklady* 10; 707–10.

Levine, E. and E. Domany (2001): "Resampling method for unsupervised estimation of cluster validity." In: *Neural Computation* 13; 2573–93.

Levine, L. and H. Crockett (1966): "Speech variation in a Piedmont community: postvocalic /r/." In: *Sociolinguistic Inquiry* 36; 204–26.

Li, B., C. Zhang, and R. Wayland (2012): "Acoustic Characteristics and Distribution of Variants of /l/ in the Nanjing Dialect." In: *Journal of Quantitative Linguistics* 19; 281–300.

Li, H. and N. Abe (1998): "Word clustering and disambiguation based on co-occurrence data." In: *ACL - COLING 17*; 749–55.

Likas, A., N. Vlassis, and J. Verbeek (2003): "The global k-means clustering algorithm." In: *Pattern Recognition* 36; 451–61.

Lin, D. and P. Pantel (2001): "Induction of semantic classes from natural language text." In: *Proceedings of the seventh ACM SIGKDD International Conference on Knowledge Discovery and Data Mining*; 317–22.

Linden, K. (2004): “Evaluation of Linguistic Features for Word Sense Disambiguation with Self-Organized Document Maps.” In: *Computers and the Humanities* 38; 417–435.

Lindquist, H. (2009): *Corpus Linguistics and the Description of English*. Edinburgh: Edinburgh University Press.

Linn, M. and R. Regal (1985): “Numerical taxonomy as a tool in dialect research.” In: *Papers from the Fifth International Conference on Methods in Dialectology*. Ed. by H. Warkentyne; 245–61.

— (1988): “Verb analysis of the Linguistic Atlas of the North-Central States: a case study in preliminary analysis of a large data set.” In: *Methods in Dialectology: Proceedings of the Sixth International Conference at University College, North wales, August 1987*. Ed. by M. Thomas; 138–54.

— (1993): “Missing data and computer mapping.” In: *Verhandlungen des Internationalen Dialektologenkongresses Bamberg 1990*. Ed. by W. Viereck; 253–67.

Liu, H. and H. Motada (2008): *Computational Methods of Feature Selection*. London: Chapman and Hall CRC.

Liu, P., D. Zhou, and N. Wu (2007): “VDBSCAN: Varied density based spatial clustering of application with noise.” In: *Proceedings of the IEEE Conference ICSSSM 2007*; 528–31.

Liu, Y. et al. (2010): “Understanding of internal clustering validation measures.” In: *2010 IEEE International Conference on Data Mining*; 911–16.

Lüdeling, A. and M. Kytö (2008): *Corpus Linguistics. An International Handbook. Vol 1*. Berlin: Mouton de Gruyter.

— (2009): *Corpus Linguistics. An International Handbook. Vol 2*. Berlin: Mouton de Gruyter.

Luhn, H. (1957): “A statistical approach to mechanized encoding and searching of literary information.” In: *IBM Journal of Research and Development* 1; 309–17.

Maaten, L. van der, E. Postma, and H. van den Herik (2009): *Dimensionality Reduction: A Comparative Review*. Tilburg: Tilburg University Technical Report TiCC-TR 2009.

Machamer, P. and M. Silberstein (2007): *The Blackwell Guide to the Philosophy of Science*. Hoboken NJ: John Wiley and Sons.

MacQueen, J. (1967): "Some methods for classification and analysis of multivariate observations." In: *Proceedings of the Fifth Berekeley Symposium*; 281–97.

Maguire, W. and A. McMahon (2011): *Analysing Variation in English*. Cambridge: Cambridge University Press.

Manning, C., P. Raghavan, and H. Schütze (2008): *Introduction to Information Retrieval*. Cambridge: Cambridge University Press.

Manning, C. and H. Schütze (1999): *Foundations of Statistical Natural Language Processing*. Cambridge MA: MIT Press.

Mark, H. and J. Workman (2005a): "Linearity in calibration: how to test for non-linearity." In: *Spectroscopy* 20(9).

— (2005b): "Linearity in calibration: quantifying nonlinearity." In: *Spectroscopy* 20(12).

— (2005c): "Linearity in calibration: the importance of nonlinearity." In: *Spectroscopy* 20(1).

— (2006): "Linearity in calibration: quantifying nonlinearity, part II." In: *Spectroscopy* 21(1).

Martin, M., J. Liermann, and H. Ney (1998): "Algorithms for bigram and trigram word clustering." In: *Speech Communication* 24; 19–37.

Martinetz, T., S. Berkovich, and R. Schulten (1993): "Neural gas network for vector quantization and its application to time-series prediction." In: *IEEE Transactions on Neural Networks* 4; 558–69.

Martinez, W., A. Martinez, and J. Solka (2011): *Exploratory Data Analysis with MATLAB*. London: CRC Press.

Mary, S. and K. Kumar (2012): "Evaluation of clustering algorithm with cluster validation metrics." In: *European Journal of Scientific Research* 69; 52–63.

Maulik, U. and S. Bandyopadhyay (2002): "Performance evaluation of some clustering algorithms and validity indices." In: *IEEE Transactions on Pattern Analysis and Machine Intelligence* 24; 1650–54.

McEnery, T. and A. Hardie (2012): *Corpus Linguistics. Method, Theory, and Practice*. Cambridge: Cambridge University Press.

McEnery, T. and A. Wilson (1996): *Corpus Linguistics. An Introduction*. Edinburgh: Edinburgh University Press.

— (2001): *Corpus Linguistics. An Introduction.* 2nd ed. Edinburgh: Edinburgh University Press.

McMahon, A. and W. Maguire (2011): "Quantitative historical dialectology." In: *Analysing Older English.* Ed. by D. Denison et al. Cambridge University Press; Cambridge.

McMahon, A. and R. McMahon (2003): "Finding families: Quantitative methods in language classification." In: *Transactions of the Philological Society* 101; 7–55.

— (2005): *Language Classification by Numbers.* Oxford: Oxford University Press.

McMahon, J. and F. Smith (1996): "Improving statistical language model performance with automatically generated word hierarchies." In: *Computational Linguistics* 22; 217–47.

— (1998): "A review of statistical language processing techniques." In: *Artificial Intelligence Review* 12; 347–91.

Mehler, A., O. Pustylnikov, and N. Diewald (2011): "Geography of social ontologies: testing a variant of the Sapir-Whorff Hypothesis in the context of Wikipedia." In: *Computer Speech and Language* 25; 716–40.

Merkl, D. (1997): "Exploration of text collections with hierarchical feature maps." In: *Principles of Data Mining and Knowledge Discovery: Proceedings of the First European Symposium, PKDD '97*; 101–111.

Merkl, D. and A. Rauber (1997a): "Alternative ways for cluster visualization in self-organizing maps." In: *Proceedings of the Workshop on Self-Organizing Maps (WSOM–97*; 106–11.

— (1997b): "Cluster Connections: A visualization technique to reveal cluster boundaries in self-organizing maps." In: *Proceedings of the 9th Italian Workshop on Neural Nets*; 22–24.

Meschenmoser, D. and S. Pröll (2012): "Fuzzy clustering of dialect maps using the empirical covariance. Grouping maps in dialect corpora based on spatial similarities." In: *International Journal of Corpus Linguistics* 17; 176–97.

Meyer, C. (2002): *English Corpus Linguistics. An Introduction.* Cambridge: Cambridge University Press.

Miles, J. and M. Shevlin (2001): *Applying Regression and Correlation.* London: Sage Publications.

Miller, G. (1971): "Empirical methods in the study of semantics." In: *Semantics: An Interdisciplinary Reader*. Ed. by D. Steinberg and L. Jakobovits. Cambridge: Cambridge University Press; 569–585.

Miller, G. and W. Charles (1991): "Contextual correlates of semantic similarity." In: *Language and Cognitive Processes* 6; 1–28.

Miller, G. and P. Nicely (1955): "An analysis of perceptual confusion among English consonants." In: *Journal of the Acoustic Society of America* 27; 338–52.

Miller, M. (1988): "Ransacking linguistic survey data with a number cruncher." In: *Methods in Dialectology: Proceedings of the Sixth International Conference at University College, North Wales, August 1987*. Ed. by M. Thomas; 464–79.

Milligan, G. (1996): "Clustering validation: results and implications for applied analyses." In: *Clustering and Classification*. Ed. by P. Arabie, L. Hubert, and G. D. Soete. Singapore: World Scientific; 341–75.

Milligan, G. and M. Cooper (1985): "An examination of procedures for determining the number of clusters in a data set." In: *Psychometrika* 50; 159–79.

— (1988): "A study of standardization of variables in cluster analysis." In: *Journal of Classification* 5; 181–204.

Milroy, L. and M. Gordon (2003): *Sociolinguistics. Method and Interpretation*. Hoboken, NJ: Blackwell.

Milroy, L. et al. (1999): "Phonological variation and change in contemporary English: evidence from Newcastle-upon-Tyne and Derby." In: *Variation and Linguistic Change in English*. Ed. by S. Conde and J. Hernandez-Compoy. Cuadernos de Filologia Ingelsa; 35–46.

Milton, J. and J. Arnold (2003): *Introduction to Probability and Statistics*. 4th ed. New York: McGraw Hill.

Mirkin, B. (2011): *Core Concepts in Data Analysis: Summarization, Correlation, and Visualization*. Berlin: Springer.

— (2013): *Clustering. A Data recovery Approach*. London: CRC Press.

Moisl, H. (2009): "Exploratory multivariate analysis." In: *Corpus Linguistics. An International Handbook. Vol 2*. Ed. by A. Lüdeling and M. Kytö. Mouton de Gruyter; 874–99.

— (2010): "Variable scaling in cluster analysis of linguistic data." In: *Corpus Linguistics and Linguistic Theory* 6; 75–103.

— (2011): "Finding the minimum document length for reliable clustering of multi-document natural language corpora." In: *Journal of Quantitative Linguistics* 18; 23–52.

— (2012): "Mapping phonetic variation in the Newcastle Electronic Corpus of Tyneside English." In: *Synergetic Linguistics: Text and Language as Dynamic Systems*. Ed. by S. Naumann et al. Wien: Praesens Verlag; 135–147.

Moisl, H. and V. Jones (2005): "Cluster analysis of the Newcastle Electronic Corpus of Tyneside English: A comparison of methods." In: *Literary and Linguistic Computing* 20; 125–146.

Moisl, H. and W. Maguire (2008): "Identifying the Main Determinants of Phonetic Variation in the Newcastle Electronic Corpus of Tyneside English." In: *Journal of Quantitative Linguistics* 15; 46–69.

Moisl, H., W. Maguire, and W. Allen (2006): "Phonetic variation in Tyneside: exploratory multivariate analysis of the Newcastle Electronic Corpus of Tyneside English." In: *Language Variation – European Perspectives*. Ed. by F. Hinskens. Amsterdam: John Benjamins; 127–42.

Montemagni, S. et al. (2013): "Synchronic patterns of Tuscan phonetic variation and diachronic change: Evidence from a dialectometric study." In: *Literary and Linguistic Computing* 28; 157–72.

Motulsky, H. and A. Christopoulos (2004): *Fitting Models to Data using Linear and Nonlinear Regression*. Oxford: Oxford University Press.

Mucha, H. (2006): "On validation of hierarchical clustering." In: *Advances in Data Analysis. Proceedings of the 30th Meeting of the Gesellschaft für Klassifikation, Dortmund, Mar. 8-10*. Ed. by R. Decker and H. Lenz; 115–22.

Mucha, H. and E. Haimerl (2004): "Automatic validation of hierarchical cluster analysis with application in dialectometry." In: *Classification-the Ubiquitous Challenge. Proceedings of the 28th Meeting of the Gesellschaft für Klassifikation, Dortmund, Mar. 9-11*. Ed. by C. Weihs and W. Gaul; 513–520.

Mukherjee, J. and S. Gries (2009): "Collostructional nativisation in New Englishes: verb-construction associations in the International Corpus of English." In: *English World-Wide* 30; 27–51.

Müller, K., B. Möbius, and D. Prescher (2000): "Inducing probabilistic syllable classes using multivariate clustering." In: *Proceedings of the 38th Annual Meeting of the Association for Computational Linguistics (ACL-00), Hong Kong*.

Munkres, J. (2000): *Topology*. 2nd. London: Pearson Education International.

Naumann, C. (1976): "Grundzüge der Sprachkartographie und ihrer Automatisierung." In: *Germanische Linguistik* 1–2.

— (1977): "Klassifikation in der automatischen Sprachkartographie." In: *Germanische Linguistik* 3–4; 181–210.

— (1982): "Klassifikation sprachlicher Sequenzen am Beispiel der computativen Bearbeitung von Wörtern für Sprachkarten." In: *Studien zur Klassifikation* 10; 132–40.

Nerbonne, J. (2003): "Linguistic Variation and Computation." In: *Proceedings of the 10th Meeting of the European Chapter of the Association for Computational Linguistics April, 2003*; 3–10.

— (2006): "Identifying Linguistic Structure in Aggregate Comparison." In: *Literary and Linguistic Computing* 21; 463–476.

— (2008): "Variation in the Aggregate: An Alternative Perspective for Variationist Linguistics." In: *Northern Voices: Essays on Old Germanic and Related Topics offered to Professor Tette Hofstra*. Ed. by K. Dekker, A. MacDonald, and H. Niebaum. Leuven: Peeters; 365–382.

— (2009): "Data-Driven Dialectology." In: *Language and Linguistics Compass* 3; 175–198.

— (2010): "Mapping Aggregate Variation." In: *Language and Space. International Handbook of Linguistic Variation. Vol. 2 Language Mapping*. Ed. by A. Lameli, R. Kehrein, and S. Rabanus. Berlin: Mouton de Gruyter; 2401–2406.

Nerbonne, J. and W. Heeringa (1997): "Measuring dialect distances phonetically." In: *Workshop on Computational Phonology, Special Interest Group of the Association for Computational Linguistics. Madrid, 1997*. Ed. by J. Coleman; 11–18.

— (2001): "Computational comparison and classification of dialects." In: *Dialectologia et Geolinguistica* 9; 69–83.

— (2007): "Geographic Distributions of Linguistic Variation Reflect Dynamics of Differentiation." In: *Roots: Linguistics in Search of its Evidential Base*. Ed. by S. Featherston and W. Sternefeld. De Gruyter Mouton; 267–297.

— (2010): "Measuring Dialect Differences." In: *Language and Space: Theories and Methods*. Ed. by J. Schmidt and P. Auer. Berlin: Mouton de Gruyter; 550–567.

Nerbonne, J., W. Heeringa, and P. Kleiweg (1999): "Edit Distance and Dialect Proximity." In: *Time Warps, String Edits and Macromolecules: The Theory and Practice of Sequence Comparison*. Ed. by D. Sankoff and J. Kruskal. Stanford: CSLI Press; v–xv.

Nerbonne, J. and E. Hinrichs (2006): "Linguistic Distances." In: *Linguistic Distances Workshop at the joint conference of International Committee on Computational Linguistics and the Association for Computational Linguistics, Sydney, July, 2006*. Ed. by J.Nerbonne and E.Hinrichs; 1–6.

Nerbonne, J. and P. Kleiweg (2007): "Toward a Dialectological Yardstick." In: *Journal of Quantitative Linguistics* 14; 148–167.

Nerbonne, J. and W. Kretzschmar (2003): "Introducing Computational Techniques in Dialectometry." In: *Computers and the Humanities* 37; 245–255.

— (2006): "Progress in Dialectometry: Toward Explanation." In: *Literary and Linguistic Computing* 21; 387–97.

Nerbonne, J. et al. (1996): "Phonetic distance between Dutch dialects." In: *CLIN VI. Papers from the sixth CLIN meeting, University of Antwerp*. Ed. by G. Durieux, W. Daelemans, and S. Gillis; 185–202.

Nerbonne, J. et al. (2008): "Projecting Dialect Differences to Geography: Bootstrap Clustering vs. Noisy Clustering." In: *Data Analysis, Machine Learning, and Applications. Proceedings of the 31st Annual Meeting of the German Classification Society*. Ed. by C. Preisach et al.; 647–654.

Nerbonne, J. et al. (2011): "Gabmap – A Web Application for Dialectology." In: *Dialectologia* II; 65–89.

Nevin, B. (2002): *The legacy of Zellig Harris: Language and information into the 21st century. Volume 1: Philosophy of science, syntax and semantics*. Amsterdam: John Benjamins.

Nevin, B. and S. Johnson (2002): *The legacy of Zellig Harris: Language and information into the 21st century. Volume 2: Mathematics and computability of language*. Amsterdam: John Benjamins.

Ng, R. and J. Han (2002): "CLARANS: A method for clustering objects for spatial data mining." In: *IEEE Transactions on Knowledge and Data Engineering* 14; 1003–16.

Nichols, J. and T. Warnow (2008): "Tutorial on computational linguistic phylogeny." In: *Language and Linguistics Compass* 5; 760–820.

O Searcoid, M. (2006): *Metric Spaces*. Berlin: Springer.

Oakes, M. (1998): *Statistics for Corpus Linguistics*. Edinburgh: Edinburgh University Press.

Oja, E. and S. Kaski (1999): *Kohonen Maps*. London: Elsevier.

O'Keefe, A., and M. McCarthy (2010): *The Routledge Handbook of Corpus Linguistics*. London: Routledge.

Oksanen, J. (2010): *Cluster analysis: tutorial with R*. URL: http://cc.oulu.fi/~jarioksa/opetus/metodi/sessio3.pdf.

Osenova, P., W. Heeringa, and J. Nerbonne (2009): "A Quantitative Analysis of Bulgarian Dialect Pronunciation." In: *Zeitschrift für slavische Philologie* 66; 425–458.

Palander, M., L. Opas-Hänninen, and F. Tweedie (2003): "Neighbours or Enemies? Competing Variants Causing Differences in Transitional Dialects." In: *Computers and the Humanities* 37; 359–372.

Pampalk, E., A. Rauber, and D. Merkl (2002): "Using Smoothed Data Histograms for Cluster Visualization in Self-Organizing Maps." In: *Artificial Neural Networks. ICANN 2002*. Ed. by J. Dorronsoro; 871–76.

Pantel, P. and D. Lin (2002): "Discovering word senses from text." In: *Proceedings of ACM SIGKDD Conference on Knowledge Discovery and Data Mining*; 613–19.

Paolillo, J. (2002): *Analyzing Linguistic Variation: Statistical Models and Methods*. Stanford: CSLI.

Parsons, L., E. Hague, and H. Liu (2004): "Subspace clustering for high-dimensional data: a review." In: *ACM SIGKDD Exploration Newsletter* 6; 90–105.

Pascual, D., F. Pla, and J. Sanchez (2010): "Cluster validation using information stability measures." In: *Pattern Recognition Letters* 31; 454–61.

Patane, G. and M. Russo (2001): "The enhanced-LBG algorithm." In: *Neural Networks* 14; 1219–37.

Patwardhan, S., S. Banerjee, and T. Pedersen (2006): "UMND1: Unsupervised Word Sense Disambiguation Using Contextual Semantic Relatedness." In: *Proceedings of SemEval-2007: 4th International Workshop on Semantic Evaluations, Prague*; 390–393.

Pedersen, T. (2006): "Unsupervised Corpus Based Methods for WSD." In: *Word Sense Disambiguation : Algorithms and Applications*. Ed. by E. Agirre and P. Edmonds. Berlin: Springer; 133–166.

— (2010): *Computational Approaches to Measuring the Similarity of Short Contexts : A Review of Applications and Methods*. University of Minnesota Supercomputing Institute Research Report UMSI 2010/118.

Peirsman, Y., D. Geeraerts, and D. Speelman (2010): "The automatic identification of lexical variation between language varieties." In: *Natural Language Engineering* 16; 469–490.

Peissig, J. and M. Tarr (2007): "Visual object recognition: Do we know more than we did 20 years ago?" In: *Annual Review of Psychology* 58; 75–96.

Pellowe, J. and V. Jones (1978): "On intonational variety in Tyneside speech." In: *Sociolinguistic Patterns of British English*. Ed. by P. Trudgill. London: Arnold.

Pellowe, J. et al. (1972): "A dynamic modelling of linguistic variation: the urban (Tyneside) linguistic survey." In: *Lingua* 30; 1–30.

Pena, J., J. Lozano, and P. Larranaga (1999): "An empirical comparison of four initialization methods for the K-means algorithm." In: *Pattern Recognition Letters* 20; 1027–40.

Pereira, F., N. Tishby, and L. Lee (1993): "Distributional clustering of English words." In: *Proceedings of the 31st Annual Meeting of the Association for Computational Linguistics (ACL '93)*; 183–90.

Petroni, F. and M. Serva (2008): "Language distance and tree reconstruction." In: *Journal of Statistical Mechanics: Theory and Experiment* P08012; 1–15.

Pham, D. and D. Karaboga (2011): *Intelligent Optimisation Techniques: Genetic Algorithms, Tabu Search, Simulated Annealing, and Neural Networks*. Berlin: Springer.

Pölzlbauer, G. (2004): "Survey and comparison of quality measures for self-organizing maps." In: *Proceedings of the Fifth Workshop on Data Analysis (WDA'04)*. Ed. by J. Paralic, G. Pölzlbauer, and A. Rauber; 67–82.

Pölzlbauer, G., A. Rauber, and M. Dittenbach (2005a): "A vector field visualization technique for self-organizing maps." In: *Proceedings of the Ninth Pacific-Asia Conference on Knowledge Discovery and Data Mining (PAKDD'05)*. Ed. by T. Ho, D. Cheung, and H. Li; 399–409.

— (2005b): "Advanced visualization techniques for self-organizing maps with graph-based methods." In: *Proceedings of the Second International Symposium on Neural Networks (ISNN'05)*. Ed. by J. Wang, X. Liao, and Z. Yi; 75–80.

— (2005c): "Graph projection techniques for self-organizing maps." In: *Proceedings of the European Symposium on Artificial Neural Networks (ESANN'05)*. Ed. by M. Verleysen; 533–538.

Popper, K. (1959): *The Logic of Scientific Discovery*. New York: Basic Books.

— (1963): *Conjectures and Refutations: The Growth of Scientific Knowledge*. London: Routledge.

Prokic, J. and J. Nerbonne (2008): "Recognising Groups among Dialects." In: *International Journal of Humanities and Arts Computing* 2; 153–72.

Prokic, J., M. Wieling, and J. Nerbonne (2009): "Multiple Sequence Alignments in Linguistics." In: *Language Technology and Resources for Cultural Heritage, Social Sciences, Humanities and Education Workshop at the 12th Meeting of the European Chapter of the Association for Computational Linguistics. Athens.* Ed. by L. Borin and P. Lendvai; 18–25.

Prokic, J. et al. (2009): "The Computational Analysis of Bulgarian Dialect Pronunciation." In: *Serdica Journal of Computing* 3; 269–298.

Pröll, S. (2013): "Detecting structures in linguistic maps – Fuzzy clustering for pattern recognition in geostatistical dialectometry." In: *Literary and Linguistic Computing* 28; 108–118.

Psillos, S. (2007): *Philosophy of Science A-Z*. Edinburgh: Edinburgh University Press.

Psillos, S. and M. Curd (2008): *The Routledge Companion to Philosophy of Science*. London: Taylor and Francis.

Putschke, W. (1977): "Automatische Sprachkartographie: Konzeption, Probleme, und Perspektiven." In: *Germanische Linguistik* 3–4; 25–40.

Raisinger, S. (2008): *Quantitative Research in Linguistics*. London: Continuum.

Ram, A. et al. (2010): "A density based algorithm for discovery density varied cluster in large spatial databases." In: *International Journal of Computer Applications* 3; nr. 6.

Rauber, A. and D. Merkl (1999): "Automatic labeling of self-organizing maps: Making a treasure-map reveal its secrets." In: *Methodologies for Knowledge Discovery and Data Mining, Proceedings of the Third Pacific-Asia Conference, PAKDD-99*; 228–237.

Rauber, A., D. Merkl, and M. Dittenbach (2002): "The Growing Hierarchical Self-Organizing Map: Exploratory Analysis of High-Dimensional Data." In: *IEEE Transactions on Neural Networks* 13; 1331–41.

Reed, D. and J. Spicer (1952): "Correlation methods of comparing dialects in a transition area." In: *Language* 28; 348–59.

Reid, M. and B. Szendroi (2005): *Geometry and Toplogy*. Cambridge: Cambridge University Press.

Rendon, E. et al. (2011): "Internal versus external cluster validation indexes." In: *International Journal of Computers and Communications* 1; 27–34.

Renfrew, C. (1987): *Archaeology and Language: The Puzzle of Indo-European Origins*. London: Cape.

— (1999): "Reflections on the archaeology of linguistic diversity." In: *The Human Inheritance: Genes, Language, and Evolution*. Ed. by B. Sykes. Oxford: Oxford University Press; 1–32.

Renfrew, C., A. McMahon, and L. Trask (2000): *Time Depth in Historical Linguistics*. Cambridge: McDonald Institute for Archaeological Research.

Renouf, A. and A. Kehoe (2009): *Corpus Linguistics: Refinements and Reassessments*. Amsterdam: Rodopi.

Reppen, R., S. Fitzmaurice, and D. Biber (2002): *Using Corpora to Explore Linguistic Variation*. Amsterdam: Benjamins.

Rexová, K., Y. Bastin, and D. Frynta (2006): "Cladistic analysis of Bantu languages: a new tree based on combined lexical and grammatical data." In: *Naturwissenschaften* 93; 189–94.

Rexová, K., D. Frynta, and J. Zrzavy (2003): "Cladistic analysis of languages: Indo-European classification based on lexicostatistical data." In: *Cladistics* 19; 120–127.

Ridley, M. (1986): *Evolution and Classification: The reformation of Cladism*. London: Longman.

Rietveld, T. and R. van Hout (1993): *Statistical Techniques for the Study of Language and Language Behaviour*. Berlin: Mouton de Gruyter.

— (2005): *Statistics in Language Research. Analysis of Variance*. Berlin: Mouton de Gruyter.

Ringe, D., T. Warnow, and A. Taylor (2002): "Indo-European and computational cladistics." In: *Transactions of the Philosophical Society* 100; 59–129.

Ritter, H., T. Martinetz, and K. Schulten (1992): *Neural computation and self -organizing maps*. London: Addison-Wesley.

Robertson, S. (1972): "Letter and reply." In: *Journal of Documentation* 28; 164–5.

— (2004): "Understanding inverse document frequency: on theoretical arguments for IDF." In: *Journal of Documentation* 60; 503–20.

Rohlf, F. (1974): "Methods of comparing classifications." In: *Annual Review of Ecology and Systematics* 5; 101–13.

Rooth, M. et al. (1998): *EM-based clustering for NLP applications*. Tech. rep. AIMS Report 4(3). IMS, Universität Stuttgart.

— (1999): "Inducing a semantically annotated lexicon via EM-based clustering." In: *Proceedings of the 37th Annual Meeting of the ACL*; 104–111.

Roth, V. and T. Lange (2004): *Feature selection in clustering problems*. Cambridge MA: MIT Press.

Roth, V. et al. (2002): "A resampling approach to cluster validation." In: *Proceedings of the International Conference on Computational Statistics, COMPTSTAT2002*. Ed. by W. Härdle and B. Rönz; 123–8.

Roweis, S. and L. Saul (2000): "Nonlinear dimensionality reduction by locally linear embedding." In: *Science* 290; 2323–26.

Rubin, G. (1970): "Computer-produced mapping of dialectal variation." In: *Language Resources and evaluation* 4; 241–46.

Ruette, T., D. Speelman, and D. Geeraerts (2013): "Measuring the lexical distance between registers in national variaties of Dutch." In: *Pluricentric Languages: Linguistic Variation and Sociocognitive Dimensions*. Ed. by A. Soares da Silva. Berlin: Mouton de Gruyter; 541–554.

Saitou, N. and M. Nei (1987): "The neighbor-joining method: A new method for reconstructing phylogenetic trees." In: *Molecular Biology and Evolution* 4; 406–425.

Sammon, J. (1969): "A Nonlinear Mapping for Data Structure Analysis." In: *IEEE Transactions on Computers* C-18(5); 401–409.

Sampson, G. and D. McCarthy (2004): *Corpus Linguistics. Readings in a widening Discipline*. London: Continuum.

Sander, J. et al. (1998): "Density-based clustering in spatial databases: the algorithm GDBSCAN and its applications." In: *Data Mining and Knowledge Discovery* 2; 169–94.

Sankoff, D. (1988): *Linguistic Variation: Models and Methods*. Waltham, Massachusetts: Academic Press.

Sankoff, D. and H. Cedergren (1971): "Some results of a sociolinguistic study of Montreal French." In: *Linguistic Diversity in Canadian Society*. Ed. by R. Darnell. Edmonton: Edmonton: Linguistic Research Inc.; 61–87.

— (1976): "The dimensionality of grammatical variation." In: *Language* 52; 163–78.

Sankoff, D. and J. Kruskal (1999): *Time Warps, String Edits and Macromolecules: The Theory and Practice of Sequence Comparison*. Stanford: CSLI Press.

Sankoff, D. and R. Lessard (1975): "Vocabulary richness: a sociolinguistic analysis." In: *Science* 190; 689–90.

Sankoff, D., R. Lessard, and N. Truong (1977): "Computational linguistics and statistics in the analysis of the Montreal French Corpus." In: *Language Resources and Evaluation* 11; 185–91.

Sankoff, D. and P. Rousseau (1974): "A method for assessing variable rule and implicational scale analyses of linguistic variation." In: *Computers in the Humanities*. Ed. by J. Mitchel. Edinburgh: Edinburgh University Press; 3–15.

Schaeffer, S. (2007): "Graph Clustering." In: *Computer Science Review* 1; 27–64.

Schikuta, E. (1996): "Grid-clustering: an efficient hierarchical clustering method for very large data sets." In: *Proceedings of the 13th International Conference in Pattern Recognition, col. 2., IEEE*; 101–5.

Schikuta, E. and M. Erhart (1997): "The BANG-clustering system: Grid-based data analysis." In: *Lecture Notes in Computer science, vol. 1280*. Ed. by X. Liu, P. Cohen, and M. Berthold. Berlin: Springer; 513–24.

Schiltz, G. (1996): "Dialectometry." In: *H. Bock and W. Polasek*. Ed. by D. Analysis, I. S. Statistical, and C. A. P. of 19th Meeting of the Gesellschaft für Klassifikation. Berlin; Springer.

Schölkopf, B., A. Smola, and K. Müller (1998): "Nonlinear component analysis as a kernel eigenvalue problem." In: *Neural Computation* 10; 1299–1319.

— (1999): "Kernel Principal Component Analysis." In: *Advances in Kernel Methods - Support Vector Learning*. Ed. by B. Schölkopf, C. Burges, and A. Smola. MIT Press; Cambridge MA.

Schütz, A. and J. Wenker (1966): "A program for the determination of lexical similarity." In: *Computation in Linguistics: a case study*. Ed. by P. Garvin and B. Spolsky. Bloomington IN: Indiana University Press; 124–45.

Schütze, H. (1992): "Dimensions of meaning." In: *Proceedings of Supercomputing '92, Minneapolis*; 787–96.

— (1995): "Distributional part-of-speech tagging." In: *EACL* 7; 141–8.

— (1998): "Automatic word sense discrimination." In: *Computational Linguistics* 24; 97–124.

Schütze, H. and M. Walsh (2011): "Half-context language models." In: *Computational Linguistics* 37; 843–65.

Seber, G. and C. Wild (2003): *Nonlinear Regression*. Hoboken NJ: Wiley Interscience.

Séguy, J. (1973a): *Atlas linguistique de la Gascogne, complément du volume VI*. Paris: Centre National de la Recherche Scientifique.

— (1973b): *Atlas linguistique de la Gascogne, volume VI*. Paris: Centre National de la Recherche Scientifique.

— (1973c): "La dialectométrie dans l'Atlas linguistique de la Gascogne." In: *Revue de Linguistique Romane* 37; 1–24.

Serdah, A. and W. Ashour (2012): "Overcoming the problem of different density-regions using the inter-connectivity and the closeness." In: *Computing and Information Systems* 16(3); 2–7.

Shackleton, R. (2007): "Phonetic variation in the traditional English dialects: a computational analysis." In: *Journal of English Linguistics* 33; 99–160.

Sharma, S. (1996): *Applied Multivariate Techniques*. Hoboken NJ: John Wiley and Sons.

Shaw, D. (1974): "Statistical analysis of dialectal boundaries." In: *Language Resources and Evaluation* 8; 173–177.

Shawe-Taylor, J. and N. Cristianini (2004): *Kernel Methods for Pattern Analysis*. Cambridge: Cambridge University Press.

Shepard, R. (1972): "Psychological representation of speech sounds." In: *Human Communication: a unified view*. Ed. by E. David and P. Denes. McGraw-Hill; London.

Shutova, E., L. Sun, and A. Korhonen (2010): "Metaphor identification using verb and noun clustering." In: *Proceedings of the 23rd International Conference on Computational Linguistics, Beijing, China*; 1002–1010.

Sims-Williams, P. (1998): "Genetics, linguistics, and prehistory: thinking big and thinking straight." In: *Antiquity* 72; 505–27.

Singhal, A., C. Buckley, and M. Mitra (1996): "Pivoted document length normalization." In: *Proceedings of the 19th ACM Conference on Research and Development in Information Retrieval (SIGIR-96)*; 21–29.

Singhal, A. et al. (1996): "Document Length Normalization." In: *Information Processing and Management* 32; 619–633.

Sneath, P. and R. Sokal (1963): *Numerical Taxonomy: The Principles and Practice of Numerical Classification*. London: Freeman.

Sokal, R. and F. Rohlf (1962): "The comparison of dendrograms by objective methods." In: *Taxon* 11; 33–40.

Spärck Jones, K. (1972): "Exhaustivity and specificity." In: *Journal of Documentation* 28; 11–21.

— (2004): "A statistical interpretation of term specificity and its application in retrieval." In: *Journal of Documentation* 60; 493–502.

Spärck Jones, K., S. Walker, and S. Robertson (2000): "A probabilistic model of information retrieval: development and comparative experiments, part 2." In: *Information Processing and Management* 36; 809–40.

Speelman, D., S. Grondelaers, and D. Geeraerts (2003): "Profile-Based Linguistic Uniformity as a Generic Method for Comparing Language Varieties." In: *Computers and the Humanities* 37; 317–337.

Spruit, M. (2006): "Measuring syntactic variation in Dutch dialects." In: *Literary and Linguistic Computing* 21; 493–506.

Spruit, M., W. Heeringa, and J. Nerbonne (2009): "Associations among Linguistic Levels." In: *Lingua* 119; 1624–40.

Stefanowitsch, A. (2010): "Cognitive linguistics meets the corpus." In: *Cognitive lingusitics: convergence and expansion*. Ed. by M. Brdar, S. Gries, and M. Fuchs. Amsterdam: John Benjamins; 257–290.

Stefanowitsch, A. and S. Gries (2006): *Corpus-based approaches to metaphor and metonymy*. Berlin: Mouton de Gruyter.

— (2009): "Corpora and grammar." In: *Corpus linguistics: an international handbook, Vol. 2*. Ed. by A. Lüdeling and M. Kytö. Berlin: Mouton de Gruyter; 933–951.

Stein, B., S. Eissen, and F. Wissbrock (2003): "On cluster validity and the information need of users." In: *International Conference on Artificial Intelligence and Applications*. Ed. by M. Hanza; 216–21.

Steinbach, M., L. Ertöz, and V. Kumar (2004): "The challenges of clustering high-dimensional data." In: *New Directions in Statistical Physics*. Ed. by L. Wille. Berlin: Springer; 273–309.

Stolz, W. (1965): "A probabilistic procedure for grouping words into phrases." In: *Language and Speech* 8; 219–35.

Stone, J. (2004): *Independent Component Analysis. A Tutorial Introduction*. Cambridge MA: MIT Press.

Strang, B. (1968): "The Tyneside Linguistic Survey." In: *Zeitschrift für Mundartforschung* 4; 788–794.

Stubbs, M. (1996): *Text and Corpus Analysis. Computer-assisted Studies of Language and Culture*. Hoboken NJ: Blackwell.

Sun, L. and A. Korhonen (2009): "Improving verb clustering with automatically acquired selectional preferences." In: *EMNLP '09: Proceedings of the 2009 Conference on Empirical Methods in Natural Language Processing*; 638–647.

— (2011): "Hierarchical Verb Clustering Using Graph Factorization." In: *EMNLP '11: Proceedings of the Conference on Empirical Methods in Natural Language Processing*; 1023–33.

Sutherland, W. (2009): *Introduction to Metric and Topological Spaces*. 2nd ed. Oxford: Oxford University Press.

Szmrecsanyi, B. (2008): "Corpus-based dialectometry: Aggregate morphosyntactic variability in British English dialects." In: *International Journal of Humanities and Arts Computing* 2; 279–296.

— (2011): "Corpus-based dialectometry: a methodological sketch." In: *Corpora* 6; 45–76.

— (2013): *Grammatical Variation in British English Dialects: A Study in Corpus-Based Dialectometry*. Cambridge: Cambridge University Press.

Szmrecsanyi, B. and B. Kortmann (2009): "The morphosyntax of varieties of English worldwide: a quantitative perspective." In: *Lingua* 119; 1643–1663.

Tabachnick, B. and L. Fidell (2007): *Using Multivariate Statistics*. London: Pearson Education.

Tabak, J. (2004): *Geometry: The Language of Space and Form*. New York: Facts on File.

Tagliamonte, S. (2006): *Analysing Sociolinguistic Variation*. Cambridge: Cambridge University Press.

Tan, P., M. Steinbach, and V. Kumar (2006): *Introduction to Data Mining*. London: Pearson Addison Wesley.

Taylor, C. (2008): "What is corpus linguistics? What the data says." In: *ICAME Journal* 32; 179–200.

Temizel, T. et al. (2007): "The effect of data set characteristics in the choice of clustering validity index type." In: *Proceedings of 22nd International Symposium on Computer and Information Sciences (ISCIS'07)*; 1–6.

Tenenbaum, J., V. deSilva, and J. Langford (2000): "A global geometric framework for nonlinear dimensionality reduction." In: *Science* 290; 2319–23.

Thomas, A. (1977): "A cumulative matching technique for computer determination of speech-areas." In: *Germanistische Linguistik* 3–4; 275–88.

— (1980): "Computer analysis of a dialectal transition belt." In: *Computers and the Humanities* 14; 241–51.

— (1988): "Methods in Dialectology." In: *Proceedings of the Sixth International Conference held at the University College of North Wales, 1987.*

Tokunaga, T., M. Iwayama, and H. Tanaka (1995): "Automatic thesaurus construction based on grammatical relation." In: *Proceedings of the 14th International Joint Conference on Artificial Intelligence (IJCAI '95), Montreal*; 1308–13.

Trudgill, P. (1974): *The Social Differentiation of English in Norwich*. Cambridge: Cambridge University Press.

Turney, P. and P. Pantel (2010): "From Frequency to Meaning: Vector Space Models of Semantics," in: *Journal of Artificial Intelligence Research* 37; 141–188.

Ueberla, J. (1995): "More Efficient Clustering of N-Grams For Statistical Language Modeling." In: *EUROSPEECH '95. Fourth European Conference on Speech Communication and Technology, Madrid*; 1257–1260.

Uiboaed, K. et al. (2013): "Variation of verbal constructions in Estonian dialects." In: *Literary and Linguistic Computing* 28; 42–62.

Ultsch, A. (2003a): "Maps for visualization of high-dimensional data spaces." In: *Proceedings of the Workshop on Self Organizing Maps, WSOM03*; 225–30.

— (2003b): *U*-Matrix: a tool to visualize cluster in high-dimensional data.* Tech. rep. 36. Department of Computer Science, University of Marburg.

Ultsch, A. and L. Herrmann (2005): "The architecture of emergent self-organizing maps to reduce projection errors." In: *Proceedings of the 13th European Symposium on Artificial Neural Networks*; 1–6.

Ultsch, A. and F. Mörchen (2005): *ESOM-Maps: tools for clustering, visualization, and classification with Emergent SOM.* Marburg: Technical Report, Data Bionics Research Group, University of Marburg.

Ultsch, A. and H. Siemon (1990): "Kohonen's self-organizing feature maps for exploratory data analysis." In: *Proceedings of the International Neural Network Conference, INNC '90*; 305–8.

Ushioda, A. (1996): "Hierarchical clustering of words and application to NLP tasks." In: *Fourth Workshop on Very Large Corpora, Association for Computational Linguistics*. Ed. by E. Ejerhed and I. Dagan; 28–41.

Valls, E., M. Wieling, and J. Nerbonne (2013): "Linguistic advergence and divergence in north-western Catalan: A dialectometric investigation of dialect leveling and border effects." In: *Literary and Linguistic Computing* 28; 119–46.

Valls, E. et al. (2012): "Applying the Levenshtein Distance to Catalan Dialects. A Brief Comparison of Two Dialectometric Approaches." In: *Verba. Anuario Galego de Filoloxía* 39; 35–61.

Van Hulle, M. (2000): *Faithful Representations and Topographic Maps.* Hoboken NJ: John Wiley and Sons.

Venna, J. and S. Kaski (2001): “Neighborhood preservation in nonlinear projection methods: an experimental study.” In: *Lecture Notes in Computer Science* 2130; 485–91.

Verleysen, M. (2003): “Learning high-dimensional data.” In: *Limitations and future trends in neural computation*. Ed. by S. Ablameyko et al. Amsterdam: IOS Press; 141–162.

Vesanto, J. (1999): “SOM-based data visualization methods.” In: *Intelligent Data Analysis* 3; 111–26.

— (2000): *Using SOM in Data Mining*. Helsinki: Helsinki University of Technology, Department of Computer Science and Engineering.

Vesanto, J. and E. Alhoniemi (2000): “Clustering of the self-organizing map.” In: *IEEE Transactions on Neural Networks* 11; 586–600.

Vidal, R. (2011): “Subspace clustering.” In: *IEEE Signal Processing Magazine* 28; 52–68.

Viereck, W. (1984): “Der Einsatz des Computers in der amerikanisch-englischen und britisch-englischen Dialektologie und Soziolinguistik.” In: *Zeitschrift für Dialecktologie und Linguistik* 51; 6–30.

— (1988): “The computerisation and quantification of linguistic data.” In: *Methods in Dialectology: Proceedings of the Sixth International Conference at University College, North Wales, August 1987*. Ed. by M. Thomas; 524–50.

Villmann, T., R. Der, and T. Martinetz (1994): “A new quantitative measure of topology preservation in Kohonen’s feature maps.” In: *Proceedings of ICNN’94, IEEE International Conference on Neural Networks*; 645–48.

Villmann, T. et al. (1997): “Topology preservation in self-organizing feature maps: exact definition and measurement.” In: *IEEE Transactions in Neural Networks* 8; 256–66.

Vriend, F. de et al. (2008): “The Dutch–German Border: Relating Linguistic, Geographic and Social Distances.” In: *International Journal of Humanities and Arts Computing* 2; 119–34.

Walde, S. Schulte im (2000): “Clustering Verbs Semantically According to their Alternation Behaviour.” In: *Proceedings of the 18th International Conference on Computational Linguistics. Saarbrücken, Germany*; 747–753.

— (2006): “Experiments on the automatic induction of German semantic verb classes.” In: *Computational Linguistics* 32; 159–94.

Walde, S. Schulte im and C. Brew (2002): "Inducing German semantic verb classes from purely syntactic subcategorisation information." In: *Proceedings of the 40th Annual Meeting of the Association for Computational Linguistics, Philadelphia*; 223–230.

Wales, K. (2006): *Northern English: A Social and Cultural History*. Cambridge: Cambridge University Press.

Wall, M., A. Rechtsteiner, and L. Rocha (2003): "Singular value decomposition and principal component analysis." In: *A Practical Approach to Microarray Data Analysis*. Ed. by D. Berrar, W. Dubitzky, and M. Granzow. Dordrecht: Kluwer; 91–109.

Walpole, R. et al. (2007): *Probability and Statistics for Engineers and Scientists*. 8th. London: Pearson Education International.

Wang, L. et al. (2009): "Automatically determining the number of clusters in unlabeled data sets." In: *IEEE Transactions on Knowledge and Data Engineering* 21; 335–50.

Wang, W., J. Yang, and R. Muntz (1997): "STING: a statistical information grid approach to spatial data mining." In: *23rd International Conference on Very Large Databases*; 186–95.

Warnow, T. (1997): "Mathematical approaches to comparative linguistics." In: *Proceedings of the National Academy of Sciences* 94; 6585–6590.

Waterman, S. (1995): "Distinguished usage." In: *Corpus Processing for Lexical Acquisition*. Ed. by B. Boguraev and J. Pustejovsky. Cambridge MA: MIT Press; 143–72.

Watt, D. and W. Allen (2003): "Illustrations of the IPA: Tyneside English." In: *Journal of the International Phonetic Association* 33; 267–71.

Watt, D. and L. Milroy (1999): "Patterns of variation in Newcastle vowels." In: *Urban Voices: Accent Studies in the British Isles*. Ed. by P. Foulkes and G. Docherty. London: Arnold; 25–46.

Watters, P. (2002): "Discriminating word senses using cluster analysis." In: *Journal of Quantitative Linguistics* 9; 77–86.

Wells, J. (1982): *Accents of English I. An Introduction*. Cambridge: Cambridge University Press.

Whitehead, C. and D. Towers (2002): *Guide to Abstract Algebra*. 2nd ed. Basingstoke: Palgrave Macmillan.

Wichmann, S. and A. Saunders (2007): "How to use typological databases in historical linguistic research." In: *Diachronica* 24; 373–404.

Wieling, M. (2012): *A Quantitative Approach to Social and Geographical Dialect Variation*. University of Groningen: Groningen Dissertations in Linguistics 103.

Wieling, M., E. Margaretha, and J. Nerbonne (2011): "Inducing Phonetic Distances from Dialect Variation." In: *CLIN Journal* 1; 109–118.

— (2012): "Inducing a Measure of Phonetic Similarity from Pronunciation Variation." In: *Journal of Phonetics* 40; 307–314.

Wieling, M. and J. Nerbonne (2009): "Bipartite Spectral Graph Partitioning to Co-Cluster Varieties and Sound Correspondences in Dialectology." In: *Text Graphs 4, Workshop at the 47th Meeting of the Association for Computational Linguistics*. Ed. by M. Choudhury et al.; 14–22.

— (2010a): "Bipartite spectral graph partitioning for clustering dialect varieties and detecting their linguistic features." In: *Computer Speech and Language* 25; 700–715.

— (2010b): "Hierarchical bipartite spectral graph partitioning to cluster dialect varieties and determine their most important linguistic features." In: *TextGraphs-5 Workshop on Graph-Based Methods for Natural Language Processing 16*; 33–41.

Wieling, M., J. Nerbonne, and H. Baayen (2011): "Quantitative Social Dialectology: Explaining Linguistic Variation Geographically and Socially." In: *PLoS ONE* 6; nr. 9.

Wieling, M., J. Prokic, and J. Nerbonne (2009): "Evaluating the Pairwise String Alignments of Pronunciations." In: *Language Technology and Resources for Cultural Heritage, Social Sciences, Humanities and Education Workshop at the 12th Meeting of the European Chapter of the Association for Computational Linguistics. Athens*. Ed. by L. Borin and P. Lendvai; 26–34.

Wieling, M., R. Shackleton, and J. Nerbonne (2013): "Analyzing phonetic variation in the traditional English dialects: Simultaneously clustering dialects and phonetic features." In: *Literary and Linguistic Computing* 28; 31–41.

Wood, G. (1972): "The computer in analysis and plotting." In: *American Speech* 47; 195–202.

Woods, A., P. Fletcher, and A. Hughes (1986): *Statistics in Language Studies*. Cambridge: Cambridge University Press.

Xu, J. and W. Croft (1998): "Corpus-based stemming using co-ocurrence of word variants." In: *ACM Transactions on Information Systems* 16; 61–81.

Xu, R. and D. Wunsch (2005): "Survey of clustering algorithms." In: *IEEE Transactions on Neural Networks* 16; 645–78.

— (2009): *Clustering*. Hoboken NJ: Wiley.

Xu, X. et al. (1998): "A distribution-based clustering algorithm for mining large spatial databases." In: *Proceedings of the Fourteenth International Conference on Data Engineering, ICDE'98*; 324–31.

Yang, Y., J. Lafferty, and A. Waibel (1996): "Word clustering with parallel spoken language corpora." In: *Proceedings of the Fourth International Conference on Spoken Language Processing, ICSLP'96*; 2364–67.

Young, S. and G. Bloothooft (1997): *Corpus-Based Methods in Language and Speech Processing*. Dordrecht: Kluwer Academic.

Yue, S. et al. (2008): "A general grid-clustering approach." In: *Pattern Recognition Letters* 29; 1372–84.

Zernik, U. (1991): "Train 1 vs train 2: tagging word sense in a corpus." In: *Lexical Acquisition: On-line Resources to Build a Lexicon*. Ed. by U. Zernik. Lawrence Erlbaum Associates; Hillsdale NJ.

Zhang, T., R. Ramakrishnan, and M. Livny (1996): "BIRCH: and efficient data clustering method for very large databases." In: *Proceedings of the ACM SIGMOD Conference on Management of Data*; 103–14.

Ziviani, N. and B. Ribeiro-Neto (1999): "Text operations." In: *Modern Information Retrieval*. Ed. by R. Baeza-Yates and B. Ribeiro-Neto. Addison-Wesley; London.

Subject index

D

E

G

H

N

O

T

W

Z